Environmental Division 2019

Held at the 2019 AIChE Annual Meeting

Orlando, Florida, USA
10 - 15 November 2019

ISBN: 978-1-7138-0533-5

Printed from e-media with permission by:

Curran Associates, Inc.
57 Morehouse Lane
Red Hook, NY 12571

Some format issues inherent in the e-media version may also appear in this print version.

Printed with permission by Curran Associates, Inc. (2020)

For permission requests, please contact AIChE
at the address below.

AIChE
120 Wall Street, FL 23
New York, NY 10005-4020

Phone: (800) 242-4363
Fax: (203) 775-5177

www.aiche.org

Additional copies of this publication are available from:

Curran Associates, Inc.
57 Morehouse Lane
Red Hook, NY 12571 USA
Phone: 845-758-0400
Fax: 845-758-2633
Email: curran@proceedings.com
Web: www.proceedings.com

TABLE OF CONTENTS

566474 Hidden Economic Impacts of Nutrient Runoff from Livestock Waste

Livestock waste may cause soil quality detriment from the in-excess nutrients and acidification, air quality degradation from ammonia and methane emissions, and water pollution issues from nutrient and pathogens runoff to the water bodies, leading to eutrophication, algal blooms, and hypoxia [1-2]. Despite the significant environmental benefits of performing waste management for these organic materials, the recovery of value-added products from livestock waste is not a current practice due to the high investment costs required and the low market incentives for the products that are recovered [3]. In addition, the deployment of waste treatment technologies is also hindered by the difficulty to attribute an economic value to the social and environmental damages algae blooms impart.

In this talk, we present a methodology to quantify the impacts associated with harmful algae blooms in water bodies. Specifically, we quantify the economic loss in property values, loss in recreational benefits (such as swimming and fishing) and water cleanup costs. We use the change in Secchi depth (change in water quality) as the basis for our calculations. For the loss in property values, the hedonic analysis done by Dodds et al. [4] reveals a strong correlation between Secchi depth and the property values. We provide a modeling framework to integrate this analysis in a decision-making setting. The algae blooms can be toxic to human health, thus forcing frequent beach closures. We use statistical models [5, 6] such as the logistic regression and negative binomial models to estimate the loss in revenue related to swimming and fishing opportunities. In areas where the affected waterbody is a source for drinking water, clean up procedures such as alum treatment are required to make the water potable. We will present a holistic approach to account for these impacts and conduct simultaneous economic and environmental analysis for waste treatment. This approach can also account for mid-point indicators such as the primary energy and water footprint associated with waste treatment technologies.

Lastly, we will present a case study that highlights the effect of including these hidden economic impacts of phosphorus runoff on the overall decision-making for treatment of livestock waste in the Upper Yahara watershed region in the State of Wisconsin. Excessive amounts of phosphorus have accumulated in this area, primarily due to livestock manure and the heavy use of agricultural fertilizers. Rain and snow melt often wash these nutrients into waterways, which lead to the blue-green algae blooms in the Yahara lakes [7]. We observe that accounting the hidden economic impacts can activate an integrated waste management and valorization market [8] and help in achieving phosphorus balance within the study area. The state, tribes, and federal regulatory agencies can use this framework to develop regulatory and non-regulatory incentives for waste treatment that can help in achieving nutrient pollution reduction targets in an environmentally and economically sustainable manner.

References:

[1] Burkholder, J., Libra, B., Weyer, P., Heathcote, S., Kolpin, D., Thorne, P. S., & Wichman, M. (2006). Impacts of waste from concentrated animal feeding operations on water quality. *Environmental health perspectives*, *115*(2), 308-312.

[2] Aguirre-Villegas, H. A., & Larson, R. A. (2017). Evaluating greenhouse gas emissions from dairy manure management practices using survey data and lifecycle tools. *Journal of cleaner production*, *143*, 169-179.

[3] Sampat, A. M., Ruiz-Mercado, G. J., & Zavala, V. M. (2018). Economic and Environmental Analysis for Advancing Sustainable Management of Livestock Waste: A Wisconsin Case Study. *ACS Sustainable Chemistry & Engineering*, *6*(5), 6018-6031.

[4] Dodds, W. K., Bouska, W. W., Eitzmann, J. L., Pilger, T. J., Pitts, K. L., Riley, A. J., ... & Thornbrugh, D. J. (2008). Eutrophication of US freshwaters: analysis of potential economic damages.

[5] Vesterinen, J., Pouta, E., Huhtala, A., & Neuvonen, M. (2010). Impacts of changes in water quality on

recreation behavior and benefits in Finland. *Journal of Environmental Management*, *91*(4), 984-994.

[6] Wisconsin Department of Natural Resources. (2012). Phosphorus reduction in Wisconsin water bodies: An economic impact analysis. Technical report, 2012.

[7] Water sustainability and climate in the Yahara Watershed. *University of Wisconsin-Center for Limnology available at https://wsc.limnology.wisc.edu/about/watershed*. (2018). [Online; accessed 23-July-2018].

[8] Sampat, A.M.; Hu, Y.; Sharara, M.; Aguirre-Villegas, H.; Ruiz-Mercado, G.; Larson, R.A.; and Zavala, V.M. (2018). Coordinated Markets for Scalable Management of Organic Waste and Derived Products. Under Review.

571451 Energy-Water Nexus in Eco-Industrial Park

Abstract:
Present day energy and water systems are tightly intertwined. Water is used in all phases of energy production and electricity generation. Energy is required to extract, convey, and deliver water of appropriate quality for diverse human uses, and then again to treat waste waters prior to their return to the environment. Historically, interactions between energy and water have been considered on a regional or technology-by-technology basis. At the national and international levels, energy-water systems have been developed, managed, and regulated independently. Water and energy are critical, mutually dependent resources—the production of energy requires large volumes of water and water infrastructure requires large amounts of energy.

An Eco-industrial park (EIP) is a community of manufacturing and service businesses seeking enhanced environmental and economic performance through collaboration in managing environmental and resource issues including energy, water, and materials. By working together, the community of businesses seeks a collective benefit that is greater than the sum of the individual benefits each company would realize if it optimized its individual performance only. The goal of an EIP is to improve economic performance of the participating companies while minimizing their environmental impact (Das, 2005; Das and Cabezas, 2018; EPRI, 2011; Lowe, 2001).

A case study pertinent to energy-water-food nexus in the agro industrial eco-complex is illustrated. This is particularly important to the theme of process integration and intensification because it also involves the production of a good (paper), the sustainable generation of energy (from bagasse – e.g., waste-to-energy), a human food (sugar), and sustainable use of water.

References:
Das, T. K. (2005). Toward Zero Discharge: Innovative Methodology and Technologies for Process Pollution Prevention, Wiley-InterScience, Hoboken, NJ.

Das, T. K. and Cabezas, H. (2018). "Tools and Concepts for Environmental Sustainability in the Food-Energy-Water Nexus: Chemical Engineering Perspective," Environmental Progress and Sustainable Energy, 37(1), 73-81.

EPRI, Electric Power Research Institute (2011). "Water Use for Electricity Generation and Other Sectors: Recent Changes (1985-2005) and Future Projections (2005-2030)". Palo Alto, CA: EPRI.

Lowe, E. A. (2001). "Eco-industrial Park Handbook for Asian Developing Countries. A Report to Asian Development Bank," Environment Department, Indigo Development, Oakland, CA.

577780 Electrochemical Processing to Capture Phosphorus from Concentrated Animal Feeding Operations Waste: Experimental and Process Simulation Investigations

Recovery of phosphates and nitrates to prevent watershed eutrophication presents a new challenge in animal waste management. In 2011, the United States applied 4.32 million short tons of inorganic P_2O_5 fertilizer with continued growth in application rate expected [1]. Beneficiated phosphate rock is the primary source of phosphorus found in fertilizer products, with an estimated 68 Gt remaining worldwide and 74% of these reserves located in Morocco [2,3]. Although peak phosphorus supply is not a major present concern, phosphorus supply chain sustainability requires further consideration to ensure food and clean water security.

The development of phosphorus recycling technologies has recently become of interest, sparking a proliferation of scientific and technical research. Reducing phosphorus loss by increasing its recycling will increase phosphorus supply chain sustainability, food security, and watershed health. Recovery of nitrogen and phosphorus via struvite ($MgNH_4PO_4 \cdot 6H_2O$) precipitation is a recycling technique which can reduce watershed eutrophication and potentially increase fertilizer supply chain sustainability. Nucleation and growth of struvite crystals is controlled by solution pH which is dependent upon soluble magnesium, ammonia, and phosphorus specie concentrations.

Ohio University (OHIO) has been investigating a modular electrochemical-based technique to indirectly induce struvite precipitation within the liquid fraction of animal waste. Struvite precipitation was performed via chronoamperometry and struvite precipitates were characterized by scanning electron microscopy with energy-dispersive X-ray spectroscopy. Key parameters controlling electrochemical struvite precipitation include: solution pH, temperature, turbulence, and the presence of competing ions (Ca^{2+}). Solution temperature greater than 20°C had negligible effects on struvite precipitation kinetics. The presence of Ca^{2+} was found to suppress phosphorus removal. The impact of operating voltage on phosphorus removal efficiency and current efficiency with time has also investigated. Further, process simulations have been developed to aid estimating operating costs for a commercial scale system integrated into a concentrated animal feeding operation (CAFO). This presentation will review recent laboratory and process simulation results.

[1] "https://www.ers.usda.gov/data-products/fertilizer-use-and-price/," USDA, 2 May 2018. [Online]. [Accessed 2018].

[2] M. Heckenmüller, D. Narita, G. Klepper, Global availability of phosphorus and its implications for global food supply: an economic overview, Kiel Institute for the World Economy, Germany, (2014).

[3] S. M. Jasinski, Phosphate rock, U.S. Geological Survey, Mineral Commodity Summaries, (2017).

[4] A. B. Ribeiro, E. P. Mateus, N. Couto, Electrokinetics across disciplines and continents, Springer International Publishing Switzerland, New York (2016).

578042 Pore-Scale Effects of Dilute Organic Macromolecules and Nanoparticles to Inhibit Moisture Loss

The retention of soil moisture is a complex and dynamic process, influenced by physical, chemical, and biological processes all interacting at the pore scale. For example, the function of microbial communities is inextricably linked with soil microstructure, such as when microbes secrete extracellular polymeric substances (EPS) to regulate soil moisture and promote plant growth. EPS regulates moisture three ways: it serves as a hydrogel (swelling during wet conditions and remaining hydrated during dry conditions), it alters soil surface properties (creating more water-repellent surfaces), and it promotes aggregation of soil particles. EPS-mediated moisture retention can be understood by systematically controlling the aqueous composition and the microscale environment simultaneously. Here, we show the combined effects of physical microstructure and EPS concentration on retention of soil moisture in different PDMS micromodels. We employed a physically complex emulated soil micromodel to evaluate the spatial distribution of water under drying pressure versus purified EPS concentration. Here, the EPS was collected from stationary-phase *Sinorhizobium meliloti* cultures and suspended at different concentrations in growth media salts or in artificial groundwater. Experimental results showed that the 15 μg/mL of EPS solution dried eight times slower than deionized water, and the 58 μg/mL of EPS solution dried 16 times slower than deionized water in identical soil micromodels while the residual saturation of 15 μg/mL and 58 μg/mL of EPS solutions were 6 ± 1% and 35 ± 21%, compared with a residual saturation of 0 for deionized water. Studies using colloidal agro-additives including graphene oxide and biochar showed the degree of oxidation might effect on moisture retention. In this project, we build a systematic platform to evaluate pore-scale effects of dilute organic macromolecules and organic nanoparticles on inhibiting moisture loss. We show that relatively low EPS concentration dramatically reduce the rate of water loss, increase steady state moisture content, and enhance the variability of moisture distribution. We anticipate this work will impact the future development of sustainable agriculture biotechnology for increasing food production while using less water.

571798 Process Intensification of the Conversion of Waste Cooking Oil to Fame Using a T- Microreactor

Fatty Acid Methyl Ester (FAME) or biodiesel is considered as an environment-friendly fuel because of its unique properties: it is non-toxic, biodegradable, and produces less greenhouse gas emissions in comparison to fossil fuels [1]. Generally, esters are produced by mechanical stirring via transesterification process in which the fatty acids (oil) is converted into ester by reacting it with alcohol in the presence of a catalyst. However, mechanical stirring has a low efficiency in overcoming the mass transfer limitation between the oil and alcohol phases resulting in lower reaction rates and higher capital production costs [2]. Recently, intensification technologies have been investigated for ester production in order to overcome this issue. Many researchers have employed different intensification processes such as the utilization of microfluidic devices, microwave, ultrasonic, and hydrodynamic cavitation reactors[3][4]. These technologies have been utilized to minimize the limitations in mass transfer between the oil and alcohol. Reducing the mass transfer resistance results in a significant reduction of the overall reaction time which significantly reduces the consumption of energy compared to the mechanical stirring process [5].

Microreactors have been extensively used for liquid-liquid multiphase systems to reduce the mass and heat transfer limitation because of the high interfacial area achieved between the phases. Ester production is almost 15 times faster in a microreactor compared to the conventional method. Numerous flow patterns, such as drop flow, slug flow, and parallel flow have been reported in the literature on ester production by using microreactors from different feedstock of virgin oil [6].

In this work, we present a study on the production of esters from waste cooking oil. In particular, we study the impact of various process parameters such as oil-to-methanol ratio, catalysts concentration, and residence time on the conversion of FAME. We specifically look at the challenges of using waste cooking oil instead of virgin oil during the transesterification process in the microreactor. To our best knowledge, there is no research available on ester production in microreactors using waste cooking oil as a feedstock.

References:

[1] Balat M. Potential alternatives to edible oils for biodiesel production – A review of current work. Energy Convers Manag 2011;52:1479–92. doi:https://doi.org/10.1016/j.enconman.2010.10.011.

[2] Marchetti JM. A summary of the available technologies for biodiesel production based on a comparison of different feedstock's properties. Process Saf Environ Prot 2012;90:157–63. doi:10.1016/j.psep.2011.06.010.

[3] Azcan N, Danisman A. Microwave assisted transesterification of rapeseed oil 2008;87:1781–8. doi:10.1016/j.fuel.2007.12.004.

[4] Gholami A, Hajinezhad A, Pourfayaz F, Ahmadi MH. The effect of hydrodynamic and ultrasonic cavitation on biodiesel production: An exergy analysis approach. Energy 2018;160:478–89. doi:10.1016/j.energy.2018.07.008.

[5] Chuah LF, Klemeš JJ, Yusup S, Bokhari A, Akbar MM. A review of cleaner intensification technologies in biodiesel production. J Clean Prod 2017;146:181–93. doi:10.1016/j.jclepro.2016.05.017.

[6] Enrique L-G, Enrique O-N, Alejandro M-C, Krishna DPN. Process Intensification of Biodiesel production using a Tubular Micro-Reactor (TMR): Experimental and Numerical Assessment. Chem Eng Commun 2017;204:467–75.

571841 Simulation and Energy Assessment of Ammonia-Containing Gas Separation and Recovery Process with Ionic Liquids

Simulation and energy assessment of ammonia-containing gas separation and recovery process with ionic liquids

Guoxiong Zhan[1,2,3], Fei Chang[1,2], Shaojuan Zeng[1], Zengxi Li[2], Xiangping Zhang[1,2]*

[1]Institute of Process Engineering, Chinese Academy of Sciences, 100190, Beijing, China

[2]School of Chemical Engineering, University of Chinese Academy of Sciences, Beijing, 100049, China

[3]Sino-Danish College, University of Chinese Academy of Sciences, Beijing, 100049, China

*Corresponding author email: xpzhang@ipe.ac.cn

Abstract: Huge amounts of toxic gases (e.g. NH_3) and greenhouse gases (e.g. CO_2) emission have caused serious environment problems. Besides, more strict laws and standards have been proposed to reduce emissions and punish the action of excess emissions in recent years. Therefore, NH_3 and CO_2 separation from industrial gas is considered as a significant and essential technology to reduce atmospheric pollution. The traditional water method is used to separate NH_3 and CO_2 from industrial gas, but it led to a lot of ammonia-nitrogen wastewater, high energy consumption, low NH_3/CO_2 separation efficiency and NH_3 recovery ratio. Therefore, there is still a great challenge to get an effective way to separate and recover NH_3 and CO_2 from tail gas. Ionic liquids (ILs) has already been applied in industrial tail separation due to extremely low volatility and good affinity with gas molecule [1]. In this work, novel ILs with good selectivity towards NH_3/CO_2 and high NH_3 absorption capacity were designed and synthesized. The thermodynamic data between NH_3/CO_2 and ILs were measured and regressed by the NRTL equation [2]. Based on those results, the separation process of ammonia/carbon dioxide-containing gas with ionic liquid has been simulated and optimized. The result indicates that the designed IL can reduce NH_3 concentration under 1000 ppm in tail gas, and the removal efficiency and recovery purity are 70% and over 97%, respectively. Then, based on the former results, the assessment of energy, economy and green degree for separation system has been calculated to figure out the integrated performance of the system under varied situations. The results can provide guidelines for developing new absorption process to meet lower NH_3 concentration emission, higher NH_3 purity and recovery ratio along with lower energy consumption.

Keywords: Ionic liquids; gas separation; NH_3; CO_2; process simulation; energy assessment

REFERENCES

[1] S. J. Zeng, X. P. Zhang, L. Bai, X. C. Zhang, H. Wang, J. J. Wang, X. Y. Liu and S. J. Zhang, *ChemRev*, 2017.117.9626-9673

[2] X. Y. Liu, T. Zhou, X. P. Zhang, S. J. Zhang, X. D. Liang, R. Gani, G. M. Kontogeorgis, *Chem Eng Sci,* 2018.192.816-828

556753 Solubility and Diffusivity of Syngas Components in a Next Generation Pre-Combustion Solvent

Solubility and Diffusivity of Syngas Components in a Next Generation Pre-Combustion Solvent

Lei Hong [1,2], *Nicholas Siefert* [1], *Robert Thompson* [1,2], *Wei Shi* [1,2], *Kevin Resnik* [1,2], *David Hopkinson* [1]

[1]*National Energy Technology Laboratory, 626 Cochrans Mill Road, Pittsburgh, PA 15236*

[2]*Leidos Research Support Team, 626 Cochrans Mill Road, Pittsburgh, PA 15236*

ABSTRACT.

A key to improve the efficiency of integrated gasification combined cycle power plants with CO_2 capture and storage is to increase the temperature at which carbon dioxide is captured and regenerated. The current baseline for pre-combustion carbon capture technology is a solvent based process, such as Selexol® (mixtures of dimethylethers of polyethelenglycol) and Rectisol® (chilled methanol). To obtain high CO_2 capture efficiency (>90%), sub-ambient conditions are required for those solvents. However, cooling the syngas to below room temperature is costly and requires energy-intensive chillers. In addition, since both Selexol and Rectisol are highly hydrophilic, this can introduce corrosion issues for equipment. The hydrophilic solvents are not ideal to be used at warm temperatures partly due to water presence in the gas stream which could be absorbed in the solvent and significantly decrease CO_2 uptake. A process based on a hydrophobic solvent which can achieve a warm CO_2 separation would improve power plant efficiency as much as 2-3% compared to those processes requiring sub-ambient temperatures (Henry Pennline, et.al., Fuel Processing Technology, 89 (2008) 897-907). The National Energy Technology Laboratory (NETL) has developed a hydrophobic solvent, i.e. disub-4PEG, which has shown great potential to replace current solvents, Selexol or Rectisol. Preliminary pilot scale tests demonstrated that disub-4PEG performs better than Selexol surrogate with both higher CO_2 uptake and higher CO_2/H_2 selectivity at the same operating conditions. Absorption of water in Selexol affects the absorption of other gases, mainly CO_2 and H_2. The computational study shows that H_2O absorption in a Selexol surrogate solvent could significantly decrease both CO_2 uptake and CO_2/H_2 solubility selectivity. In contrast, for disub-4PEG, water in the fuel gas stream will have minimal effects on CO_2 solubility and CO_2/H_2 solubility selectivity due to very small amounts of water absorption in the hydrophobic disub-4PEG solvent. Finally, the solubility and diffusivity of different gases including H_2, CO_2, N_2 and CH_4, which are the main components of syngas, will be presented, which are essential for a reliable process design.

567864 Carbon Capture and Storage (CCS): The Way Forward

Carbon capture and storage (CCS) is expected to play an important role in meeting climate change mitigation targets (IPCC, 2018), providing a least cost pathway to decarbonise different sectors. There are a variety of potential applications for CCS, these include delivering low carbon heat, power and fuels, decarbonising industry and net removal of atmospheric CO_2 (*i.e.*, negative emissions). This paper presents the current state-of-the-art technologies for CO_2 capture, transport, utilisation and storage from a multi-scale perspective, starting from the global level and moving to the molecular scale (based on work in Bui et al. (2018)). We provide insight on how to move the discipline forward, highlighting key research challenges that should be addressed over the course of the next decade, giving a balanced perspective on the scientific, policy and commercial priorities.

Role of CCS and negative emissions technologies

There is a broad consensus that CCS has reached the technical maturity required for specific application (*i.e.*, power sector and industry). There are presently 23 commercial-scale CCS facilities worldwide in operation or under construction, which permanently sequester up to 37 million tonnes of CO_2 per annum (Global CCS Institute, 2019). However, IPCC integrated assessment modelling (IAM) shows that scenarios limiting global warming to 2°C requires CO_2 removal rates of 10 Gt_{CO2}/yr by 2050 and 25 Gt_{CO2}/yr by 2100. When considering a 1.5°C limit, there are no feasible scenarios without negative emissions. It is clear current CCS deployment rates do not reflect projected level required to meet emission reduction targets. In many scenarios, the dominant negative emissions technology (NET) is bioenergy with CCS (BECCS), which is likely due to the lack of other NET options within IAMs. Thus, incorporating alternative NETs into IAMs is a key research priority (Bui et al., 2018).

BECCS is currently deployed for industrial applications in 5 plants worldwide. The estimated technical potential of BECCS is a CO_2 removal rate of between 3–20 Gt_{CO2}/year. The resources (*e.g.*, land, water) required to deliver BECCS is case-specific, key factors include feedstock type, region specific constraints and decisions made along the biomass supply chain. Direct capture of CO_2 from air (DAC) is technically possible, with two demonstration plants currently operating in Switzerland (Climeworks) and Canada (Carbon Engineering). There are technical and economic challenges associated with the very dilute CO_2 concentration of air, *i.e.*, high energy requirements, significantly higher capture costs. Therefore, there is a need to address policy questions around regulating and incentivising negative emissions. Furthermore, in the absence of a mature CCS industry, attempting large-scale deployment of BECCS or DACS will be challenging.

CCS technology improvements

For the CCS technology development to accelerate, technology benchmarks need to be updated, preferably with current industrial best practice. Comparing new sorbent materials against obsolete benchmarks (e.g., 30 wt% MEA) is potentially limiting progress. Importantly, experimental testing should study these materials under conditions representative of the "real world", *e.g.*, high CO_2 partial pressure for desorption, presence of contaminants. With an immense number of potential materials for CCS, pilot-scale testing of all of them is not practicable. A more efficient approach to screen materials is the use of high throughput modelling and simulation approaches, which combine molecular- and process- scale information. Future designs of next-generation CO_2 capture sorbents (e.g., ionic liquids, MOFs) need to consider the cost aspects in combination with reduced energy requirements and higher capacity.

The metrics used to assess CCS technology improvement can vary. The CCS community tends to focus on reducing the cost to capture CO_2. On the other hand, the owners of the CO_2-emitting facilities prioritise minimising the cost of their low carbon product, *e.g.*, electricity. It is important to recognise that the majority of "CCS costs" are associated with increases in capital cost, as opposed to operating cost. Whilst continued focus on improving thermodynamics is helpful, priority must be given to research which promises reduction in capital costs. Rather than just minimising the cost of capturing CO_2, R&D initiatives aimed towards "improving" CCS should take a whole systems approach. Thus, shifting the focus towards reducing the cost per unit of decarbonised product (*e.g.*, steel, cement, power) and how the decarbonised process will compete in the market (*i.e.*, what will be displaced by what). As the energy system becomes more diverse, not all energy technologies provide the same services. Therefore, metrics such as the "system value" is a more holistic approach to evaluate and compare low-carbon energy technologies (as illustrated in Figure 1). Distinct from intermittent renewables, coal/gas/biomass-CCS offers reliable dispatchable power, thereby providing energy system resilience and flexible operability.

CCS commercialisation and political considerations

Large-scale deployment CCS is needed for deep decarbonisation. The technical elements of CCS are well-understood. The financial and commercial models to enable CCS deployment are becoming increasingly clear. There is substantial evidence of the economy-wide GDP and employment benefits associated with CCS deployment. However, public acceptability and understanding of the impact on the political economy are at an early stage. Some governments provide generous subsidies to low-carbon technologies such as offshore wind and nuclear power, these are similar in scale as what would be required for CCS. Unlike nuclear power or onshore wind, there are no strong opponents, but neither are there advocates willing to lobby strongly for CCS. Thus, shifts in incentives and regulations are needed to enable changing of interests and economics, which will eventually lead to large-scale deployment of CCS.

References

Bui, M., Adjiman, C. S., Bardow, A., Anthony, E. J., Boston, A., Brown, S., Fennell, P. S., Fuss, S., Galindo, A., Hackett, L. A., Hallett, J. P., Herzog, H. J., Jackson, G., Kemper, J., Krevor, S., Maitland, G. C., Matuszewski, M., Metcalfe, I. S., Petit, C., Puxty, G., Reimer, J., Reiner, D. M., Rubin, E. S., Scott, S. A., Shah, N., Smit, B., Trusler, J. P. M., Webley, P., Wilcox, J. & Mac Dowell, N. (2018). Carbon capture and storage (CCS): the way forward. *Energy & Environmental Science*, 11 (5), 1062-1176.

Global CCS Institute (2019). *CCS Facilities Database*. https://www.globalccsinstitute.com/resources/ccs-database-public/ [accessed 4/2/2019].

IPCC (2018). Global warming of 1.5°C. Switzerland: *Intergovernmental Panel on Climate Change (IPCC)*, https://www.ipcc.ch/sr15/.

570004 Corrosion Analysis of RTI's Non-Aqueous Solvent for Carbon Capture

RTI International is developing a novel non-aqueous solvent (eCO_2Sol) based process for post-combustion CO_2 capture from coal-fired power plant flue gas. The process has the potential of substantially reducing the thermal energy required for solvent regeneration compared to the NETL baseline Case 12. RTI has already shown the reduction of thermal energy using NAS in a Bench-scale Gas Absorption System (BsGAS) at RTI with simulated flue gas and a pilot-scale unit at SINTEF's Tiller facility in Trondheim, Norway with coal-derived flue gas. Further NAS testing at the National Carbon Capture Center (NCCC) was used to investigate solvent degradation, corrosion, and emissions using the Slipstream Solvent Test Unit (SSTU) with long-term exposure to coal-derived flue gas from Southern Company's Plant Gaston in Wilsonville, Alabama to reduce the scale-up risks for this technology. Testing was conducted during May-August 2018, with 580 hours of time on stream with coal-derived flue gas from the power plant. The post-combustion test facilities at NCCC include the 0.5 MWe Pilot-Scale Solvent Test Unit (PSTU), the 50 KW Slipstream Solvent Test Unit (SSTU), and other customer supplied test units.

The capital cost of CO_2 capture plants is one of the largest drivers in the cost of CO_2 captured. As a result, efforts to reduce the capital costs are one of the most effective ways to bring down the cost of a technology. Most capture systems are currently built with stainless steel due to the corrosive nature of MEA and other aqueous solvents, but cheaper alternatives including carbon steel or plastic liners may be suitable for solvents like eCO_2Sol. This paper will focus on the corrosion analysis of the eCO_2Sol completed during testing at NCCC.

Corrosion coupons were inserted in the piping downstream of the absorber and reboiler that each included carbon steel and stainless steel as well as polypropylene at the absorber outlet. The coupons were removed from the system at the end of the testing and were analyzed for weight change and any surface defects. It was observed that the eCO_2Sol (0.015 mm/yr) exhibited extremely lower corrosion rates, ~100x less than aqueous solvents (MEA – 50 mm/yr, Gunasekaran et al., 2013) for carbon steel due to the lower water content of the solvent. The corrosion rate can be correlated with the conductivity of the solvent. The corrosion rate for stainless steel was about one order of magnitude lower than for the carbon steel. Comparisons with other values reported in literature showed power law correlations fit well between the conductivity of solvents and the corrosion rate.

Samples of the solvent were routinely taken at various points in the system during steady state operation to analyze the system operation and detect any changes in the solvent. Samples were analyzed for amine, water, and CO_2 content from both the absorber and desorber side of the system. In addition, the concentrations of metals in the samples were analyzed using inductively coupled plasma mass spectrometry (ICPMS) for metal content. Chromium (Cr), iron (Fe), nickel (Ni), manganese (Mn), and cobalt (Co) metals are present in the SSTU materials of construction and are not present in the flue gas in significant quantities. The presence of these metal ions shows that the occurrence of corrosion in the process equipment. Fe, Cr, and Ni ions were significantly lower than MEA in the tests. The major metal components present in MEA are two orders of magnitude higher than the metals measured in eCO_2Sol. For instance, a maximum of ~100 ppm Fe was seen in the MEA campaign, whereas ~1 ppm Fe was observed in the eCO_2Sol campaign.

Fischer et al. (2017) suggested that a linear correlation exists between the rate of corrosion and the rate of formate ion production. The corrosion rate increases with an increase in formate ion concentration. The measured corrosion rates of eCO_2Sol and others from literature were compared and a power law correlation was fit to the data of multiple solvents.

Long-term corrosion studies are also underway for the eCO_2Sol and will be compared to the data collected at NCCC. The low corrosion rate of the eCO_2Sol indicates that CO_2 capture plants could use low-cost materials of construction, lowering the capital cost significantly and reducing the cost of CO_2 capture. A technoeconomic analysis will also be included to show the reduction in cost of capture from the use of cheaper materials.

574691 Catalytic Oxygen Removal from Pressurized Oxy-Combustion Flue Gases Using a Reverse Flow Fixed-Bed Reactor

Pressurized oxy-combustion (OC) is regarded as an energy-efficient alternative to atmospheric OC. To utilize the captured CO_2 from pressurized OC for enhanced oil recovery or store it in geological formations, the CO_2 stream must meet the required purity specifications. One major impurity in OC flue gases is residual O_2 (e.g., 3 vol%). The O_2 limit is typically required to be below 100 ppmv to lessen corrosion effects on pipelines, avoid reactions with oil and overheating injection points in oil wells, and mitigate the oxidation of caprock species in sequestration applications.

In this study, a catalytic approach was investigated to remove O_2 impurity in CO_2 streams from pressurized OC systems. The flue gas entered a laboratory high-pressure reverse flow fixed-bed (RFFB) reactor, where O_2 residue was removed through a catalytic reduction reaction with injected natural gas (CH_4). With the reverse-flow reactor operation, an autothermal operation (wherein the heat generated by the CH_4 oxidation reaction was sufficient to maintain essentially complete conversion) was achieved without preheating the feed gas. Non-precious metal catalysts synthesized using wet chemistry methods were investigated in the experiments.

This presentation will first provide an overview of the OC technology and RFFB reactor process for catalytic O_2 removal. Then, the focus will be on the test results of O_2 removal using the RFFB reactor system with a simulated 15 bar OC flue gas at different temperature, O_2/CH_4, and space velocity conditions. The performances of the selected catalysts and the effects of the RFFB operating conditions will be compared and discussed to provide insights to the process development.

576281 Optimal Techno-Economic Design and Operation of a Fixed Bed System Using Micro-Encapsulated Solvent for CO_2 Capture

A number of high viscosity and phase change solvents have strong potential to reduce the penalty for CO_2 capture due to their high working capacity and low regeneration energy. However, due to hydrodynamic limitations in conventional towers and issues in transporting the solvent between the absorber and regenerator, these solvents can only be used under low working capacity thus greatly limiting their potential. Furthermore, a number of potential CO_2 capture solvents are also high corrosive. These solvents need to be highly diluted, which leads to considerable reduction in their working capacities. Microencapsulation of solvents is a new technology where the solvent is encapsulated within a polymer membrane. Vericella et al., (2015) have reported encapsulation of carbonate solutions encapsulated in a silicone membrane Micro-encapsulated carbon sorbents (MECS) can not only lead to realization of the high potential of a large family of solvents, but they also offer large surface area that can reduce the mass transfer and heat transfer resistances.

To realize the full potential of this novel technology, optimal contactor technology should be designed and selected with due consideration of thermodynamics, mass transfer, heat transfer, reaction kinetics, and hydrodynamics. Raksajati et al. (2017), performed a high level assessment of a few reactor configurations for microencapsulated MEA as the solvent. Hornbostel et al. (2019) developed a simplified carbonate capsule model to estimate the sizes and energy penalties of both fixed bed and fluidized bed absorbers. However, the authors assumed isothermal conditions and made simplifying assumptions about the regeneration process such as the assumption of a constant heat of reaction and fixing the regeneration time to 10 minutes, etc. The authors also did not account for water flux across the capsule shell, which can significantly affect the performance of the MECS. Detailed rate-based modeling of fixed bed reactors for MECS and optimal techno-economic design of these systems are currently not available in the existing literature.

In this work, a comprehensive model of the MECS system is first developed with rigorous chemistry models and thermodynamic models where the electrolyte non-random two-liquid (NRTL) model is used for the vapor-liquid equilibrium. Rate parameters are optimally estimated using the dynamic experimental data obtained under varying temperature and pressure conditions. A first-principles, non-isothermal, dynamic model of a fixed bed contactor containing microencapsulated sodium carbonate as the solvent is developed. For regeneration, both indirect heating by using an embedded heat exchanger and direct heating through steam injection are evaluated. Finally, the cyclic temperature swing absorption (TSA) process is modeled. Due to the strong trade-off between the capital and operating costs of this system and due to large number of decision variables, a mathematical programming problem is solved to obtain the optimal economic performance. However, since the fixed bed process is a cyclic process, a dynamic optimization problem needs to be solved.

Dynamic optimization of the fixed bed system is performed by using a direct transcription method for minimizing the equivalent annual operating cost (EAOC). Two different reactor materials-concrete and carbon steel- are evaluated. Due to the high heat content of the solvent and the capsule present during the regeneration stage, it is absolutely critical to recover the heat. Various configurations and working fluids are considered in the nonlinear programming problem to obtain the optimal design of the fixed bed system as well as its operating conditions.

References

1. Vericella, J. J. et al. "Encapsulated liquid sorbents for carbon dioxide capture", Nature Communications, 6:6124, 1-7, 2015.
2. Anggit Raksajati, Minh T. Ho, and Dianne E. Wiley, "Techno-economic Evaluation of CO2 Capture from Flue Gases Using Encapsulated Solvent", Industrial & Engineering Chemistry Research 1604-1620, 56 (6), 2017.

3. K.Hornbostel, D.Nguyen, W.Bourcier, J.Knipe, M.Worthington, S.McCoy, J.Stolaroff, Packed and fluidized bed absorber modeling for carbon capture with micro-encapsulated sodium carbonate solution, Applied Energy, 1192-1204, 2019

576359 Techno-Economic Analysis of Post-Combustion Carbon Capture with a 3rd Generation Water-Lean Solvent

Aqueous amines offer one of the most competitive, mature and scalable technology platforms for post-combustion carbon capture. However, the high costs associated with both energy penalties and the necessary capital investments limit its economic viability. The significant energy required to regenerate aqueous amines is due to the high water content. The reason is that the water content increased the condenser duty and this, in turn, increases the reboiler duty. Therefore water-lean solvents have the potential to significantly reduce the reboiler duty. CO_2-Binding Organic Liquids (CO_2BOLs) comprise a class of water-lean solvents being developed at PNNL. These solvents have shown considerable promise towards reducing the carbon capture energy consumption by reducing the solvent water content. The energetic benefits of CO_2BOLs can be further realized through advanced process configurations, which have been intensively studied for aqueous amines, but not yet for water-lean solvents. Indeed, due to the physical and thermodynamic property differences between aqueous and water-lean solvents, viable configurations for aqueous amines may not be optimal for water-lean solvents. Further, expensive construction materials, such as stainless steels, which are required to prevent corrosion for many aqueous amines, may not be required for CO_2BOLs due to their weaker corrosiveness and unique wetting properties. This opens up the option of replacing stainless steel packing with plastics, which would result in a considerable saving of capital cost. With the above motivations, equivalent work calculations and techno-economic analysis were conducted to quantitatively evaluate different process configurations for post-combustion carbon capture using a 3rd generation CO_2BOL. In this presentation, we identify the advantages and limitations of this CO_2BOLs solvent, and the opportunities to reduce the cost of carbon capture.

564642 Kinetic Parameter Estimation from Experimental Laboratory Data for Modeling Reactions Occurring in a Groundwater Permeable Reactive Barrier

Permeable reactive barrier (PRB) technology is used to treat contaminated groundwater in situ in a groundwater aquifer. Groundwater contaminants passing through the PRB are treated by reactions occurring within the unit or immediately downgradient of it. A biowall is a type of PRB that contains bark mulch or other agricultural waste material, which serves as an electron donor to promote biodegradation processes such as reductive dechlorination for trichloroethylene degradation. The biodegradation of bark mulch, and the anaerobic conditions resulting from said biodegradation, can lead to the presence of organic carbon and sulfide, the latter arising from biological reduction of sulfate, in the groundwater downgradient of a PRB. The presence of these dissolved species can be deleterious to the environment if the treated groundwater discharges to a nearby surface water receptor.

The research work presented here focuses on the effluent concentrations of dissolved species from a laboratory column operated in a manner to mimic a groundwater biowall. The data from this laboratory column is used to estimate the kinetic parameters of a time-dependent one-dimensional fate and transport model of the biowall and downgradient groundwater. The model was developed by applying mass balance equations over a control volume of the aquifer between a biowall and a surface water body receptor. MATLAB was used to solve the family of partial differential equations for a timescale extending to approximately 4 years, using appropriate initial / boundary condition values. The species incorporated into the model include glucose (surrogate for carbon source upon degradation of solid carbonaceous material), lactic acid (surrogate for fatty acids), the electron donor sulfate and the product of sulfate reduction, namely sulfide. The agreement between model results and the laboratory column data is presented.

568766 Pyrolytic Remediation of Petroleum-Contaminated Soil in a Pilot-Scale Rotary Kiln Reactor: Reaction Mechanisms, Soil Transformations and Treatment Intensity Tradeoffs

Scope and Significance

The vast majority (about 98%) of all oil spills occur on land. Terrestrial oil spills from pipelines and fixed facilities are usually of moderate size (100-1,000 m^3) but release 10-25 million gallons of petroleum products in the environment every year. Because biodegradation is extremely slow in the anaerobic zone of soils, land spills pose long term threats to groundwater quality. Current remediation methods for petroleum-contaminated soils are either relatively slow or have unintended consequences in the form of soil damage and high-energy usage. Furthermore, some treatment processes such as aerobic bioremediation can activate toxic hydrocarbons and transform them to more noxious byproducts such as PAH derivatives, raising the possibility of meeting regulatory cleanup goals without achieving full soil detoxification. Clearly, there is a pressing need for more efficient and *sustainable remediation*of oil-contaminated soils.

Pyrolysis is receiving increasing attention as an on-site remediation approach because of its potential to rapidly and reliably remove total petroleum hydrocarbons (TPH) with lower energy requirements and better post-treatment soil fertility than other *ex situ*thermal remediation approaches. However, soil fertility restoration is not consistently achieved by thermal treatment methods, underscoring the need for mechanistic insight into how treatment conditions affect soil detoxification and ability to support plant growth.

Our initial studies on pyrolytic treatment focused on small-scale batch tests to assess TPH removal and soil fertility restoration, as well as to understand the fundamental reaction mechanisms behind these outcomes. After desorption of light hydrocarbons between 150-350°C, pyrolysis reactions dominate in the range of 400-500°C, releasing gaseous products and forming a solid char. Plant growth studies showed favorable soil fertility metrics in soils that were pyrolyzed at 420°C for 3 h, suggesting the potential for improved ecosystem restoration following remediation compared to traditional thermal technologies. While these laboratory-scale experiments were instrumental in demonstrating the proof of concept for pyrolysis and in informing process design, pilot-scale studies are needed to delineate the merits and limitations of pursuing multiple treatment objectives (e.g., TPH removal compliance, soil detoxification and fertility restoration) as a function of pyrolysis conditions (e.g., temperature, soil residence time and soil oil content). Ensuring reliable remediation also requires special attention to residual PAHs in the treated soil or its leachate, since these priority pollutants are not only present in crude oil, but *could also be formed during thermal treatment*.

Pyrolysis Experiments in a Continuous Reactor, Soil Detoxification and Treatment Intensity Tradeoffs

We report here results from the first pilot-scale study of pyrolytic soil treatment performed in a continuous rotary kiln reactor to determine how the processing conditions affect not only TPH and PAH removal efficiency, but also detoxification efficacy and the ability of the treated soil to support plant growth.

Two contaminated soils with different oil contents were pyrolyzed in a continuously-fed rotary kiln reactor. A total of 15 pyrolysis runs were carried out with temperatures ranging between 370 and 470°C, 15-60 min residence times and solid flow rates between 28 and 7 lb/h. The 7-inch diameter rotary kiln had four 24-inch electrically-heated zones that were independently controlled to (a) raise the solids to the desired temperature in the first zone, and (b) maintain the reactor temperature constant (within 5-10°C) in zones 2, 3 and 4. For the purpose of this study, the solids residence time was the time the solids were exposed to the pyrolysis temperature in zones 2-4. Pure nitrogen in countercurrent flow was used to "sweep" the desorbing hydrocarbons and pyrolysis products.

TPH and PAH concentrations and key agronomic parameters of the treated soils were measured for each set of operating conditions. To quantify the effectiveness of pyrolysis in reducing potential toxicity to humans (e.g., through inhalation of contaminated dust or ingestion of impacted groundwater), the cytotoxicity of soil extracts was tested by the MTT colorimetric assay with a human bronchial epithelial cell line (BEAS-2B). Plant growth studies were also conducted to characterize potential tradeoffs between pyrolytic treatment intensity, remediation efficiency and soil fertility restoration.

Treatment at 420°C with only 15-min residence time resulted in high removal efficiencies for both total petroleum hydrocarbons (TPH) (99.9%) and polycyclic aromatic hydrocarbons (PAHs) (94.5%), and restored fertility to clean soil levels (i.e., *Lactuca sativa* biomass dry weight yield after 21 days increased from 3.0±0.3 mg for contaminated soil to 8.8±1.1 mg for treated soil, which is almost equal to the 9.0±0.7 mg yield for uncontaminated soil). Viability assays with a human bronchial epithelial cell line showed that pyrolytic treatment effectively achieved detoxification of contaminated soils.

As expected, TPH and PAH removal efficiencies increased with increasing treatment intensity (i.e., higher temperatures and longer residence times). However, higher treatment intensities decreased soil fertility. While pyrolytic treatment at 420°C for 30 min restored soil fertility to 98% of the clean soil level, treatment at 470°C for 15- or 30-min reduced soil fertility to 51% and 39% of the clean soil level, which was only marginally higher that the fertility of the contaminated soil (33% of clean soil fertility). This is a strong indication that that there exists an optimum system-specific treatment intensity for fertility restoration.

Overall, this study highlights tradeoffs between pyrolytic treatment intensity, hydrocarbon removal efficiency and fertility restoration while informing the design, optimization, and operation of large-scale pyrolytic systems to efficiently remediate crude-oil contaminated soils.

577849 Nitrate Reduction By Catalytic Hydrogenation: Controlling Hydrogen Delivery with Nano-Enabled Polymeric Hollow Fibers

As a result of anthropogenic nitrogen fertilizer inputs, nitrate concentrations in surface and ground waters have dramatically increased during the last century. To minimize the adverse health impacts of nitrate, the World Health Organization has set a recommended maximum contaminant level (MCL) of 50 mg/L NO_3^- (~11 mg/L as N-NO_3^-) in drinking water. Catalytic hydrogenation emerges as abiotic treatment with potential application as point-of-use treatment to reduce nitrate to innocuous nitrogen. Past research of our group identified bimetallic nanocatalysts of In-Pd as promising catalyst due to its high catalytic activity and close to 100% selectivity towards nitrogen gas. However hydrogen delivery remains as technology barrier for implementation. This work explores the nano-enabling of polymeric hollow fibers with In-Pd catalysts and their use in continuous flow reactors for nitrate remediation. The delivery control of hydrogen through hollow fibers overcome the risks associated to direct hydrogen bubbling reactors and diminishes physical-foot print needs. Different coating and nanoparticle attachment technologies are evaluated under long-term performance in terms of (i) nitrate reduction catalytic activity, (ii) selectivity towards nitrogen, and (iii) stability of coatings from leaching.

582684 Time for a Paradigm Shift: Toward an Engineering that Accounts for People and the Planet

Engineering has made tremendous contributions to enhancing human well-being, but it has also contributed to the degradation of ecosystems that are essential for sustaining our well-being. For the last few centuries, engineering has developed with a paradigm that ignores the role of ecosystems and the behavior of people. With this paradigm, engineering aims to dominate nature, and considers ecosystems to be an infinite source and sink. Engineering also usually assumes that improvements in technological efficiency will translate into societal benefits. It does not consider the fact that efficiency may encourage consumption that may nullify the societal benefits of greater efficiency. This talk will argue that the current paradigm of engineering is a barrier to its sustainability. Therefore, for engineering to make positive contributions to sustainable development, its paradigm needs to shift toward explicitly accounting for the role of people and the planet. Rather than trying to dominate nature, engineering needs to learn from it, account for its role, and respect its limits. Rather than assuming full adoption and proper use of technologies, engineering needs to consider how the behavior of individuals and markets will determine the overall impact of the technology. Such a paradigm shift requires expansion of the engineering boundary to include ecological and economic systems. Sustainable engineering brings together industrial ecology, sustainability science, and systems engineering, and is well-suited to meeting this challenge. Recent work is showing how this paradigm shift can result in innovative solutions that can often improve profit, environmental quality, and human well-being. It presents many challenges and opportunities to chemical engineers for transdisciplinary work toward sustainable development.

582685 Community-Engaged Environmental Measurements

The emergence of commodity sensors is changing the way we think about pollution measurements, and it offers a great opportunity to engage citizens in understanding and addressing environmental challenges. We have been partnering with community members and students across Utah's Salt Lake Valley as citizen scientists who work with an interdisciplinary team of researchers to create a distributed online network of approximately 100 low-cost, air-quality sensors. This region periodically experiences some of the worst particulate matter (PM) pollution episodes in the country. Our network provides dynamic neighborhood-scale PM estimates, and it has captured dramatic differences in PM concentration resulting from persistent cold air pools, fireworks, wildfires, and dust events that would not be evident from the regulatory monitors alone. I will highlight how our team addresses community engagement as well as data quality, integration, and visualization. Through outreach efforts to local schools, we also engaged students as citizen scientists through activities that: (1) allow them to build their own light-scattering based PM sensor from building blocks and simple electronics; (2) train them on how to be good sensor hosts; and (3) help them make sense of real-world data. Citizen science poses a number of challenges, and I will describe some of these and strategies to address them.

588861 The Chemical Engineering Profession and Its Infinite Possibilities

Tayo Femi-Fowode is a passionate Chemical Engineer with over 10 years of corporate and leadership experience. He is currently a Business Integration Leader for GE Power's Monitoring and Diagnostic Center in Atlanta, GA. In his role, Tayo drives digital integration of 5000 monitored turbines, generators and other equipment to minimize downtime, optimize plant performance and save operating cost for more than 950 power plants in 75 countries and serving 350 million people worldwide. Before joining GE Power in 2011, he was Project Manager in the asset management division of Rolls-Royce Energy.

A second-generation Nigerian-American, Tayo holds an MBA from Columbia Business School with concentration in Strategy and Finance. He also holds a Master of Science and a Bachelor of Science *(magna cum laude)* degrees in Chemical Engineering from Columbia University and Prairie View A&M University respectively.

588862 Direct contact membrane distillation (DCMD) membrane distillation techniques for desalination

Direct contact membrane distillation (DCMD) is one of the simplest and most widely used membrane distillation techniques for desalination. It is a thermally driven separation process and is sought to achieve high water vapor flux while being thermally efficient. Typically, a hydrophobic porous membrane is used and water vapor from the salt solution is allowed to pass through the pores and condense on the other side of the membrane. Investigation on other structures of membranes is carried out. Membranes include those in the form of an integral composite, in which the structures of PVDF membranes are changed, and a stacked composite, in which different membranes are stacked on top of each other (e.g. a hydrophobic ePTFE membrane over a hydrophilic PVDF membrane, and a hydrophobic PVDF membrane over another hydrophobic PVDF membrane). These membranes were found to yield a higher water vapor flux than the conventional single hydrophobic membrane. Membranes are also characterized using a variety of characterization techniques: Contact angle, liquid entry pressure (LEP), bubble-point pressure, scanning electron microscopy (SEM), and atomic force microscopy (AFM).

588863 My Career Path

My career thus far has been shaped by many factors learned early in life. It started in my hometown of Yazoo City, Mississippi. It was there that I had the job that I consider to be my first engineering experience which was working with my dad at Gainwell Tire Services. While there, I discovered that I had a fondness for hands-on work and gravitated toward working closely with people as part of a team. This was instrumental in not only my early career development but shaped my life as a whole. From there, while in high school, I worked at our local construction company, Parker Builders. During this time, I developed a strong partiality for repairing process equipment. Also, a sense of work ethic was formed as there were several days where I worked on a hot tin roof for hours. After graduation, I headed to college at Louisiana Tech University. While there, I was a part of the football team while pursuing my Chemical Engineering degree. This conceivably impossible task was made possible largely due to the work ethic learned early in life. It also helped further developed my team-membership skills.

Later, I transferred to Southern Arkansas University. It was here that I not only compounded upon my team-membership skills but developed into a team leader as a captain of the football team. While there I took on several internships which included stints in the chemical, aerospace, energy, and mechanical fields. One of those internships was Lanxess Solutions in El Dorado, Arkansas. The skills learned early in life helped me very much in many of the projects given during this time and were good enough to be hired full time as a Shift Leader upon graduating college. In this role, I developed many technical skills and gained an understanding of the operations side of the plant. After some time there, I went to New Orleans, Louisiana to work at Galata Chemicals as a Process Engineer. While at Galata, I was able to further develop technical skills as well as learn the safety aspect that is imperative in a plant setting. After spending a year and a half there, I was offered a Process Engineer II position at Oxea Chemicals, my current role. Thus far I have learned even more about the chemical field and continue to grow professionally with the organizational training offered.

588932 Sustainable Energy Drivers: A Perspective from a Career in the Nuclear Sector

Dr. Harold T. Conner, Jr. has a 52-year career in the nuclear sector, having determined at an early age that he wanted a career in chemical engineering. Along the way, he blazed an impressive path by achieving a lot of firsts: the first African American student in the University of Tennessee-Martin Engineering School, the first cooperative engineering student at the University, and the first African American to receive bachelor's and master's degrees in chemical engineering from the University of Tennessee-Knoxville. He worked for more than 30 years in Oak Ridge, working in almost every facility at the Oak Ridge K-25 Site, a uranium enrichment facility that would be closed down in the mid-1980s and then undergo cleanup. Conner became site manager of the K-25 Site and was then named vice president of Environmental Management and Enrichment Facilities. He would leave Oak Ridge and worked at various DOE complexes across the nation until returning to Oak Ridge to help with the cleanup of the K-25 Site, which is now called the East Tennessee Technology Park (ETTP). Working for UCOR, an AECOM led partnership with Jacobs, his expertise has significantly helped the company work toward achieving Vision 2020, DOE's plan to complete ETTP cleanup by 2020. The site is being transformed into a private sector industrial park, national park, and conservation area. Conner is now semi-retired but serves as a consultant for UCOR.

588933 A Chemical Engineer in the Energy Journey

Dr. Quinta Nwanosike Warren, PE, PMP is a chemical engineer and an energy professional. She is an entrepreneur and has worked in industry, academia, and government. She holds a PhD in Chemical & BioMolecular Engineering from the Georgia Institute of Technology, and a Bachelors in Chemical Engineering from the Pennsylvania State University. She is a Project Management Professional, and a registered Professional Engineer.

Dr. Warren previously worked for ConocoPhillips for on CO2 capture research, and as a reservoir engineer for heavy oil and tight gas assets. She has worked with the US Department of Energy, learning about and contributing to domestic and international policy around carbon management and power generation. She also worked for the Millennium Challenge Corporation (MCC), a US government agency that works on international development. At MCC, she provided energy expertise and contributed to the development of energy projects.

Dr. Warren owns and runs a consulting company called Energy Research Consulting (www.EngrRC.com) which provides energy and water solutions and strategies. The company's projects include creating the STEM-focused Dakar American University of Science and Technology (www.DAUST.org) in Senegal, and designing a water treatment system for a mushroom farming initiative in Nigeria. Dr. Warren recently published a textbook titled "A Practical Guide to Oil & Gas Resource Characterization for Geologists and Reservoir Engineers."

574929 Understanding Formation of Nitrogen Heterocycles during Catalytic Hydrothermal Liquefaction

Hydrothermal liquefaction (HTL) is a promising technology for conversion of wet wastes into fuel precursor oils. In this work, we studied catalytic-HTL of food waste using a series of acid and base catalysts. While base catalysts are generally more effective at increasing oil yield and decreasing its oxygen content, acidic catalysts are especially effective at removing nitrogen content. Removing nitrogen from the bio-oil is an important challenge for the industrial application of the process. Heterocyclic nitrogen compounds are especially problematic, as they are difficult to remove using hydrodenitrogenation. Moreover, in the context of fixing nitrogen, HTL has potential for re-utilization of fixed nitrogen, reducing the need for costly and energy intensive nitrogen reduction processes. Unfortunately, the complex reaction mixture resulting from catalytic HTL of food waste and other realistic feed streams is too complicated to understand mechanisms. Computational studies can overcome this challenge, providing a guide for targeted experimental work and forming the basis of science-based reactor models. To further simplify, we studied homogeneous formation of heterocyclic under HTL conditions. In this work, a reaction network that involves interactions between reducing sugars, amino acids, and dienes is chosen as a basis for the formation of undesirable nitrogen heterocycles. Specific attention is given to aza-Diels-Alder cycloadditions and Maillard reactions, as both of these are potential pathways to formation of organo-nitrogen compounds. Quantum chemical calculations are carried in Gaussian 16 on reactants, intermediates, and products using the CBS-QB3 method. Conformers of each structure are identified by scanning the potential energy surface. Transition states of each pathway are determined and validated using intrinsic reaction coordinate calculations. These results are used in CanTherm to calculate thermochemistry and rate coefficients while treating hindered rotors. These results suggest new mechanisms for formation of undesirable nitrogen heterocycles that can guide selection of HTL processes and conditions to minimize their formation.

574908 Biochar and Activated Carbons Derived from Food Waste: Properties and Applications

Biochar and activated carbons derived from food waste: properties and applications

Lei Yu [a], David P. Gamliel [b], Katherine Saltzgiver [a], Brianna Markunas [a] and Julia A. Valla [a]

[a] Department of Chemical and Biomolecular Engineering, University of Connecticut, Storrs 191 Auditorium Road, Unit 3222, Storrs, CT 06269-4602, United States

[b] Physical Sciences Incorporated, 20 New England Business Center Rd. Andover, MA, 01810, United States

Food waste can be considered as a useful, renewable resource for energy and/or materials production. In this study, biochars were prepared via pyrolysis of food waste and then used as precursors for production of activated carbons via physical and chemical activation. Pyrolysis and activation conditions were varied to optimize the properties of biochar and activated carbons, respectively. Activated carbons were characterized by N_2 sorption-desorption, scanning electron microscope (SEM), energy-dispersive X-ray spectroscopy (EDX), elemental analysis, inductively coupled plasma atomic emission spectroscopy (ICP-OES), X-ray powder diffraction (XRD), ^{13}C nuclear magnetic resonance (NMR), Fourier-transform infrared spectroscopy (FTIR) and Raman spectroscopy. Adsorption experiments of aromatic hydrocarbons were conducted to evaluate the feasibility of using the biochars and activated carbons derived from food waste for the adsorption of aromatic hydrocarbons (benzene and naphthalene) from water. The activated carbon with the highest adsorption capacity was prepared via steam activation at 950 °C for 1 h. The BET surface area, micropore volume and total pore volume of the activated carbon was 745 m^2/g, 0.185 cm^3/g and 0.594 cm^3/g, respectively. The benzene and naphthalene sorption capacity was 460 mg/g and 150 mg/g, respectively. The physicochemical properties and adsorption capacities of the activated carbons prepared from food waste were also compared to the corresponding properties and capacities of the activated carbons derived from miscanthus, as well as commercial activated carbons. Our study revealed that activated carbons prepared by physical activation of biochars derived from food waste resources are very promising candidates for water purification with capacities for aromatic hydrocarbons similar to the ones observed using commercial activated carbons. The results are presented in the Figure 1 below.

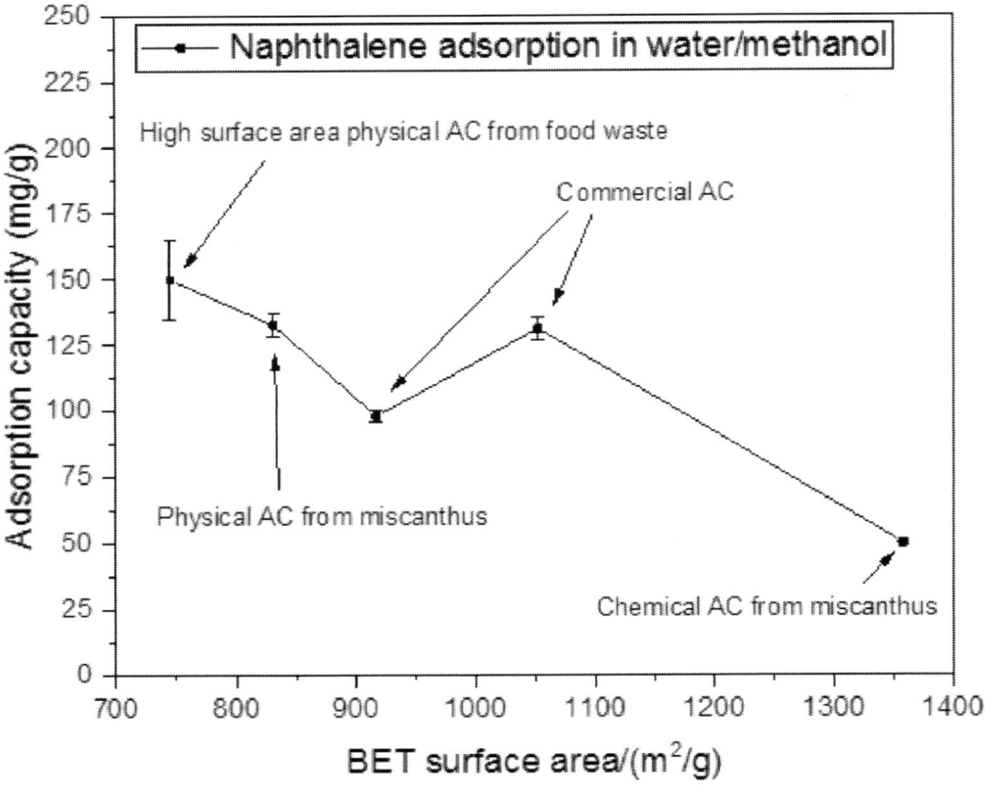

Figure 1: Adsorption capacity of naphthalene in activated carbons derived from various resources

566272 A Framework for Assessing Solid Waste Energy Recovery Systems for Small-Scale on-Site Applications

The generation of solid waste, which includes wood, paper, plastics, organic wastes (e.g., food wastes) and other debris can be problematic for isolated, small-scale communities. While several known technologies, such as pyrolysis or gasification, can be employed for on-site applications, there is no waste-to-energy technology that can accept all types of solid waste. At present, small-scale communities must decide which technology to use (if any) based on their unique site and situation. The purpose of this study is to develop a framework for assessing available technologies for the destruction of solid wastes with the simultaneous production of energy for small-scale operations (i.e., for populations of 100 to 1,000 people). To assess the feasibility of several existing technologies, we first surveyed several available technologies (by contacting company representatives and/or conducting on-line research) to characterize the technology and obtain the following performance metrics: rate of waste treated, type of waste treated, pre-processing requirements, transportability, number of operators, physical footprint requirements, support requirements (fuel, electricity), permitting and military testing & evaluation, environmental control limitations, safety, operator skills and training, specialty equipment, operational and down time, production of heat & electricity, production of residuals, noise & odor, and air emissions. Second, we developed a framework for assessing characteristics of each technology considering the typical solid waste generation rates (based on characteristics from U.S. military forward operating bases) for population sizes ranging from 100 to 1,000 people. Within the framework, individual metrics where binned into three overarching categories: "waste inputs", "support requirements", and "products / residuals". Each technology was assigned a baseline score considering each individual metric and the binned categories for comparison. Depending on the need, the framework further allowed for weighted analysis of each binned category. The developed framework was employed to assess several pyrolysis or gasification technologies for the U.S. military, but could be modified and employed for numerous small-scale on-site waste-to-energy applications.

576722 Microplastic Environmental Effect on the Aquatic Ecosystem of the Thermaic Gulf

The aim of this study is to quantify the abundance of floating microplastics in sediments and surface water of Thermaic Gulf, Greece and to identify the spatial distribution of the microplastic particles and their polymeric characterization. Moreover, the frequency of occurrence of ingested microplastics in marine biota also was examined. Microplastics are now found worldwide in all aquatic compartments such as surface water, water column, and sediments (Ivar do Sul and Costa, 2014). Recently, it has been estimated that 5 trillion microparticles are floating globally, while up to 890,000 plastic particles/ km^2 were predicted in the Mediterranean surface waters. (Eriksen et al., 2014).

Thermaic Gulf is a gulf constituting the northwest corner of the Aegean Sea. The city of Thessaloniki is at its northeastern tip, and it is bounded by Pieria and Imathia on the west and the Chalkidiki peninsula on the east, with Cape Kassandra at the southeasternmost corner. It is about 100 km long. Water surface samples were realized during the winter (February–March 2018). A total of 25 samples were collected at 500 m, 5 km and 10 km distance from the coast. Surface samples were collected using neuston net with cod end (300 µm mesh size) equipped with a flowmeter and towed on the water surface at 2–3 knots for 20 min. At the end of the samplings, the nets were rinsed thoroughly from the outside to ensure that both plankton and debris were washed into the end of the net and to prevent any contamination by rinsing water. Organic matter attached to the surface of microplastics was removed to allow clear identification of the type of plastics, using a solution of 30% H_2O_2. For sediment sampling, boxcorer grabs were used and to ensure representativeness of the samples, at least five sample replicates (5 m between replicates) were collected. Density separation of microplastics were carried out with a Munich Plastic Sediment Separator. After separation and purification, target microplastics were sorted from the remaining matrix. Large plastics were sorted out directly, while smaller-sized ones needed further observation under a dissection microscope. For plastic particle analysis, samples were observed under a stereo-microscope with a LED light and measured with micrometer ocular lens. Plastic particles were characterized and classified by colour, size and shape. The particles were dried at room temperature and weighted in analytical balance. Additionally, the polymer fingerprint was detected using Fourier Transform Infrared (FT-IR) spectroscopy.

A total of 15 sea bass was sampled from Thermaic Gulf. The fish were collected during the winter of 2018. The total length (TL) and total wet weight (TWW) of fish were recorded. The fish were stored at −20 °C prior microplastic detection. Also, the tissues from sea bass were analyzed.

A total of 1200 microplastics have been isolated from all 25 surface water samples, with a range of 13 to 520 total particles per sample. The minimum concentration was 13752 particles/km^2 and the maximum was 420513 particles/km^2. Regarding the sediments the minimum concentration was 0.02 MPs / g dry sediment and the maximum was 0.1 MPs/g dry sediment. Analysis of the size distribution of microplastics in surface samples showed that the most abundant size class is 1–2.5 mm both in the whole dataset. This finding suggests a potentially higher hazard or zooplankton organisms (Fossi et al., 2012). Plastic fragments were found to be primarily isolated plastic particles. Polyethylene (PE) and Polypropylene (PP) were the most abundant chemical species, in good agreement with the literature where these polymers account for the majority of the plastic particles floating in Mediterranean Sea. The results showed a higher abundance of microplastics at 5-10 km distance from the coast. Microplastics were also detected in in the guts, skin and muscles of sea bass collected at an amount that ranged from 0.08 g^{-1} to 1.1 g^{-1}.

This study reports the results of a monitoring campaign of microplastics in three different media. These data represent the first attempt to present the current state of microplastics in the Thermaic Gulf and to establish the baseline level for a future study. Being a Thermaic Gulf an important port within the European countries and because of its morphological features generate important quantities of waste originating from the South European counties so the evaluation of the presence of plastics is mandatory to apply strategies to reduce the quantity of microplastics in Thermaic Gulf.

REFERENCES

ERIKSEN, M., LEBRETON, L. C. M., CARSON, H. S., THIEL, M., MOORE, C. J., BORERRO, J. C., GALGANI, F., RYAN, P. G. & REISSER, J. 2014. Plastic Pollution in the World's Oceans: More than 5 Trillion Plastic Pieces Weighing over 250,000 Tons Afloat at Sea. *PLOS ONE,* 9**,** e111913.

FOSSI, M. C., PANTI, C., GUERRANTI, C., COPPOLA, D., GIANNETTI, M., MARSILI, L. & MINUTOLI, R. 2012. Are baleen whales exposed to the threat of microplastics? A case study of the Mediterranean fin whale (Balaenoptera physalus). *Marine Pollution Bulletin,* 64**,** 2374-2379.

IVAR DO SUL, J. A. & COSTA, M. F. 2014. The present and future of microplastic pollution in the marine environment. *Environmental Pollution,* 185**,** 352-364.

564081 Study and Optimization of the Supply Chain of the Recollection and Harnessing of Used Cooking Oil in the City of Bogotá

Currently, waste disposal is a major challenge for most mega-populated cities worldwide. Particularly in underdeveloped and developing countries, the lack of regulations and law enforcement, together with a generalized unconscious behavior regarding waste management had generated a large variety of impacts. Among the different urban residues, used cooking oil (UCO) is a highly problematic waste that is generally disposed through the sewage system. As a consequence of this improper disposal, there is blockage of sewage pipes, which in turn generates flooding during rainy seasons, proliferation of vectors (rats, cockroaches, bacteria, etc.), and bad odors while decomposing. In some cases, the waste oil reaches surface or underground waters generating a major ecological impact.

All these problems have been increasingly experienced in the capital city of Colombia, Bogota. The city and its surrounding metropolitan area concentrate nearly 20% of the country's population (~ 10 million), and it is expected to grow at 2.8% annually in the coming years. Then, the increasing population will boost impacts associated to UCOs mishandling. Nowadays the city expends nearly 6 to 8 million dollar per year on corrective sewage maintenance related to UCOs mismanagement.

In order to reduce UCO disposal problem, major cooking oil companies in Colombia have started responsible consumption programs to promote UCOs recollection and valorization. Currently, UCOs are collected to be used as feedstock for soaps and biodiesel production. However, as these are low-added value commodities, sustainability of the recollections and exploitation chain is challenging. Thus, all the constituent steps of the collection, treatment and transformation processes must be optimized to ensure sustainability. Particularly, supply chain optimization is a necessary task to accomplish during the sustainability assessment.

In this work, the analysis and optimization of a UCO recollection program, implemented in the city of Bogota by a major fat and oils, was accomplished. An analysis of available UCOs volumes, source locations, recollections frequencies and routes, and the corresponding impacts (e.g. fuel consumption) was carried out. Subsequently, the optimization of the supply chain was focused on maximizing recollection volumes while minimizing time, costs and the associated impacts (i.e. environmental). Thus, the supply chain was modelled using a Travelling Salesman Problem like formulation with combinatorial Genetic Algorithm as solution algorithm for the route optimization. The formulation is based on a graph where each collection location was represented by a node and the route is formed by changing the paths between them. Restrictions were incorporated to take into account shift times, truck capacities, and recollections times. The optimization problem was formulated and solved in Python.

572109 Economic Feasibility of Evaporation-Based Liquid Digestate Treatment for Biogas Plants

Biogas plants (BGP) are important renewable energy sources and they contribute to sustainable development. In BGP, organic waste is used to produce biogas which is either utilized for on-site electricity generation or is upgraded to biomethane and then distributed to a natural gas network or used as a fuel (compressed biomethane - CBG).

The core technology is the anaerobic digestion unit. Its development has currently achieved maturity. However, the current frontier of biogas technologies research is focused on liquid digestate processing which supports circular economy of biogas industry.

Evaporation-based systems are very suitable for further liquid digestate treatment especially when storage and transportation are associated with significant costs for operators. It reduces liquid digestate volume by thickening and at the same time preserves its fertilizing potential. Further, it can produce clean water suitable for discharge into the environment, i.e. is able to prevent ammonia from entering the outgoing freshwater stream.

On the other hand, such treatment unit has higher investment cost and reasonable payback period is not always achievable. It depends on many parameters such size of the plant, liquid digestate production, transportation and storage cost, waste heat availability, thickening-related parameters, etc.

This contribution provides detailed sensitivity analysis to various parameters in order to show what is important to consider when making decision about an investment into an evaporation-based liquid digestate treatment system. Since BGPs are costly, supporting tools are frequently applied by governments. These include investment support, higher feed-in tariffs for electricity, CHP incentives, etc. The conditions provided by such tools are also considered.

Tools supporting the biogas plant operation and investments are crucial. Reasonability of investment into an evaporation-based system for liquid digestate significantly decrease without them. However it is not the necessary condition. Beside the investment cost, the parameters that play key role are transportation cost, dry matter content in liquid digestate and cost of chemicals. Less important parameters, but still significant, are application cost of the liquid digestate, energy consumption of an evaporation unit, dry matter content in thickened concentrate and how much waste heat is available. For a specific combination of values of these parameters, the evaporation-based system for liquid digestate treatment is feasible.

Strategies for improvement of anaerobic digestion from agricultural waste: a critical review

Safenaz Shaaban, Doaa Bassuney, Mennat Allah Mohamed, and Mahmoud Nasr*

Sanitary Engineering Department, Faculty of Engineering, Alexandria University, 21544, Alexandria, Egypt

*Corresponding author: Dr. Mahmoud Nasr

Tele. (+20) 100 63 90 400

Email: mahmmoudsaid@gmail.com; mahmoud-nasr@alexu.edu.eg;

https://orcid.org/0000-0001-5115-135X

Abstract

Agricultural wastes, also known as agro-wastes, refer to the residual products discarded from various farming practices. Agro-wastes consist of multiple organic and non-biodegradable substances, and they include large quantities of crop residues such as rice straw, sugarcane bagasse, wheat shell, soybean hulls, and corn stalk. Agricultural wastes also comprise a wide range of animal residues such as livestock feces and urine, poultry litter and excreta, and carcasses (or parts) of animals, as well as, agro-industrial byproducts such as sugarcane bagasse ash (a byproduct of sugar factories), cassava bagasse, soy hulls, and sugar-beet pulp. Moreover, agricultural wastes contain nitrogen and phosphorus species released from crop fertilization, as well as, toxic elements due to the utilization of pesticides (herbicides and insecticides). Furthermore, agricultural wastes include lignocellulosic biomass such as wood and straw residues, which is considered a renewable resource for the 2nd generation of biofuels production. The fractions of cellulose, hemicellulose, and lignin vary among agricultural wastes (e.g., corn stover, rice husk, wheat straw, and sweet sorghum stalk). Based on the nature of farming systems, agro-wastes can be collected in the forms of solid, liquid, or slurry. The annual quantity of agricultural wastes has been estimated as 998 million tonnes, in which the organic fraction can reach up to 80% of the total solid wastes.

Anaerobic digestion process

Anaerobic digestion has found successful applications as an environmentally friendly approach for the conversion of agro-wastes into valuable products and renewable fuels (Shaaban and Nasr, 2020). The input agro-wastes (i.e., as a substrate) include wet organic substances, animal slurries, biosolids, and livestock manure (pig and cattle). The anaerobic bioconversion process reduces the volume and mass of agro-wastes, and the residual matter after digestion can be used as a soil amendment and as secondary fertilizer compounds (Mthethwa et al., 2019). In anaerobic digestion, biodegradable materials are stabilized and broken down by the action of microbial organisms in an oxygen-deprived environment. The microbial community includes fermentative bacteria, acetogens, and methanogens. The biological degradation reactions are complex, consecutive, and parallel, as they undergo multiple steps of hydrolysis, acidogenesis, acetogenesis, and methanogenesis. During these processes, the waste feedstock is decomposed into simple chemical components such as sugars, which are converted to organic acids. Finally, the organic acids are converted to biogas, *viz*., about 55–75% CH_4, CO_2, and traces of other gases. The percentage of CH_4 varies according to the operating temperature and the amount of organic fraction in the feed substrate. These steps can be summarized as follows:

In the hydrolysis step, hydrolytic bacteria convert the complex organic compounds (e.g., proteins, lipids, and carbohydrates) to monomers (low-molecular-weight matters) including amino acids, long chain fatty acids, and simple sugars.

In the acidogenesis phase, acid bacteria stimulate the fermentation process, producing intermediary products such as volatile fatty acids (VFAs), alcohols, hydrogen, and carbon dioxide (Mthethwa et al., 2018).

In the acetogenesis process, acid-forming bacteria or acetogens utilize VFAs to obtain acetic acid, carbon dioxide, and hydrogen.

In the methanogenesis phase, methanogenic bacteria transform the intermediary products to CH_4 gas via two pathways:

Pathway-I: Methanogenic bacteria, converting CH_3COOH to CH_4 and CO_2.

Pathway-II: Methanogenic bacteria, converting CO_2 and H_2 to CH_4 and H_2O.

Anaerobic digestion is sustainable management of agricultural waste as it converts the organic matter into renewable energy that can replace conventional fossil fuels. Anaerobic digestion has the advantage of producing biogas and fertilizer, leading to minimizing the net energy input of the system. Moreover, the amount of waste sludge yielded from the anaerobic digestion system is lower than that of the aerobic treatment mechanisms. However, anaerobic digestion requires a high level of investment for a commercial or large scale application. Moreover, the environmental conditions should be strictly optimized to avoid the accumulation of volatile fatty acids and the inhibition of bacterial growth. Additionally, the fluctuation of input variables (e.g., flowrate, organic load, and chemical composition) provides a negative impact on the efficiency, control, and stability of the process. Furthermore, a lack of knowledge about the activities of living organisms and the phenomena occurring in the bioprocess offers serious challenges for the wide implementation of anaerobic digestion (Kumari et al., 2017).

Anaerobic digestion of agricultural wastes is influenced by various factors, which can be summarized as follows:

Carbon to nitrogen (C/N) ratio

A C/N ratio of 20 – 30 has been recognized to provide the optimum condition for most anaerobic digestion applications. The unbalanced C/N ratio results in the accumulation of ammonia, leading to the inhibition of microbial activities. Accordingly, different types of chemical components are mixed with the feedstock wastes to provide the favorable C/N ratio that can improve gas productivity.

Organic loading rate

Organic loading rate (OLR) refers to the mass of volatile solids (VS) or chemical oxygen demand (COD) per cubic meter of reactor's volume per day. OLR is used to estimate the ability of a digester to convert the feedstock waste into biogas and other end products. The high OLR requires large digester's size, leading to an increase in the initial cost of investment.

Medium pH

pH is an important environmental factor in the anaerobic digestion process, as it affects the growth rate of microorganisms and the biological conversion pathways (Nasr et al., 2015). The optimum pH range for the biomethanation processes has been reported as 6.6 – 7.0.

Alkalinity

The accumulation of volatile fatty acids in the digester during the acidogenesis phase can lead to the deactivation of methanogenic bacteria. Hence, alkalinity (e.g., ability to resist acidification via the addition of calcium carbonate) is an essential factor in the anaerobic digestion process as it safeguards the buffer capacity and stability of the digester.

Inoculum (Seeding)

A stable methanogenic community is required for the efficient degradation of organic compounds to produce biogas. Hence, the anaerobic digestion system should be enriched with sufficient quantities of micro-organisms. The acclimatization stage allows the bacterial cells to adapt to the new environmental conditions (Nasr et al., 2015).

Substrate pretreatment

Agricultural wastes may include a heterogeneous complex of biodegradables and non-biodegradables. Accordingly, the input substrate for anaerobic digestion should be subjected to a pretreatment process (e.g., physicochemical, and biological treatments) (Reddy et al., 2017). This stage is used to (a) form high amounts of readily biodegradable substrates, (b) reduce the formation of inhibitory products, (c) minimize operational costs (e.g., energy demands), and (d) avoid the loss of sugars.

Other factors

Anaerobic digestion of agricultural wastes and metabolic conditions for microorganisms' growth are also influenced by several factors including activators (e.g., enzymatic additives), retention time, and temperature regime (mesophilic and thermophilic) (Nasr et al., 2013).

Because agricultural wastes are abundant (available), low cost, and renewable, they can be used as a valuable input for bioenergy generation. Hence, this study presents a comprehensive review on the application of anaerobic digestion for biogas production from agricultural wastes. The review illustrates the optimum operating conditions (e.g., culture temperature, medium pH, nutrients availability, organic load, and digestion time) that maximize biogas productivity and improve organic biodegradability. The reactor's configurations, viz., suspended bed reactor, fixed bed reactor, fluidized bed reactor, two-stage processes, and hybrid reactors employed for the enhancement of digester stability are discussed. The types of agricultural wastes such as lignocellulose biomass, livestock manure, crop residues, poultry litter, and cattle wastes used for biomethane production are illustrated. These objectives are represented regarding recent literature studies. Recommendations for future works are also suggested.

Acknowledgements

The authors would like to acknowledge Nasr Academy for Environment Protection

References

Kumari, S., Nasr, M. & Kumar, S., 2017. Technological advances in biohydrogen production from microalgae. *Algal Biofuels: Recent Advances and Future Prospects,* pp. 347-360.

Mthethwa, N., Nasr, M., Bux, F. & Kumari, S., 2018. Utilization of *Pistia stratiotes* (aquatic weed) for fermentative biohydrogen: Electron-equivalent balance, stoichiometry, and cost estimation. *International Journal of Hydrogen Energy,* 43(17), pp. 8243-8255.

Mthethwa, N.P., Nasr, M., Kiambi, S.L., Bux, F., Kumari, S. 2019. Biohydrogen fermentation from *Pistia stratiotes* (aquatic weed) using mixed and pure bacterial cultures. *International Journal of Hydrogen Energy,* 44(33), pp. 17720-17731.

Nasr, M., Tawfik, A., Ookawara, S. & Suzuki, M., 2013. Biological hydrogen production from starch wastewater using a novel up-flow anaerobic staged reactor. *BioResources,* 8(4), pp. 4951-4968.

Nasr, M., Tawfik, A., Ookawara, S., Suzuki, M., Kumari, S., Bux, F., 2015. Continuous biohydrogen production from starch wastewater via sequential dark-photo fermentation with emphasize on maghemite nanoparticles. *Journal of Industrial and Engineering Chemistry,* Volume 21, pp. 500-506.

Nasr, M., Tawfik, A., Suzuki, M. & Ookawara, S., 2015. Mathematical modeling of bio-hydrogen production from starch wastewater via up-flow anaerobic staged reactor. *Desalination and Water Treatment,* 54(1), pp. 50-58.

Reddy, K., Nasr, M., Kumari, S., Kumar, S., Gupta, S.K., Enitan, A.M., Bux, F. 2017. Biohydrogen production from sugarcane bagasse hydrolysate: effects of pH, S/X, Fe^{2+}, and magnetite nanoparticles. *Environmental Science and Pollution Research,* 24(9), pp. 8790-8804.

Shaaban, S. & Nasr, M., 2020. Toward Three R's Agricultural Waste in MENA: Reduce, Reuse, and Recycle. *In: Negm A., Shareef N. (eds) Waste Management in MENA Regions. Springer Water. Springer, Cham,* pp. 337-353.

570267 Towards a Comprehensive Decision-Support Framework for Sustainable Manufacturing

From the sustainability science point of view, industrial sustainability is a triple-bottom-line based, discretized system state transition problem, which requires continuous sustainability assessment and decision analysis in order to derive best possible decisions for implementation However, put sustainability in practice is not a trivial task and it needs a systems approach with a suitable framework that enables decision makers to integrate sustainability thinking into their business strategies successfully.

To provide a fundamental definition for sustainability, we defined the sustainability as vector function [1]. Using this definition, the sustainability status of a system and its transition toward sustainability can be measured, monitored, and interpreted. As sustainability is a continuous process, we formulated the problem as a multi stage decision problem in [2]. To achieve the sustainability goals, we formulated the sustainability problem as a multi-objective control problem and proposed a decision support system in [3] with focusing more on tactical layer.

In this paper, we propose a new comprehensive decision support framework for manufacturing sustainability enhancement that is an extension to our recent work in [3]. This two-layer (strategic and tactical) framework is designed based on the sustainability fundamentals and control theory. The two layer that are executed at different time scales are integrated to implement sustainability development strategies successfully. A model predictive control (MPC) strategy is used at the strategic control layer to find and update optimal sustainability strategic plans continuously (e.g. at each year) for archiving short-to-long term sustainability goals with considering external information changes (e.g. new regulations, supplies and demands) provided by external monitoring component and internal information provided by sustainability assessor at the tactical layer. The tactical layer, on the other hand, is responsible for implementing sustainability strategic plans successfully by minimizing the projects' implementation costs and time. The components of each layer, implementation mechanism, and methods are presented. The proposed framework provides a structured step-by-step guide to industry for achieving economic, environmental, and social sustainability goals with considering uncertainties related to external and internal parameters that affect the business. A case study on biodiesel manufacturing is illustrated to demonstrate methodological efficacy.

References:

[1]Moradi-Aliabadi, M. and Y. Huang, "Vector-Based Sustainability Analytics: A Methodological Study on System Transition toward Sustainability," Industrial and Engineering Chemistry Research, 55(12), 3239-3252, 2016.

[2] Moradi-Aliabadi, M. and Y. Huang, "Multistage Optimization for Chemical Process Sustainability Enhancement under Uncertainty," ACS Sustainable Chemistry and Engineering, 4(11), 6133-6143, 2016.

[3] Moradi-Aliabadi, M. and Y. Huang, "Decision Support for Enhancement of Manufacturing Sustainability: A Hierarchical Control Approach," ACS Sustainable Chem. Eng., 6(4), 4809–4820, 2018.

---

* All correspondence should be addressed to Prof. Yinlun Huang (Phone: 313-577-3771; E-mail: yhuang@wayne.edu).

577723 Evaluating Dynamic Resilience of Intermittent and Uncontrollable Techno-Ecological Systems

Engineering sciences have made a constant effort towards the goal of sustainable development. In particular, chemical engineering has developed multiple techniques like heat and mass integration, waste reduction and eco-industrial parks aimed at reducing the ecological impact. Using the framework of Techno-Ecological Synergy (TES), Bakshi and co-workers [1, 2, 3] have incorporated ecological systems in process and supply chain design space to obtain novel ecologically and economically win-win solution. These novel systems aim to improve sustainability by matching the ecological goods or service utilization rate with its regeneration rate.

Unlike technological systems, the dynamics of ecosystems are not "controllable". In order to ensure sustainability of TES system, the design and operation should be "viable" over long periods of time i.e. the design should be resilient. The term "resilience" has been defined across multiple disciplines with varying definition. Resilience can be used to describe a system's resistance capacity, coping capacity, adaptive preference or its transformative capacity to avoid irreversible changes while sustaining large shocks [4].

In this work, we employ the mathematical framework of viability theory [5] to analyze the viable operational domain of TES systems. We employ various dynamic resilience metrics to analyze and design control strategy for ensuring a sustainable operation of the novel TES systems. We compare the resilience of TES systems with that of conventional technocentric systems. This study helps us integrate the intermittent and uncontrollable dynamics of ecosystem with the control of technological system's paradigm and design "resilient" sustainable systems.

References

[1] Bhavik R. Bakshi. Toward Sustainable Chemical Engineering: The Role of Process Systems Engineering. *Annual Reviews in Chemical and Biomolecular Engineering*, Accepted, 2019.

[2] Utkarsh Shah and Bhavik R. Bakshi. Accounting for nature's intermittency and growth while mitigating no 2 emissions by technoecological synergistic design-application to a chloralkali process. *Journal of Advanced Manufacturing and Processing*, 0(0):e10013, Mar 28, 2019.

[3] Tapajyoti Ghosh, Xinyu Liu, and Bhavik R Bakshi. Including ecosystem services in sustainable process design across multiple spatial scales. In *Computer Aided Chemical Engineering*, volume 44, pages 1837–1842. Elsevier, 2018.

[4] Christophe Béné and Luc Doyen. From resistance to transformation: a generic metric of resilience through viability. *Earth's Future*, 6(7):979–996, 2018.

[5] Jean-Pierre Aubin. *Viability theory*. Springer Science & Business Media, 2009.

574791 A Multi-Layer Process Control Framework for Sustainability: Application to Biomass/Coal Co-Gasification System

Industry, government, and society have begun to shift from economic stand-alone focus to the inclusion of sustainability in the decision-making process. This shift is due to the adverse environmental impact and unsustainable development caused by human activities, including chemical industry releases. As a result, process systems engineering (PSE) approaches have been developed, predominantly for incorporating sustainability into chemical process design and optimization. However, the development of sustainability-oriented control schemes is still scarce, especially when compared to the number of available efforts on steady-state sustainability assessment and optimization. Previous work has shown that process control plays a critical role in the transient sustainability and emission performance, especially for systems with different operating conditions (e.g., steady-state multiplicity [1]). The main challenges that have been mentioned in previous literature are: 1) existence of multiple objectives for optimization involving social, economic and environmental issues (e.g., pollution prevention and abatement); and 2) difficulty of integrating the sustainability indicators into a process control framework [2].

To address these challenges, a novel multi-layer process control framework is proposed to drive the system to a sustainable operating point that is defined using a multi-objective optimization algorithm. In particular, a sustainable Model Predictive Control (MPC) scheme is formulated based on dimensionless sustainability performance indicators from the U.S. EPA's GREENSCOPE (Gauging Reaction Effectiveness for the ENvironmental Sustainability of Chemistries with a Multi-Objective Process Evaluator) tool [3]. Such indicators are associated with the process state variables and can capture the sustainability information of the current process condition, including economic, environmental and social aspects. The selected sustainability indicators are employed as constraints in the designed controller to maintain the process operation within a pre-defined sustainable zone, where the performance indicator values are higher than the desired thresholds. To characterize this sustainable control zone, dynamic radar plots are developed to visualize the multidimensional sustainability indicators during a transient period.

The developed method is illustrated via a biomass/coal co-gasification process for syngas production with the end goal of methanol manufacturing. For this application, the whole process model is developed in Aspen Hysys based on existing literature information [4-5]. With the established models in Hysys and a link for data communication between Hysys and MATLAB, a multi-objective optimization problem is solved to maximize profit and optimize the process sustainability performance (e.g., environmental release and resource use minimization), by employing a genetic algorithm-based approach developed in MATLAB. A reduced-order model for the process is also derived in MATLAB for control implementation purposes. In this presentation, the details on the application results of this multi-layer control framework for improving sustainability are discussed, focusing on the process dynamic sustainability performance. These results demonstrate the capability of the proposed sustainable control scheme for maintaining the process within sustainable zones during transient periods.

References:

1. Li S, Mirlekar G, Ruiz-Mercado GJ, Lima FV. Development of chemical process design and control for sustainability. Processes. 2016;4(3):23.
2. Daoutidis P, Zachar M, Jogwar SS. Sustainability and process control: A survey and perspective. J. Process Control. 2016;44:184–206.
3. Ruiz-Mercado GJ, Gonzalez MA, Smith RL. Sustainability indicators for chemical processes: III. biodiesel case study. Ind. Eng. Chem. Res. 2013;52(20):6747–60.
4. Li S, Feliachi Y, Agbleze S, Ruiz-Mercado GJ, Smith RL, Meyer DE, Gonzalez MA, Lima FV. A process systems framework for rapid generation of life cycle inventories for pollution control and sustainability evaluation. Clean Technol. Environ. Policy. 2018;20: 1543-1561.
5. Robinson PJ, Luyben WL. Simple dynamic gasifier model that runs in Aspen Dynamics. Ind. Eng. Chem. Res. 2008;7784–92.

567732 Process Engineering Inspired Comparison of Sustainability and Resilience Specific Features of Natural and Human-Built Process Systems

From chemical and biological engineering points of views, sustainability and resilience means long term planning, design and control of robust large scale processes. In the past decades mankind built "artificial" processes against the principles of the natural (e.g. biological) processes, to achieve more profit in a short time horizon. In order to keep balance in the space of finite resources and reservoirs, we have to pay more attention to the computer assisted engineering of large water-food-energy- ecosystem related process networks.

The structure and functionalities of the complex natural and human-built processes are more sophisticated, than the networks in sense of network science, on the one hand. However, the case-specific local functionalities of the individual building blocks in these structures are less complex, than the various specific mathematical methods, on the other. As an intermediate solution, we have developed Programmable Structures that can be generated from two functional meta-prototypes and from the declaration of the process network. The meta-prototypes and the generated process structures are prepared for the semantically distinguished, but syntactically uniform representation of the "*model specific conservation law based, additive*" measures, and of the "*over-writable*" signals.

Applying this methodology, we studied many different process systems from cellular signaling, through technological systems to the ecosystem involved agricultural process networks. The results from the analysis of these quite different processes, studied by a unified approach, oriented our theoretical interest toward model based comparison of sustainability- and resilience-specific features of natural and human-built processes.

According to the preliminary investigations on this comparison, it became obvious that we can learn useful (chemical and biological engineering based) sustainability related design and control principles. The most important lessons can be summarized as follows:

1. Both natural and human-built processes are determined by "model specific conservation law based" stoichiometric balance processes, but the natural processes are inherently self-determined, while human-built ones are determined by short term decisions. As a consequence of the self-determinedness, natural processes prefer local solutions, in contrary to the more and more global character of human-built processes.
2. Another common feature is that the primary (controlling) information is carried by conservation law based vehicle processes in both natural and artificial processes. However, in natural processes a direct feedback exists between the signal and the vehicle process. In the lack of this feedback, human generated information processes are often characterized by the (sometimes useful, sometimes harmful) far reaching effects. The basic architecture of natural processes is characterized by the mutual cooperative control feedback between the functionally connected neighbors. However, the artificial processes are controlled by the hierarchically organized human decisions.
3. The secondary (evaluating) information appears only in the human-built processes, in form of objectives (i.e. interests). Accordingly, the evaluation feedback in the artificial processes are manifested in the various forms of usually hierarchical optimization. Nevertheless, consciously designed cooperative evaluation feedback amongst the functionally connected neighbors could also be applied in the human-built process architectures. Engineering designed and controlled cooperation means that the functionally connected neighbors must tend to evolve mutually suboptimal, compromise functioning.

Programmable Structure of process models helped to learn the clear separation of the chemistry invented "model specific conservation laws based" additive measures from the over-writable signs. This, combined with biology invented cooperative feedback, suggested design and operation principles for sustainable complex process systems, as follows:

- to consider long term balances of the model specific conservation laws based measures;

- to understand "information process" as a sub-process of "conservation process" that consumes less measures, but effects more on the functioning of the complementing sub-processes and the environment, than *'vice versa'*; and

- to understand cooperative process as a mutual (control or evaluation) feedback between the functionally connected neighbors.

The existence and utilization of these principles will be illustrated by some recently studied examples for biosystem, ecosystem, (bio)technological and agro-environmental process networks.

578082 An Input-Output Framework to Assess the Importance of Insect Pollinators to US Industry Sectors: A Case Study of Apple Production

Ecosystem services remain highly critical and largely undervalued in industrial communities. Although the supply of these services is often considered boundless, many ecosystem services, including insect-mediated pollination, have suffered severe degradation and face overconsumption beyond rate of replenishment. While insects provide other valuable ecosystem services beyond pollination, both wild and managed species of bees and other pollinating insects are instrumental in yield and/or quality of approximately 75 pollination-dependent crops in the United States of America. In forty-five years, production of pollination-dependent crops has increased more rapidly (50-62%) than the global population of managed bees (45%) while simultaneously, wild flying insects in all regions (including protected natural lands) have declined up to and exceeding 75 percent over only the last twenty-seven years, both in biomass and with decreasing biodiversity. We determine that insect-mediated pollination of crops accounts for over 30 billion US dollars of pollination-dependent crop production alone making the ecosystem service a highly valuable asset to the economic welfare of the agricultural sector. In addition, there are related agricultural sectors that rely upon pollination-dependent crop production (fertilizers, pesticides, equipment) as well as non-agricultural industry sectors (pharmaceutical, fuel, real estate) that share complex linkages with crop production and agricultural sectors, leading to intricate indirect dependence upon pollination service mediated by both commercially-managed and wild species of bees and other insects throughout the economy.

We present an input-output (I-O) modeling based framework for quantifying the economic value of industry sectors attributable to insect-mediated pollination. Of pollination-dependent crops, each has a specific dependence upon insect-mediated pollination service. Using available field data, we first reconcile and update estimates of pollinator dependence coefficients for individual crops while explicitly quantifying uncertainty. Using publicly available data from the USDA (United States Department of Agriculture), NASS (National Agricultural Statistics Service), and reconciled pollinator dependence coefficients, we establish bounds on economic value of agricultural crops attributable to insect pollinators. This information is then used in conjunction with the 2012 Benchmark I-O account data to quantify the economic dependence of non-agricultural sectors on insect-mediated pollination. We focus specifically on the apple farming industry and recreate the apple farming supply chain. We disaggregate the fruit and nut farming sector to represent the apple farming sector with greater detail.

While the direct economic value of apples attributable to insect pollinators is 3 billion dollars, indirect economic dependence of non- agricultural sectors on insect pollinators is equally as significant, though lacking previous quantification. In addition, we further identify the industry sectors that are most vulnerable to loss of insect pollinators. This model allows for specificity in analyses of indirect economic impacts of pollination service mediated by insects by providing a framework for the incorporation of crop-specific dependence coefficients into sector expenditures. The results are compelling and highlight the need for better understanding the role of insect pollinators in life cycles of agricultural and non-agricultural products and processes.

567111 Rapid Estimation of Release Inventories and Evaluation: A U.S. Metal Coating Industry Case Study

Industrial facilities might release many chemical substances during the production of value-added products. These chemical releases may cause damage to a receptor, either occupational and non-occupational population, and the environment. Therefore, to determine whether a chemical poses a considerable risk to human health and the environment, several tasks must be completed. These include determining the releases to the environment from conditions of use. Therefore, this work aims to develop a process-oriented, decision-making, and learning-from-data framework which integrates: (a) estimation of chemical releases from an industrial facility taking into consideration the type of processes, the condition of use, and the physical-chemical properties of a substance; (b) environmental sustainability indicators based on U.S. Environmental Protection Agency (USEPA)'s GREENSCOPE; (c) a decision-making procedure based on fuzzy techniques to include stakeholders' opinions; and (d) an aggregation method to incorporate the existing trade-off between the USEPA's GREENSCOPE environmental indicators. A case of study for the U.S. metal coating industry is employed for applying the proposed framework as a tool for supporting environmental decision-making efforts and chemical release minimization. In addition, this framework may be suitable for chemical risk evaluation and environmental sustainability decision-making for industrial systems.

The views expressed in this abstract are those of the authors and do not necessarily represent the views or policies of the U.S. Environmental Protection Agency.

562191 A Multi-Objective Optimization Model to Plan City-Scale Water Systems with Economic and Environmental Objectives: A Case Study in Santiago, Chile

Climate Change has become an important issue in several regions worldwide since human behavior is conditioned by the effects of Global Warming on Earth. A direct effect is water scarcity, considered a global risk by the World Economic Forum, affecting two thirds of the world's population [1]. To address this issue, long term planification of large scale systems has become a key matter. Therefore, water management and planning under uncertainty has been a priority area of research aiming at improving large scale regional water systems. In this context, optimization techniques can be a valuable tool in the water resources management in order to redesign regional water systems, and this effort constitutes the main focus of this project.

In this work, we develop a novel superstructure to model the water system of a city. The problem is formulated as a Multi-Objective Mixed-integer Nonlinear Programming to determine the optimal configuration of a regional water system. The model is formulated to decide (i) the installation of new treatment plants, (ii) the actualization of the existing ones, and (iii) the connections within the new integration network. These changes in the water network allow to recycle and reuse water. The objective functions to minimize are: (i) Global Warming Potential, (ii) water usage from the source, and (iii) total cost of the water system. These functions integrate the economic and environmental impacts simultaneously in the model, in order to redesign the existing water system. New plants can achieve four water qualities: (i) freshwater; (ii) drinking water; (iii) irrigation water; and (iv) discharge water. The main novelties of this work are the large scale orientation of the formulation and the integration of economic and environmental objectives in the planning of a city-scale water system.

A case study is proposed in order to create a solution linked to reality, without losing adaptability to other contexts worldwide: the model is formulated inspired in Santiago, capital of Chile, with a population of six million people [2]. Santiago is the political, economic and institutional center of Chile and has been signed as a hydric-stress zone [3]. The participants of the network are classified into consumers (urban, industrial, and agricultural); sources (superficial and groundwater); distribution and collection nodes; treatment plants; and final disposal sinks.

First, a simplified problem is formulated and solved using Goal Programming strategy. The result shows a modification of the present network: the new system has no discharge to rivers. Furthermore, drinking and irrigation water demand can be supplied through new recycling plants with a direct connection. The solution does not recycle water to the source (freshwater quality). The proposed network is composed by large and small scale plants. The connection of these treatments with consumption and distribution/collection nodes reduces the transport cost.

Finally, the formulation of this problem is oriented to address large scale water systems with special focus on real regional water networks. The model is proposed to provide orientation to policy makers so as to plan future changes in city-scale water networks. The next stage in this project is to include uncertainty in water availability, demand, and other effects of Climate Change on the network.

References

[1] ALBRITTON-JONSSON, F., BREWER, J., FROMER, N., TRENTMANN, F. Introduction: Scarcity in the Modern World: History, Politics, Society and Sustainability, 1800–2075. 2019.

[2] NATIONAL STATISTICS INSTITUTE. CHILEAN GOVERNMENT. Primera entrega de resultados definitivos CENSO 2017: cantidad de personas por sexo y edad. 2018.

[3] VALDÉS-PINEDA, R., PIZARRO, R., GARCÍA-CHEVESICH, P., VALDÉS, J.B., OLIVARES, C., VERA, M., BALOCCHI, F., PÉREZ, F., VALLEJOS, C., FUENTES,R., ABARZA, A., HELWIG, B. Water governance in Chile: Availavility, management and climate change. Journal of Hydrology, 519, Part C: 2538-2567, Nov. 2014.

575085 Evaluating Microwave-Synthesized Copper Oxide Nanoparticles from Copper Chloride with Antimicrobial Properties Using Life Cycle Assessment Techniques

The use of copper-based nanoparticles may serve as an alternative to precious metal-based nanoparticles. A life cycle assessment (LCA) based evaluation for synthesized copper oxide nanoparticles (CuONPs) with broad-based antimicrobial properties was carried out. The synthesis process analyzed used glucose and starch as the reducing and stabilizing agents respectively, and microwave energy in efforts to reduce environmental impacts. The LCA techniques followed ISO 14040 procedures and was carried out using GaBi 6.0 software on a declared unit of one kilogram of copper to produce CuONPs approximately 2 nm in diameter, and nearly spherical in shape, with a minimum process conversion rate of 90%. Energy grids in both Texas and California were investigated. The highest impacts relative to a United States normalization were predominantly acidification potential (AP), fossil fuel potential (FFP) and ecotoxicity potential (EcoP). The major process contribution in each of these impact categories was due to electricity usage for the microwave reaction. The next highest contribution was due to mining of copper and the least impactful were the use of starch and glucose. Using starch and glucose in the process and replacing precious metals with copper appear to reduce both environmental and economic impacts.

575184 Towards Sustainable Energy Mixes: Incorporating Planetary Boundaries into Energy Systems Models

Planetary boundaries identify a safe space for humanity to evolve that would prevent the potential occurrence of deleterious nonlinear events highly detrimental to the Earth. Current energy systems are designed considering mainly economic criteria and bounds on greenhouse gas emissions. This narrow scope can lead to unsustainable energy mixes. We present an approach to incorporate planetary boundaries into energy systems models. Taking the United States as a testbed, we show that the Paris Agreement solution results in transgressing five out of eight planetary boundaries. Due to the stringent planetary boundary on biogeochemical nitrogen flow, we find no mix capable of satisfying all planetary boundaries concurrently. Though the mix that meets seven out of eight planetary boundaries concurrently has a negative impact on one of the climate change planetary boundaries and ocean acidification due to the deployment of bio-energy with carbon capture and storage, i.e., a negative emission technology, it doubles the cost in contrast to the least cost energy mix. Incorporating planetary boundaries into energy systems models are necessary to determine the resilience of the power sector towards meeting the Earth's ecological capacity, and consequently design sustainable environmental policies.

581347 Chemical Engineering Innovations for a Renewable Economy

Chemical engineering evolved in the twentieth century with rise in the use of fossil resources and myriad of products derived from then. An array of these products required great innovations and made huge impact on the human race, and such examples include urea, transportation fuels, polymers, water purification chemicals, etc. Going forward as we make transition from fossil resources to renewable resources such as solar energy, wind and sustainably available biomass, we again have great opportunities for innovation that would not only make great impact but redefine chemical engineering. My talk will focus on my own experiences with challenges and opportunities presented to us as chemical engineers. I will share an overall perspective on the use of biomass as sustainable source of carbon, harvesting solar energy in conjunction with growing food without any land constraint, and creating solution processed solar cells that utilize chemical engineering principles along with knowledge from other engineering disciplines.

581350 Engineering Materials for Clinical Application

Dr. Schmidt's research has resulted in numerous patents that are licensed to companies, including AxoGen, Alafair Biosciences, Smith and Nephew, and Siluria Technologies. Comments for this panel will focus on peripheral nerve applications and one key success that is the foundation for the Avance Nerve Graft from AxoGen. In addition, along the way to uncovering strategies for nerve regeneration, Dr. Schmidt and her team also discovered a biomaterials processing approach that led to a separate application of protecting tendon's after surgery, resulting in the VersaWrap Tendon Protector from Alafair Biosciences, a start-up company in Austin, Texas.

Nerve Repair: Dr. Schmidt's technology for processing decellularized nerve tissue was licensed to AxoGen in Alachua, Florida (http://www.axogeninc.com) for the AVANCE™ nerve graft. Many tens of thousands (~100,000) of AxoGen's AVANCE™ grafts have been implanted in patients at hundreds of hospitals globally.

Post-surgical Wound Management: Using a novel in situ crystallization processing technique, Dr. Schmidt's team created mechanically robust hyaluronic acid-based hydrogel films that can be easily handled, sutured, and delivered laparoscopically. This research is the basis of Alafair Biosciences, a start-up company in Austin, TX (http://www.alafairbiosciences.com), which is run by Dr. Schmidt's PhD graduate and a neurosurgeon collaborator. Alafair's first product, VersaWrap™ Tendon Protector, has been implanted in more than 500 patients since December 2017.

554686 Evaluation of Chemical Additives for Glycolate Destruction in High-Level Radioactive Waste Condensate in the Defense Waste Processing Facility

Savannah River National Laboratory (SRNL) has completed a work scope in support of the Savannah River Remediation (SRR) Nitric-Glycolic flowsheet. Glycolic acid will replace formic acid as a reducing agent in the pre-processing of High Level Waste sludge performed at the Defense Waste Processing Facility (DWPF). Glycolic acid will perform the same function as formic acid, namely reduction of mercury and adjustment of feed rheology and melter oxidation/reduction potential, while significantly reducing the potential for hydrogen generation in DWPF processing. Development testing has demonstrated glycolic acid virtually eliminates hydrogen production in the pre-processing steps.

The Nitric-Glycolic flowsheet significantly improves DWPF's ability to address Savannah River Site's key challenge - the incorporation of effluent received from the Salt Waste Processing Facility (SWPF). SWPF will deliver significant effluent volume to DWPF, resulting in a concurrent increase in DWPF effluent returned to the Concentration, Storage and Transfer Facilities (CSTF). This work scope demonstrates that glycolate can be destroyed under the Nitric-Glycolic flowsheet prior to returning the DWPF recycle stream to CSTF.

To avoid potential flammability issues due to thermolysis of glycolate in CSTF, SRR tasked SRNL to quantify and mitigate glycolate returns via DWPF recycle stream. The development of a strategy for glycolic mitigation initiated with a system's engineering workshop. Various chemical and/or physical solutions for how to and where to destroy glycolate were considered – consistent with DWPF operational capabilities and process requirements. The workshop identified chemical oxidation of glycolate within the DWPF Recycle Collection Tank (RCT)[1] as the most promising option with sodium permanganate and Fenton's reagent (metallic ion catalyzed hydrogen peroxide) as the two most promising oxidants.

SRNL between Fenton's reagent and permanganate could be made. A test matrix was generated to evaluate the impact of:

- Glycolate concentration,
- Oxidant type (sodium permanganate or Fenton's reagent),
- Oxidant strike time (rate of delivery),
- Oxidant stoichiometry,
- RCT solution temperature, and
- RCT solution pH.

Testing conditions were selected to reflect DWPF process capability and requirements. Non-prototypic concentrations of process organics were added to the process simulant to challenge the methodology and better support the down select. Tests were conducted with discrete liquid sampling events and online offgas analysis for data generation.

For the test conditions performed and data obtained, three conclusions are drawn.

- Sodium permanganate was clearly superior to Fenton's reagent as an oxidizing agent for destruction of glycolate in the RCT. Permanganate strikes consistently reduced glycolate concentrations from nominally 125-250 mg/L to below the target threshold of 50 mg/L. Corresponding strikes with Fenton's reagent did not successfully mitigate glycolate.
- Successful permanganate mitigations of glycolate were demonstrated across broad range of RCT process conditions. These conditions include the following.
    1. RCT solution starting pH: 3-13

2. RCT solution temperature: 15-50 °C
3. Oxidant strike time: 10 minutes to 2 hours
4. Glycolate concentration: 125-250 mg/L (with exploratory testing up to 2000 mg/L)
5. Oxidant addition amount: 100-150% stoichiometric requirement for all organics

SRNL, in conjunction with SRR, is continuing the glycolate mitigation development effort. Additional testing include protocol testing intended to clearly define strike parameters for engineering and operational implementation, off-normal testing for remediating foam-over conditions and testing at larger scale or with tank waste. Results from the testing will be discussed.

[1] The RCT collects offgas condensate from pre-processing, vitrification, and other unit operations performed in DWPF and is the singular return vessel delivering recycle effluent back to CSTF.

568856 Tank Closure Cesium Removal at Savannah River Site

High Level Waste (HLW) at the Savannah River Site (SRS) is currently stored in aging underground storage tanks. This waste is a complex mixture of insoluble solids (sludge) and soluble salts in an alkaline solution. Continued long-term storage of these radioactive wastes poses an environmental risk. The salt waste is currently treated to remove cesium by an interim solvent extraction process prior to disposal as grout in the Saltstone Disposal Facility. This interim process will be replaced with the Salt Waste Processing Facility (SWPF) to increase the salt waste treatment capacity. Treatment and disposal of the salt waste is the critical path to waste retrieval, risk reduction and closure of old style tanks at SRS.

Savannah River Remediation (SRR) is developing a supplemental at-tank ion exchange capability for the removal of cesium from liquid salt waste. Building on the conceptual development of the Small Column Ion Exchange (SCIX) process, the experience of modular commercial nuclear plant decontamination, and following the disaster response associated with Fukushima, SRR is deploying a modular at-tank treatment system for the selective removal of cesium from bulk salt waste effectively and efficiently. This Tank Closure Cesium Removal (TCCR) initiative will provide a supplemental treatment capability to existing and future salt processing and improve confidence in supporting the desired acceleration of waste retrieval and tank closure efforts.

The TCCR system configuration is consists of modular process structures located near Tank 10H and 11H. Salt waste will flow through a pre-filter to reject any solid material from entering the treatment system and be returned to Tank 10H. Clarified salt waste will then flow from the pre-filter into the first stage ion exchange column. The ion exchange media utilized will be highly selective for the removal of cesium and allow most other soluble constituents to pass through the media. Additional ion exchange columns are available and can be configured in series to improve decontamination factor, protect against flow "channeling" in the column, guard against media breakthrough, and allow for more continuous operation of the system. Decontaminated (cesium removed) effluent would flow from the ion exchange columns and sent to Tank 11H prior to being transferred to Saltstone for disposal.

Once the ion exchange media in a column becomes loaded with cesium to the extent practical ("spent"), the column with media will be removed from the system and replaced with a new ion exchange column loaded with fresh media. The spent column will be temporarily stored in an Interim Safe Storage (ISS) location within the SRS Tank Farm Facility. This ISS concept avoids the initial need for back-end process facilities (deferring the cost of these facilities) and allows for identification and evaluation of potential future disposal alternatives.

In addition to accelerating waste retrieval and tank closure activities, demonstration of this capability has several other benefits. Deployment of commercially available technologies with minor modification can expand Department of Energy (DOE) utilization of commercially supplied treatment processes. This "at-tank" treatment system could be modified to evaluate and demonstrate removal of other radioactive constituents (strontium, technetium, iodine, etc.) of bulk waste. The modular aspect of this capability can allow for additional units to be delivered and deployed at SRS, as well as, provide supplemental treatment capabilities for waste treatment at other DOE sites.

577672 Large Cemented Gibbsite Agglomerates in Alkaline Nuclear Waste at the Hanford Site and the Impacts to Remediation

Jet-sluicing has been the primary method of moving nuclear waste sludge between storage tanks at sites such as Hanford in the United States. Recent experience at an area of Hanford comprising twelve, 530,000-gallon underground tanks (C Farm), with waste that had aged in place for decades, showed that jet sluicing left a small heel of waste in the tanks. This study sought to determine why this material was left behind, focusing on Hanford tank 241-C-109. The waste heel in this tank was sampled and then analyzed by X-ray diffraction and scanning electron microscopy (SEM) with energy dispersive spectroscopy. Large gibbsite [$Al(OH)_3$] particles were found to be the major component in the waste heel. Individual, primary gibbsite particles were to 20 to 100 microns in diameter but were cemented together to form agglomerates as big as seven centimeters in diameter that were difficult to crush with a laboratory mortar and pestle. SEM showed that either amorphous aluminum hydroxide or gibbsite itself was the cementing agent. While large cemented gibbsite nodules have been reported in acid soils in nature, this study shows that similarly large gibbsite agglomerates can form in alkaline environments when aged for sufficiently long time, here for over 40 years prior to waste retrieval. Fluid dynamics calculations indicate that these cemented particles would be difficult to mobilize with standard jet slurry technologies. Even if successfully mobilized, the particles would be difficult to keep suspended and would not fit into transfer lines that are commonly 5-cm pipe diameter. Thus, these large gibbsite particles explain some of the difficulty in suspending aged alkaline nuclear waste with standard sluicing technologies.

571866 Processing of Actinide Materials for Reuse/Disposal

The US Department of Energy (DOE) manages materials that contain a range of actinide isotopes that were produced from the 1960s through the 1980s by irradiating targets in production reactors to produce heavy isotopes and enriching them for DOE programmatic use, scientific research, and industrial and medical applications. Because the production reactors and enrichment facilities have been shut down, many of these unique materials will never be produced again. Oak Ridge National Laboratory (ORNL) enriched many of the materials and has since used them in DOE's center for production, storage, and distribution of heavy isotopes (uranium through californium) for the heavy-element research program. As a result, ORNL has accumulated an inventory of rare actinide isotopes that are being held for reuse because they have potential intrinsic value to DOE, even though the materials have no currently defined use. Steps are being taken to evaluate the inventory, recover the useful materials, and distribute them to programmatic users. If no programmatic need for the materials is found, the materials are processed for transuranic waste disposal. This paper describes the processes that are being used to recover or disposition the materials.

563598 In-Situ Raman and ATR-FTIR Spectroscopy for Monitoring of Molecular Species in Low-Activity Nuclear Waste

A significant portion of the nuclear waste stored at the Hanford site is low-activity waste, primarily made up of non-radioactive components such as sodium salts. The DOE has initiated the Direct Feed Low-Activity Waste (DFLAW) project to expedite the clean-up of low-activity waste. Under DFLAW, a portion of the liquid low-activity waste will be preprocessed to reduce cesium levels and solid content and directly sent for vitrification. The complexity of the waste and its hazardous nature make the analysis challenging, thus reliable measurements of the composition and solid content are needed to facilitate processing.

In our present research, we are exploring use of in-situ monitoring tools, such as ATR-FTIR and Raman spectroscopy, to collect high-dimensional measurements of the waste. The data is collected on simulated waste mixtures that represent different DFLAW locations at Hanford. Our approach includes collecting training sets for the simulated mixtures that are used to build regression models. Next, we use the models to predict the composition of unknown mixtures. Our goal is to use the experimental findings to compare the performance of Raman and ATR-FTIR probes and build a robust model that will generate the composition of the multicomponent waste system.

Our research has shown that using in-situ ATR-FTIR and Raman spectroscopy, coupled with data-driven models can be used to monitor key species in simulated nuclear waste mixtures. This approach has the potential to reduce or eliminate the need for offline sample collection, which will minimize employee exposure to hazardous waste, while improving sample turn-around times.

558836 Ultra-Microporous Materials for Efficient Removal of Krypton from Nuclear Reprocessing Facilities

Reprocessing of used nuclear fuel (UNF) is a prerequisite for any future expansion of civilian nuclear power. Major challenges during the reprocessing of nucler fuel is the release of volatile radio nuclides that must be captured and subsequently stored for an extended period of time before being released into atmosphere. While methods such as cryogenic distillation and fluorocarbon based absorption have been proposed and tested, solid state adsorbents in particular permanently porous Metal-Organic Frameworks (MOFs) are considered better alternatives in terms of cost and engineering control to capture these volatile radionuclide including ^{85}Kr (t1/2 = 10.8 years). However, because of presence of more polarizable Xe during the reprocessing, adsorbent materials including MOFs are found to be more Xe- philic than Kr. To address this challenge, a two-bed breakthrough method is being developed herein which leads to enhanced Kr adsorption and separation of ^{85}Kr during nuclear reprocessing. The extent of enhancement of Kr adsorption depends on the nature of adsorbing materials. Here, we evaluate the Kr adsorption capability of a series of hybrid ultra-microporous materials, namely SIFSIX-3-M (M= Zn, Cu, Ni, Co or Fe) for Kr adsorption and separation performance. The Fe-analogue in particular shows excellent Kr adsorption than the current benchmark materials: CC3, Ag-mordenite, NiMOF-74 and SBMOF-1, under simulated nuclear reprocessing off-gas conditions. The higher Kr adsorption and separation efficacy was attributed to the precise pore diameter of the Fe-analogue for such separation. All materials were systematically studied for gamma irradiation stability, which reveals that the metal center in these isostructural materials plays a crucial role in their irradiation stability. Zn and Ni analogues show phase change at 1KGy, while SIFSIX-3-Fe shows phase change at 3KGy. SIFSIX-3-Co structure was stable up to 10KGy before it undergoes phase change while SIFSIX-3-Cu was the most stable analogue. The phase change of all materials were interpreted using PXRD. These results attest the potential of these materials for Kr removal under the nuclear reprocessing conditions.

567054 The Migration Law of Nuclides in Steam Reforming Reaction of Spent Radioactive Resins

The treatment of spent radioactive resins in the nuclear industry is important due to its great potential hazard to human beings and environments. The two keys during the disposal of the resins are volume reduction ratio and leaching resistance. Considering these two aspects, the steam reforming technology is a better method for resin disposal with high volume reduction ratio (>6) and radionuclides retention efficiency (>99.99%). During the reaction, nuclides are released when the spent resins are pyrolyzed into light hydrocarbons, sulfides, carbon oxides and a little residue at 650-750°C. Meanwhile, the nuclides are contacted with silicon-aluminum compounds and structure directing agents in the reactor to form specific mineral, which can immobilize the nuclides in its structure. The key to the reaction is that the nuclides are released efficiently from the spent resins with large volume, and then immobilized by minerals with small volume. Therefore, it is of great significance to study the migration law of nuclides to regulate the immobilization efficiency. In this work, the migration process of nuclides in the pyrolysis of resins under the steam reforming reaction conditions is investigated in the fixed bed reactor. The pyrolysis products obtained from different reaction times and temperatures are analyzed by SEM-EDX to observe the morphologies and nuclides distribution. The form of nuclides are studied by XRD, and the quantification of nuclides are obtained by ICP-MS. It is found that different migration paths of nuclides occur in different pyrolysis stages of resins, and a migration model is proposed to describe the phenomenon. Most of the nuclides are retained in the residue with the form of oxides and inorganic salts.

Keywords: Migration law; Nuclides; Steam reforming reaction

574095 Abiotic Degradation with Reactive Iron Mineral Coatings in Anaerobic Sediment

Abiotic dehalogenation with reactive iron minerals plays an important role in the natural attenuation of chlorinated organic solvents. These reactive iron minerals are most significant in redox transition zones. However, there are a number of challenges in collecting anaerobic cores representative of the subsurface environment. In this study, an anaerobic sediment core was collected from an industrial site with 1,4-dichlorobenzene and 4-chloroaniline plumes. With the redox condition preserved, a number of approaches including sequential extraction were applied to quantify reactive iron mineral coatings. Results revealed an abundance of siderite, pyrite, mackinawite, magnetite, and iron oxyhydroxides in three redox transition zones. Experiments were carried out to evaluate abiotic dehalogenation of 1,4-dichlorobenzene. The reaction rate followed a second order rate expression that was a function of the 1,4-dichlorobenzene and the reactive mineral coatings present in the sediment; for the three transition zones rate, constants for the dominant reactive mineral coatings were obtained. The time to achieve 90% removal ranged from 50 to 200 days for the zones studied. To degrade 4-chloroanaline, persulfate was applied with the reduced iron mineral coatings in the sediments. Persulfate forms $SO_4^{-\cdot}$, a radical, and a strong and effective oxidizing agent. Degradation rates revealed first-order rate constants with 90% removals requiring 9 and 22 hours. From these studies, the degradation pathways and models for target constituents of concern can be used to enhance these processes in contaminated subsurface systems.

579690 Nanoparticles for Energy and Environment Applications – Wastewater Treatment

Oil extraction involves thermal processes that use large amounts of water which also results in large amounts of oil sands process-affected water (OSPW). Because of the zero-discharge policy, this water cannot be discharged back in rivers or lakes so it is stored in tailing ponds. One of the challenges faced with produced water and interfered with their effective water management is particle settling. Without the introduction of a process or material, it would take these particles thousands of years to settle. Despite the fact that many existing processes are promising for fine particle settling, they lack several properties that can be improved by nanoparticle technology.

In this study, nanoparticles are prepared and modified to promote interaction with suspended particles in the wastewater. The prepared material was fully characterized using FTIR, TGA, BET, XRD and SEM. Nanoparticles provide promising results for the enhancement of fine particle settling making it an environmentally friendly and cost-effective process.

Nanoparticles is a new innovative technology with many applications for energy and environment sustainability. They are effective in enhancing fine particles settling in tailing ponds that would otherwise remain suspended for decades. They also promote dewatering in alignment to the Canadian government emphasis on active management of oil sands tailings and land reclamation.

573928 Laboratory Investigation of Combined Chemical/Biological Treatment of Hexavalent Chromium, Carbon Tetrachloride and Trichloroethylene in Aqueous System

Among the hundreds of chromium contaminated sites in the United States are Hanford and Savannah River Sites, which are under the purview of the U.S. Department of Energy (DOE). At the Hanford Site, which is the most contaminated nuclear site in the United States, aquifer and groundwater contamination represents a complex mixture of inorganics including hexavalent chromium , Cr(VI), radionuclides and nitrates, and organic contaminants, e.g., trichloroethylene (TCE) and carbon tetrachloride (CCl_4).

Conventional treatment of chromium contamination focuses on immobilization of the highly toxic and water-soluble form Cr(VI), by reducing it to the relatively less toxic and less mobile trivalent form, Cr(III) while organic chlorinated compounds and nitrates can also be reduced using ferrous sulfate. Both types of contaminants (inorganic and organic) also respond to biodegradation. In this paper, results will be presented from a two-step comprehensive approach: (i) to chemically reduce hexavalent chromium, carbon tetrachloride and trichloroethylene using ferrous sulfate and zero valent iron (ZVI) amendments, which leads to nontoxic compounds, but also results in oxidation of ferrous to ferric iron. (ii) In the second step, ferric iron is regenerated to ferrous through biotic transformation using metal-reducing bacteria, *Shewanella oneidensis* MR-1. The microorganisms support biodegradation of contaminants, and convert ferric to ferrous iron in the process, leading to a self-containing sustainable process. The results can be used in designing an at a hazardous waste site.

571321 Rapid Dechlorination of Polychlorinated Biphenyls (PCBs) in an Engineered System

Polychlorinated biphenyls (PCBs) are posing great environmental problems due to their toxicity. Removal of PCB in polluted sites are challenged by the high physical/chemical treatment costs, while microbial dichlorination of PCBs are largely impeded by difficulties in growing PCB dechlorinators to a high cell abundance. In this study, we accelerated the onset of PCB dechlorination by pre-cultivating functional microbes on alternative electron acceptors in an engineered system. Extensive dechlorination of Aroclor 1260 was achieved within 1 week by culture CG4 pre-grown tetrachloroethene (PCE). Compared to control cultures fed solely with Aroclor 1260, the PCB dechlorination rate was up to 30 times greater in cultures pre-cultivated with alternative electron acceptors. However, when culture CG4 was simultaneously exposed to PCE, and PCBs, PCB dechlorination was not observed until complete depletion of PCE. Expression of the bifunctional reductive dehalogenase (RDase) gene *pcbA4* was inhibited by the presence of both substrates and continued with only one substrate in the medium. Utilization of chloroethenes and PCBs as substrates by culture CG4 was found to be largely sequential rather than concurrent, suggesting that PCE is a preferred substrate for the RDase responsible for PCE and PCB dechlorination.

570464 Adsorption on NDMA on Biomass-Derived Activated Carbons in Lab-Scale Columns

The purpose of this research is to test the ability of biochars made from local agroforestry and forestry wastes, to adsorb N-nitrosodimethylamine (NDMA) at very low concentrations out of water. According to the EPA, NDMA is a probable human carcinogen found in ground and drinking water, especially in ground water near rocket testing sites that used unsymmetrical dimethylhydrazine fuel. The treatment standards for NDMA typically range from 1 to a few ng/L (parts per trillion), which creates analytical challenges for monitoring and development of remediation methods.

In this study, biochars produced from pecan shells and local forestry waste were tested in a lab-scale column for adsorption of NDMA. Concentrations of carbon-14-labeled NDMA in well water obtained near a NASA rocket-test site were measured at regular time intervals over the course of the flow through the biochar-filled column for several adsorbents. Liquid scintillation counting was used for quantification to enable faster and most consistent results with smaller sample sizes. Data from the breakthrough curves was used to calculate column design parameters to estimate the cost of adsorption treatment.

562651 Transforming Plastic Waste into Microporous Adsorbents for Environmental Remediation

By 2050 plastics will outweigh fish if 8 million tonnes of plastics continue to flow into oceans each year and damage marine ecosystems at an annual cost of ~ £9.9 bn. This can be overcome with better plastic recycling technologies. Existing plastic recycling methods are focused on energy-intensive mechanical and chemical methods. In this work, we employed the facile method of solvent-recycling to transform plastic waste into microporous materials with high specific surface areas of up to 1000 m^2/g. The specific surface areas and morphologies of these microporous recyclates could be tailored to suit various environmental remediation applications using different combinations of reaction conditions such as reagent type, ratios and temperature, benefitting carbon capture, dye removal and inorganic toxicants sequestration from water. The adsorption capacities of these recyclate materials were comparable to state-of-the-art nanomaterials such as metal organic frameworks (MOFs), polymers synthesized from expensive fine substrates and commercial adsorbents. Our approach simultaneously benefits the living environment by tackling plastic pollution via the transformation of plastic waste into adsorbents that can potentially be employed to sequester a variety of environmental pollutants – greenhouse gases, organic and inorganic molecules.

569836 Study of Foam Injection Technology for Soil Remediation Contaminated with Hydrocarbons

Soil flushing and soil washing are two remediation techniques preferred for cleaning soil contaminated with hydrocarbons due to their technical simplicity and cost. The efficiency of these techniques when injecting only water may be moderate due to the interaction of the contaminants with the soil components (organic, inorganic). The efficiency of soil flushing techniques may be increased with the use of surfactants. Surfactants decrease the interfacial tension between oil and water leading to the drag of contaminants in the flushing solution; also, they may interact with the soil surface, detaching the adsorbed hydrocarbon components by wettability exchange mechanisms.

The use of surfactants seem a good alternative to increase the efficiency of soil remediation processes, however, the injection of surfactant solutions may lead to different problems such as dragging of contaminant to clean sites even to water reservoirs causing secondary pollution. In the same line, surfactants adsorb to the minerals in the soil causing modification of the natural chemistry of the soil avoiding the natural recomposition of the system. Different strategies have been proposed: the implementation of surfactant solutions at low concentration, the use of bio-surfactants, and recently the injection of foams. Foams are colloid comprising on air trapped in a water matrix, this colloidal system can carry very low amount of surfactant in a very huge volume and possess a high viscosity, which can give control to the flooding process. In this work, we present a review on the latest technologies of foam injection for soil remediation, emphasizing the key variables to consider during soil flooding techniques. In the same line, we present the results we have obtained with a formulation comprising a mixture of synthetic surfactant (SDS) and a bio-surfactant in a test for removing hydrocarbons from soil, foam stability and its relation to hydrocarbon removal.

556406 Facile Ball Milling Method for Developing Surface Functional Groups on Carbon-Based Adsorbents and Their Roles in the Removal of Heavy Metals in Aqueous Solution

Water pollution by heavy metals poses a serious threat to both human health and the environment due to their toxic properties and tend to accumulate in living organisms. Carbon-based adsorbents such as graphene and its derivatives, carbon nanotubes, activated carbon, and biochar are often used to remove heavy metals from aqueous solutions. This work provides a comprehensive review of recent research on various carbon adsorbents in terms of the governing removal mechanisms of their surface functional groups and the associated removal behaviors and performances to heavy metals in aqueous solutions. It then categorizes various synthesis methods that are commonly used to functionalized carbon adsorbent for enhanced heavy metals removal. Facile ball milling method is successfully developed in this study to introduce additional heteroatoms, primarily oxygen and nitrogen onto biochar surfaces for enhanced material and sorptive properties to heavy metals in aqueous solutions. Ball milled biochar had greater internal and external specific surface area, larger pore volume, nano-size properties, stronger negative zeta potential for both O and N functionalizing methods. Faster adsorption kinetics and 6 -10 times enhanced Ni (II) adsorption were displayed. Ball milling thus has great potential to increase the efficiency of environmentally friendly biochar for various environmental applications.

571889 Environmentally-Responsible Conversions from Gasification Waste to Energy Storage Devices: Synthesis of Waste-Derived Nanomaterials and Environmental Evaluation

Waste gasification technology is considered as a sustainable solution to reduce waste volumes and recover energy by producing syngas. However, the leftover of gasification become another waste which needs combustion and landfill. Herein, with the concept of waste-to-resource, an environmentally-responsible pathway has been developed to extract vanadium from a gasification waste, carbon soot, and further converted into vanadium pentoxide. In this study, V_2O_5 nanosheets were synthesized from the leachate of gasification waste, which show comparable specific capacitance as the one from commercial V_2O_5. A flexible supercapacitor was designed by using waste-derived nanosheet and carbon nanoparticle as active materials. The whole process successfully synthesizes V_2O_5 from carbon soot and explores the potential application of carbon waste. Compared to the traditional process of V_2O_5 production, this process demonstrated significantly lower negative environmental impact in terms of human health, ecosystem diversity and resource availability, in a ReCiPe 2016 environmental impact assessment. Moreover, the interesting physical and chemical properties of carbon soot raise the possibility of utilizing carbon soot as a readily-available resource in a large scope of application of carbon materials.

AMELIORATED PHENOL UPTAKE FROM PAPER MILL EFFLUENT ONTO SPENT BREWER'S *SACCHAROMYCES CEREVISIAE* IN THE PRESENCE OF CETYLTRIMETHYL AMMONIUM BROMIDE

Natarajan Saravani, V.S.B. Engineering College, Karur, Tamilnadu,
M.Manimozhi, School of Electrical Engineering, Vellore Institute of Technology, Vellore,
**M. Arulmozhi, Bharathidasan Institute of technology (BIT) Campus, Anna University, Tiruchirappalli, Tamilnadu,*
Subburayalu Ramesh, Annamalai University, Chidambaram, Tamilnadu.

Abstract

The prime purpose of this research is to evaluate the influence of foam separation, an adsorptive bubble separation technique, employed as a finer treatment technique after biosorption. Cetyl Trimethyl Ammonium Bromide (CTAB) played the multiple roles of foam production, phenol elimination, and microbial collection and hence acts as a percentage promoter. The process has its own merits of lower capital investment, less operating and maintenance cost, less power consumption, relatively faster ad eco friendly technique. The effect of several working parameters, such as initial feed concentration, equilibrium time, biosorbent dose, pH, liquid pool height, surfactant concentration, and air flow rate, were experimentally explored. The higher floatability of *Saccharomyces cerevisiae* was obtained in a maximum time of 10 min. Among the models tested, Fowler- Guggenheim model (R^2=0.9703) was the most adequate. The experimental data fitted pseudo second order kinetics with a kinetic constant of $5.5425X10^{-5}hr^{-1}$. Fourier transform infrared spectroscopy (FTIR) spectroscopy revealed the active participation of imidazole in the biosorptive foam separation process. Electro kinetic measurements were carried out to determine the iso electric point (IEP) of the *Saccharomyces cerevisiae* cells. The zeta potential profile of the baker's yeast cell was affected by the presence of phenol at different pH values. The recovery of phenol loaded biomass and final traces of phenol by flotation were found to be 99.75%.

Corresponding author
*Dr. M. Arulmozhi, Professor and Head, Department of Petrochemical Technology, Bharathidasan Institute of technology (BIT) Campus, Anna University, Tiruchirapalli-620 024, Tamilnadu, India.
E-mail id: arulmozhisenthilkumar@gmail.com Mobile: +919364467846
Phone: +91 431 2407946 Fax: +91 431-2407333

Keywords: Cetyl Trimethyl Ammonium Bromide; Fowler- Guggenheim isotherm; Isoelectric point (IEP);*Saccharomyces cerevisiae*; Scanning Electron Microscopy; Zeta potential.

Abbreviations: CTAB, cetyl trimethyl ammonium bromide; 4-AAP, 4-amino anti pyrine; IEP, Isoelectric point; FTIR, Fourier transform infrared spectroscopy; SEM, Scanning Electron Micro

1. Introduction

A massive industrial growth has occurred wide over the globe in the past few decades. The growth is so immense that the environment has entirely changed from what it had been before to what it is today!!!(Virendra Kumar et al, 2014, Abdelwahab et al 2013). Pollution being the major rationale for this drastic change has its consequences in air, water, soil etc. Water, being a chief natural resource, is used in nearly all applications. Paper making, a hefty industry, uses a large extent of water, and consequently the discharge of wastewater is also massive (Jacob Varughese Vadakenat et al, 2014). Pollution due to paper and pulp industry has been rated among the twelve top pollutants. Therefore the proper treatment before the release to the environment is essential. One chief drawbacks of the direct discharge is the incidence of persistent phenol. Higher concentration of phenol lead to multiple disadvantages in both aquatic and human respiratory systems, and may even cause skin disorders eventually leading to cancer. These compounds once discharged have the ability to cause genetic changes to the exposed organisms (Kaizar Hossaina et al, 2015, Bandyopadhyay et al 1998, Swamy AVVS & Syamala Devi K, 2012).

Industries encloses very high phenol concentration; e.g., plastic manufacturing (600-2000ppm), petrochemical manufacturing (50-600ppm) and stocking production (6000ppm). Phenol an aromatic pollutant is present in the wastewater of numerous industries. Effluent standard of phenolic wastewaters is less than 1 ppm for discharging to surface waters. Phenolic compounds are toxic even at low levels and pose a threat to the biosphere because of their recalcitrant nature. Phenol vapors acts as air pollutant and precipitate from the air into the soil. The need to remove these contaminations has led to the development of new technologies that emphasize the detoxification and destruction of the pollutants rather than the conventional approach of disposal (Girish CR &Ramachandra Murty V 2012).

Conventional methods of phenol wastewater treatment include both physical and chemical. These methods include chlorination, ozonation, solvent extraction, incineration, chemical oxidation, membrane process, coagulation, flocculation, adsorption, ion exchange, reverse osmosis, electrolysis, etc (Wei et al 2007, Zijun Xiao et al 2012, Prpich*et al.,* 2005). But these methods led to secondary problems due to formation of toxic materials such as cyanates, chlorinated phenols, hydrocarbons, etc. Solvent extraction poses a risk of contamination of treated water by the solvent & high cost of solvent is yet another disadvantage. The solvents used for phenol recovery are benzene, isopropyl ethyl and butyl acetate. Activated carbon (frequently used adsorbent) is disposed by incineration generating many new compounds such as dioxins and furans. Chemical oxidation operates at high temperature and high pressure. The realization of the high cost of phenol waste treatment by physico-chemical methods has opened the door for the investigation of new methods. The requirement to remove these contaminants has led to the development of new technologies for the fine treatment of the phenolic compounds. Biological treatment, an alternative method is versatile, inexpensive, receiving the most attention due to its environmentally friendly approach & high biodegradation abilities.Thus treatment of phenolic wastewaters is necessary and biological methods are the most appropriate techniques due to mineralization of toxic organic compounds and inexpensiveness (Prpich and Daugulis *et al.,* 2005).

Biosorption is a surface phenomenon, where physical or chemical interactions between phenol molecules and the functional group present in the microbial cell wall plays an important role. The structure of a microbial cell wall is much intricate, thus allowing utmost options for biosorption to occur. Many sorbents are being in use including a range of bacteria, algae fungi etc. (Aditi Banerjee et al, 2010, Yang CF & Lee CM 2007, Gurusamy Annadurai et al 2002, Kumar et al. 2005, Juan & Han-Qing 2006, Jiang Yan et al 2006, Juan Wu & Han-Qing Yu 2008, Jegan et al. 2010, Aditi Banerjee et al 2010, Yu Zhang et al 2011, Aditi Banerjee et al. 2011 Riham Surkatti & Muftah H El-Naas 2014, Neetu Singh &Chandrajit Balomajumder 2016) Natarajan Saravani (2016) recognized that *Micrococcus lylae,* a Gram positive bacteria, is an efficient biosorbent for phenol removal. My Linh Nguyen & Ruey-Shin Juang (2015) investigated the ability of chitosan beads modified with histidine and *Saccharomyces cerevisiae* using glutaraldehyde as a cross linking agent (SC-HIS-CCBs) for the biosorption of phenol. They performed batch biosorption experiments using the SC-HIS-CCBs at a pH (2.0 - 7.0), and initial phenol concentration (25 -250 mg/L). The biosorption capacity of phenol was 28.4 mg/g at a pH 2.0 and 313 K.

The purpose of this study is to investigate the biosorptive foam separation capabilities of *Saccharomyces cerevisiae* to treat paper and pulp mill industry effluent. The uptake and accumulation of hazardous organics by microorganism offers an economical, practical and efficient alternative to existing methods. CTAB a cationic surfactant efficiently fascinates anionic phenolate. Isotherm and Kinetic studies was conceded out by scheming which curve fits the most. About ten models for isotherm and five models for kinetic fit were analysed.The qualitative aspects of infrared spectroscopy is one of the most prevailing attributes of this diverse and multipurpose analytical technique. Over the years, much has been published in conditions of the fundamental absorption frequencies which are the key to unlocking the structure–spectral relationships of the associated molecular vibrations. FTIR is widely used to study about the functional groups present in the outer surface of the cell. In flotation as well as in solid/liquid separation the surface properties of the minerals, i.e. the specific surface charge is of essential importance. In the phenol-water interface electrical forces are also acting. Charged solid particle is surrounded by liquid ions of opposite charge to that of particle surface. As the solid particle moves through the liquid, a liquid layer surrounding the particle will move together with it. Consequently the ionized particle will carry a thin liquid shell making with it an integral whole. The potential difference between the external surface of the shell and the liquid as a whole is called electro kinetic potential or zeta potential. A zeta-potential study is used in colloid chemistry for observing the behavior of dispersive systems in liquids & characterizes the electrical double layer on the solid liquid interface, a very important fact in flotation and flocculation processes. Surface morphology was characterized using Scanning Electron Microscope

2. Materials and methods

2.1. Materials and instruments

Phenolic effluent was collected from nearby paper mill industry and its physicochemical parameters were analyzed. Sodium hydroxide flakes (98%), Hydrochloric acid (HCl), Cetyl Trimethyl Ammonium Bromide (CTAB) was bought from SDFCL (India). Reagents used in 4-amino anti pyrine (4-AAP) procedure for spectrophotometric determination (4-AAP, K_3Fe (CN)

$_6$, K_2HPO_4, KH_2PO_4 and ammonium hydroxide) were of analytical grade and supplied by Merck (Germany).

Dried yeast cells *Saccharomyces cerevisiae,* procured from a local brewery were ground to fine powder using an agate pestle and mortar in order to increase the surface area of the bio mass. It is then stored in a desiccator for future use. Dried biomass was used as an absorbent.

Sartorius basic meter PB-11 (Germany) was used for measuring pH of feed solution and UV-Vis spectrophotometer Shimadzu-UV-2450 (Japan) used for measuring absorbance of the phenolic solution.

2.2 Biosorption experiments

The biosorption experiments were carried out in a series of 250 mL Erlenmeyer flasks containing 50 mL phenol. The phenol biosorption performance at different concentration varying from 100-250ppm was investigated. 0.10g biosorbent was transferred into flasks with an initial phenol concentration ranging from 100-250ppm. Additional experiments were carried out with various biosorbent dosages ranging from 0.05-0.2 g. The pH studies were carried out for different acidic and basic conditions (6, 7, 8, and 9). All the flasks were sealed with cotton stoppers and shaken in a rotary shaker with a speed of 120 rpm. Samples of the flasks were separately taken at appropriate times to observe residual phenol using 4-amino anti pyrine method.

2.3 Biosorptive foam separation experiments

The biosorptive foam separation process scheme is shown in (Fig.1 & Fig.2).The biosorptive foam separation column is a cylindrical glass tube (1000mm in height and 160mm diameter) with a curved opening & a sparger of pore size 40 microns, fixed at the bottom, functioning as a gas distributor. Airflow is metered using a rotameter. Foamate was collected continuously at the top of the column. The liquid obtained by breaking the foam and the residual solution were analyzed using a UV spectrophotometer at a wavelength of 510 nm for phenol concentration. Feed solutions with different surfactant concentrations were used.

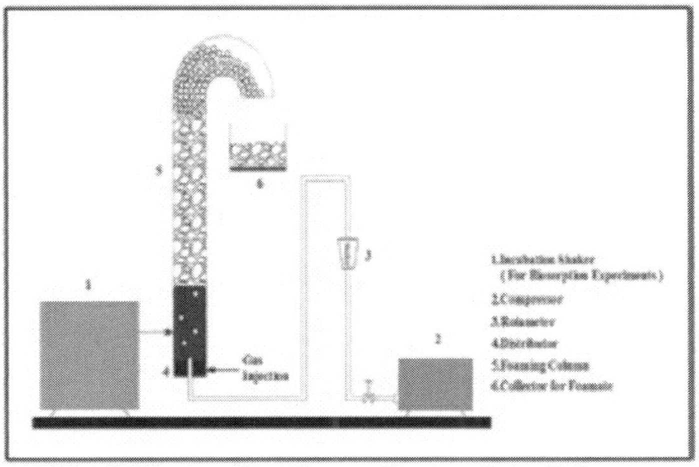

Fig.1 Biosorptive Foam Separation Column

The surfactant to be used in the process must be capable of providing electrostatic interaction with the sample to be analyzed. Hence, the surfactant used here is of cationic Cetyl trimethyl ammonium bromide (CTAB) to attract anionic phenolate.

A known volume of the phenol biosorbed solution, having predetermined concentration of phenol was taken into the foam column. 0.1 % (w/v) Surfactant was added. Air was then passed through the phenol biosorbed solution at a controlled flow rate. After allowing drainage for some specific time, the foam and the residual liquid were separated. The liquid obtained by breaking the foam and the residual pool liquid was then analysed to determine the phenol concentrations using UV Spectrophotometer. The above procedure was repeated for various surfactant concentration, flow rates and height of liquid pool height (Natarajan Saravani et al, 2016).

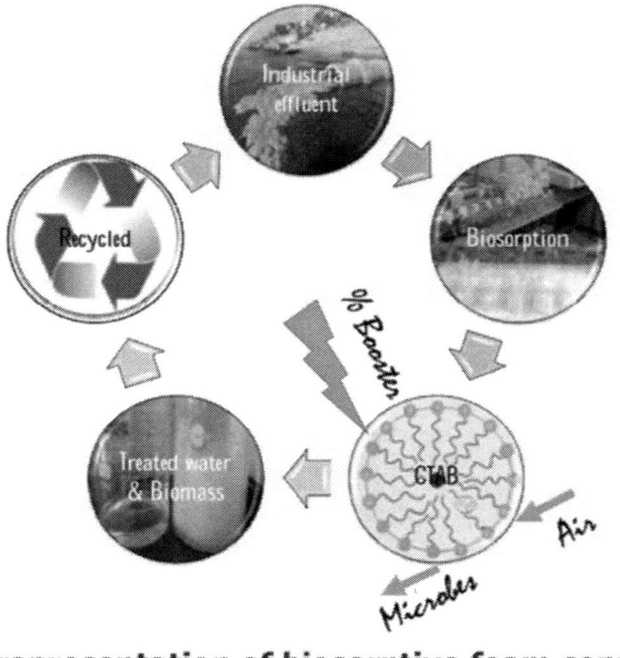

Fig.2 Schematic representation of biosorptive foam separation process

2.4 Phenol estimation

The residual phenol concentration in the culture medium was estimated using 4-aminoantipyrene method based on rapid condensation with 4-aminoantipyrine, followed by oxidation with potassium ferricyanide under alkaline conditions to give a red colour product. The concentration of phenol was calculated from the absorbance values at 510 nm using standards. Percentage removal and enrichment ratio were determined by Eq. (1), and (2):

$$\text{Removal of phenol (\%)} = \frac{C_o - C_f}{C_o} \times 100 \quad (1)$$

$$\text{Enrichment ratio} = \frac{C_{cf}}{C_o} \quad (2)$$

Where C_o and C_f are the phenol concentration (mg/L) in the influent and effluent streams respectively, and C_{cf} is phenol concentration in the collapsed foam solution (mg/L), (Natarajan Saravani et al, 2016).

2.5 Characterization experiments

The FTIR spectra of the *Saccharomyces cerevisiae* were recorded on a Perkin Elmer spectrum version 10.03.09, instrument model spectrum two and instrument serial number 92626.

The zeta (ζ) potential of the *Saccharomyces cerevisiae* cells were performed as a function of pH, for both fresh and after biosorptive foam separation using the micro electrophoretic apparatus Zeta Plus (Brookhaven Instruments Corporation, USA).

Scanning electron microscopic images were caught and hence the morphology was characterized using VEGA3 SBH version 4.1.20.3 model, and the instrument serial number VG98912771N.

3. Results & Discussion

3.1 Effects of feed concentration, equilibrium time, biosorbent dose, and pH on biosorption

Batch Experiments was ensued at room temperature, using various concentrations ranging from 100 to 250 mg/l to find the equilibrium time and optimum concentration of phenolic content in pulp and paper mill effluent. The conditions smeared to precede with the batch experiments were; speed of agitation 120 rpm, neutral pH and 0.1g dosage in 50 ml. With an increase in agitation time from 0 to 60 hr, the percentage removal reached up to 99.0%. With the increase in phenol concentration ranging from 100-250 mg/l, the percentage removal increased till 99.0% for 100 mg/l in 60 hrs. There was a rapid adsorption comprehended in the start, and becomes bit slower while the equilibrium state was achieved. Equilibrium was attained before the expected time and it was also agreed by other authors. Initial concentration difference provides the driving force. Higher the difference between the aqueous and solid phase, there is an increase in the rate of adsorbate molecule in bulk solution that passes through to reach the adsorbent surface (Juan et al, 2006).

The biosorbent dosage effects were tested for a range of 0.05g-0.20g. The results for the entire range of dose 0.05g-0.20g examined showed a very similar trend of very rapid removal at the initial, followed by a final stability and hence equilibrium. An increase of biosorbent dosage from 0.05g-0.20g leads an increase from 33.1 % to 99.0 % of removal of phenol. This increase in phenol loading capacity for a corresponding increase in dosage may be due to the accessibility of greater amounts of active sites of the sorbent remaining unsaturated. Utmost removal in the pulp and paper effluent was accomplished at a reduced dose of 0.1 g. A further increase shows decreased response due to the binding of almost all the phenol to the sorbent and establishment of equilibrium. That is, the available phenol molecules are inadequate to cover all the exchangeable sites on the sorbent, resulting in a lower phenol uptake (N. Saravani and M. Arulmozhi, 2016).

With varying pH ranging from acidic to base (6-9), pH optimization experiments were carried out using the optimum concentration (100 mg/l), dose of 0.1 g to find out the optimum. The pH is an important parameter that affects phenol solution chemistry and the surface characteristics of *Saccharomyces cerevisiae*. Response remains almost constant throughout the range of pH examined. According to the results obtained it was seen that both acidic and basic pH attain constant equilibrium and errand biosorption, yet biosorption of phenolic compound from pulp and paper was most favored at neutral pH 7. At acidic pH, the cell wall of *Saccharomyces cerevisiae* (biosorbent) was closely related with Hydronium ions competing for negative active sites and, it was caused due to the repulsive force that restricts the approach of phenolate ions. Amide groups and imidazole are positively charged, when protonated and play a vital role in phenolics adsorption. Moreover the amino and carboxyl groups, nitrogen and oxygen of the peptide bonds may also participate in the scene, which may mainly depends on the degree of displacement of protons, determined by the pH (Saifuddin, N. and Raziah, 2007). The maximum equilibrium biosorptive capacity of phenol compounds in pulp and paper effluent was achieved at pH 7. When the pH was increased further the degree of protonation decreases and hence the percentage removal decreases.

3.2 Effects of liquid pool height, surfactant concentration, and air flow rate on biosorptive foam separation performance

Biosorptive foam separation is a hybrid procedure, which in this context was used to enhance phenolics removal in the treatment of the paper and pulp industry effluent. Percentage removal of phenol in the effluent was noted to be 99.0% at the end of biosorption. Considering the aspect that even minute traces of phenol could be harmful, the next stage involves the removal of phenol loaded biomass and the minimal traces of phenol left behind, by the addition of a cationic surfactant. The exhausted cells and the final traces of phenol were eradicated along with the foam produced during the process. The influence of flotation parameters on percentage removal and enrichment ratio were studied and optimized during the process.

The effect of liquid pool height on the percentage removal of phenol and enrichment ratio using Cetyl Trimethyl Ammonium Bromide was studied at a surfactant concentration of 0.2 (w/v %) and an airflow rate of 4.5 lpm. As the liquid pool height increases from 10 to 30 cm the percentage removal of phenol also increases from 37.5 to 50.4%. Similarly the enrichment ratio also increased from 9.0 to 13.5. This consequence is mainly due to the concept that as the residence time of the bubbles is high, when the liquid pool height is high. The enrichment of the phenol on the surface of the bubble is also high until it reaches a stage of equilibrium. When the height is increased beyond 30 cm the percentage and the enrichment ratio decreases to 48.35% and 5.61 respectively. This is because the equilibrium has been already reached and further addition had no effect (Arulmozhi et al 2013).

The effect of Cetyl Trimethyl Ammonium Bromide on spent aqueous phenol under constant conditions of liquid pool height of 30 cm and air flow rate of 4.5 lpm & a variety of surfactant concentrations ranging from 0.1 to 0.4 (w/v %) of water were studied. The ability of surfactants in the treatment of effluents has been reported by Afzal Shah et al (Afzal Shah et al, 2016). The recovery of phenol using CTAB as the surfactant was found high due to its cationic nature as phenol is slightly acidic and anionic. As the concentration of the surfactant increased the percentage removal of phenol was also found to increase from 9.8 to 50.92%

and the enrichment ratio decreased from 10.78 to 9.73 respectively. On further increase of the surfactant concentration from 0.3 to 0.4% the percentage removal of phenol was found to decrease from 23.2 to 22.85% with an increase in the enrichment ratio from 11.96 to 12.27, in agreement to the report by Arulmozhi et al (Arulmozhi et al, 2013). The increase in the percentage removal was due to the fact that the effluent containing phenol binds to the cationic surfactant and removed to the foam. Once the level of saturation is attained the percentage of phenol decreased as surplus micelles were formed with high surfactant concentration (on increasing beyond 0.2 (w/v %).

The effect of air flow rate on enrichment ratio and percentage removal of the phenol containing effluent is at an optimum conditions liquid pool height of 30 cm and surfactant concentration of 0.2 (w/v %). As the air flow rate increases from 4.5 to 5 lpm, the percentage removal increased from 50.0% to 60.64% and the enrichment ratio increased from 1.73 to 4.85. This occurred due to the basis that initially at low flow rates, the quantity of phenol to be transported to the foam bubble surfaces are more. But the residence time of the bubble in the column is low which in turn results in more water content on the foam thereby enhancing the removal of phenol in the suspension. On further increase of air flow rate to 5.5 and 6, it was found that the percentage removal of phenol decreased to 27.8 and 25.0% and the enrichment ratio reduced to 3.6 to 1.5 due to higher air flow the bubbles reached the top of the column causing removal of surfactant rather than the trace level phenol. These results were in accordance with the results by Arulmozhi et al (Arulmozhi et al 2010).

3.3. Isotherm studies, Fowler Guggenheim isotherm model

In order to provide quantitative information on the adsorption capacity of the sorbent, experiments were carried out under the experimental conditions of 30^{0}C, 2 g/L of the sorbent and phenolic effluent of different initial concentration (100-250 mg/L). These flasks were placed on a rotating incubation shaker with a speed of 120 rpm till equilibrium. The amount adsorbed at equilibrium w_e is calculated by the Eq. (3)

$$q_e = \frac{(Co - Ce)V}{m} \quad (3)$$

Where C_o is the initial phenol concentration, V is the volume of solution (L) and m is the dry weight of the biosorbent (g) Equilibrium data were fitted using ten different isotherm models, namely Langmuir, Freundlich, Tempkin, Redlich Peterson, BET, Dubinin Radushkevich, Elovich, Halsey, Harkins Jura, and Fowler Guggenheim isotherm models. Results showed that Fowler Guggenheim isotherm fitted the data well with an R^2 value of 0.9703 (Fig. 2).

Fowler- Guggenheim isotherm model equation explains the lateral interactions of the adsorbed molecules. The equation is expressed as Eq. (4);

$$\frac{\theta}{C_o} = K_{FG}\,(1-\theta)^{n_{FG}} \quad (4)$$

And the linearized form is given by Eq. (5)

$$ln\,\frac{C_e(1-\theta)}{\theta} = -lnK_{FG} + \frac{2W\theta}{RT} \quad (5)$$

Where K_{FG} is Fowler- Guggenheim equilibrium constant (L mg^{-1})

n_{FG} is model exponent

$\theta = 1 - \frac{C_e}{C_o}$ is the degree of surface coverage.

W is the interaction between the adsorbed molecules (kJ mol^{-1})

R is the gas constant (8.314 J $mol^{-1} K^{-1}$)
T is the absolute temperature (K)

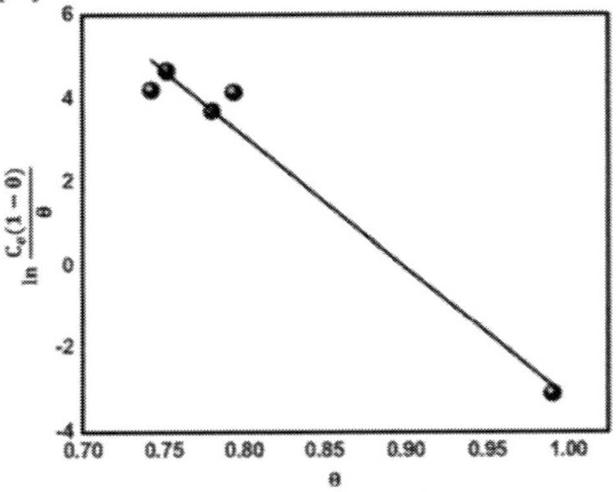

Fig. 2 Linearized Fowler Guggenheim isotherm model for phenol removal (temperature of 30°C, dose of 0.1g/50 mlbiosorbent, agitation speed of 200 rpm and pH 7).

3.4. Biosorption kinetics & Pseudo second order model

Biosorption kinetics was studied at 30^0 C and the rate was monitored till a contact time of sixty hours and the results were compared to theoretical models. Adsorbed amount of phenol at any time q_t is calculated by Eq. (6)

$$q_t = \frac{(Co-Ct)V}{m} \quad (6)$$

Where q_t is the amount of biosorption at time t (mg/g), C_o is the initial concentration of the phenolic compound in the solution (mg/L), C_t is the concentration of the phenolic compound in the solution at time t (mg/L), V is the volume of solution (L) and m is the mass of the biosorbent (g).

Pseudo first order, Pseudo second order, Elovich, Intraparticle diffusion and Mass transfer diffusion models (Natarajan Saravani et al, 2016) were tested in this study and best fit shown in Fig. (3). Results showed that the Pseudo second order model fitted the data well. The pseudo-second-order based on adsorption equilibrium capacity is expressed in Eq. (7) and linear form of pseudo second order is given in Eq. (8)

$$\frac{dq_t}{dt} = K_{2ad}(q_e - q_t)^2 \quad (7)$$

$$\frac{t}{q_t} = \frac{1}{K_{2ad}q_e^2} + \frac{t}{q_e} \quad (8)$$

Where K_{2ad} Pseudo-second-order rate constant (1/mol.hr).
A linear plot of t/q_t vs. t gives a slope of the equilibrium adsorption capacity q_e

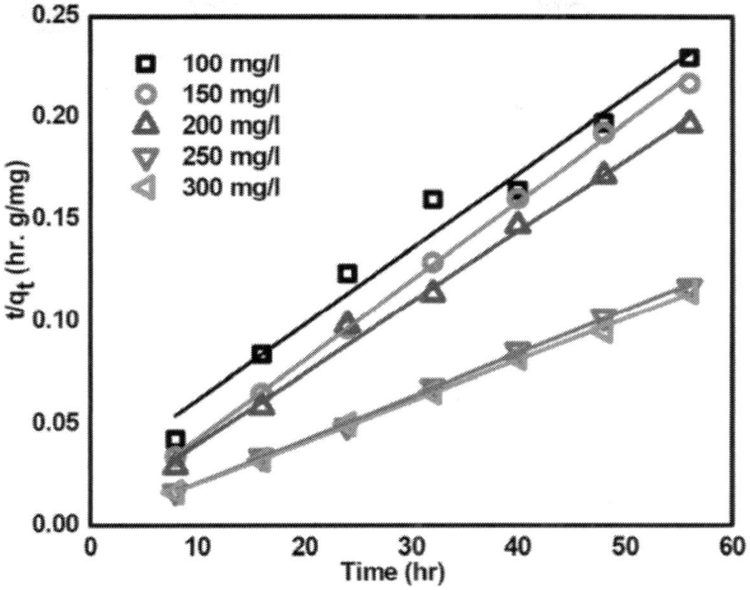

Fig. 3 Linearized Pseudo second-order model of phenol removal (temperature of 30°C, dose of 0.1g/50 mlbiosorbent, agitation speed of 200 rpm and pH 7).

3.5. Fourier transform infrared (FTIR) analysis

The FTIR spectra of raw, phenol and, phenol CTAB treated yeast in the range of 4000–400 cm^{-1} were taken to acquire information on the nature of the cell wall and phenol interaction, were shown in Fig.4. Chemical characteristics of the samples were analyzed by FTIR spectrometry after the samples were prepared as KBr discs. FTIR spectrum of the fresh biosorbent exhibited broad and strong absorption bands around 3200-3500 cm^{-1} indicating the presence of –OH stretching vibrations. Other peaks around 1691.27cm^{-1}(amide band), 1335.46 cm^{-1} (aromatic nitro compound NO_2 symmetric stretching; imidazole) and 1213.97 cm^{-1} (C–H stretching) have also been observed. Peaks around 724.14 cm^{-1} (NH bending), 523.58 cm^{-1} (-CH stretching of aromatic ring) and 438.73cm^{-1} (C–X stretching) have also been observed. The phenol treated yeast showed a broad peak at around 3600–3000 cm^{-1}, which could be attributed to sorption of the phenol onto the yeast. A shift in peak from 1691.27cm^{-1} to 1690.30 cm^{-1} and 1335.46 cm^{-1} to 1334.50 cm^{-1} is an indicative of active participation of amide band and the aromatic nitro compound NO_2 symmetric stretching (imidazole) in phenol adsorption. The peaks at 1213.97 cm^{-1} can be designated as the C-H stretching in carboxyl or amine groups. Other peaks around 724.14 cm^{-1}(NH bending), 473.44 cm^{-1} (S-S stretching) and 432.94 cm^{-1} (C–X stretching) have also been observed. The phenol CTAB treated yeast showed a new peak at 3743.15 cm^{-1} confirming the presence –OH stretching and bending and also showed a broad peak at around 3600–3000 cm^{-1}, which could be attributed to sorption of the phenol onto the yeast. A peak at 1688.37 cm^{-1} confirms the presence of amide band. A peak at 1621.84 cm^{-1} confirms the presence of intermolecular β structure of amide band. A peak at 1334.5 cm^{-1}(aromatic nitro compound NO_2 symmetric stretching). The peak at 1219.97 cm^{-1} is indicative of aliphatic C-N stretching. Other peaks around 720.28 cm^{-1}(NH bending), 472.47 cm^{-1} (S-S stretching) and 435.83 cm^{-1} (C–X stretching) have also been observed (Stuart).

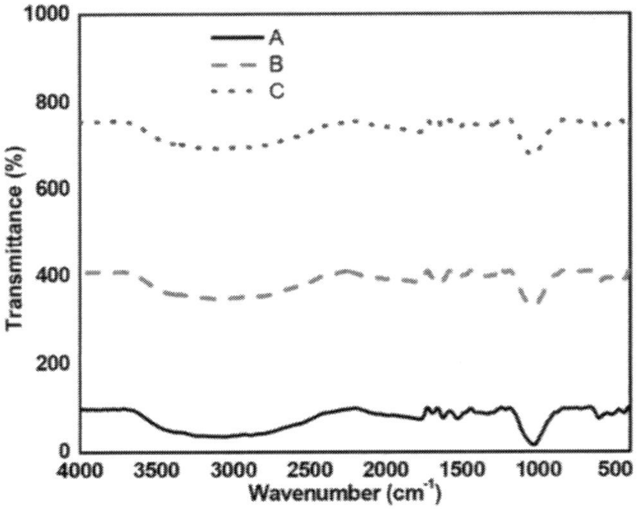

Fig. 4 FTIR spectra of *Saccharomyces cerevisiae*(A) fresh biosorbent (B) after biosorption and (C) after biosorptive foam separation,

3.6. Electro-kinetic studies

This instrument determines the electrophoretic mobility and converts to ζ potential using the Smoluchowski approximation (Natarajan Saravani et al, 2016). The reported values are an average of 10 measurements. The ζ potential measurements provide information regarding the surface charge of fresh and phenol CTAB loaded cells (Fig. 5).The pH values were adjusted with the addition of NaOH and HCl. Electrophoretic mobility measurements were also performed and the respective ζ potential values were calculated, in order to examine the possible biosorption mechanism. The electro kinetic study provides preliminary information regarding the type of surfactant to be added for the separation of fine particles by flotation. The experimental negative ζ values, right through the pH values examined, shows an isoelectric point (IEP) at a pH of around 2, proved the application of the positive surfactant. This renders hydrophobicity and hence, floatability. The hydrophobicity of yeast mainly depends on the presence of various polysaccharides, proteins, lipids, and carboxylate, amines, and imidazole, phosphate, sulfhydryl, sulfate and hydroxyl group moieties. When examined alone, the zeta potential values going on decreases till a pH of 7 and the charge reverses after that, which corresponds to maximum sorption capacity. The surface charge of the non-living biomass becomes less negative till the neutral pH and becomes more negative towards basic region in the presence of cationic surfactant. At its extreme end, the suspension is quite unstable. Microbial walls are anionic due to the presence of carboxyl, hydroxyl, phosphoryl and other negatively charged sites. The phenol binding properties of yeast are largely due to positively charged amino and imidazole groups, which are part of the cell wall structure. Due to this fixed cation content, they may exhibit high sorption capacities (Saifuddin, N. and Raziah, 2007). From the above discussions it is evident that the zeta-potential changes according to changes in the solution pH, and zeta-potential measurements makes it possible to follow the changes. After biosorptive foam separation the zeta potential of biomass becomes less negative. This could be attributed to the electrostatic interaction between the biomass, CTAB and phenolate ions. This effect is caused by the dissociation of the functional groups at high pH values, which will promote an electrostatic attraction toward negatively charged phenolate ions in the solution.

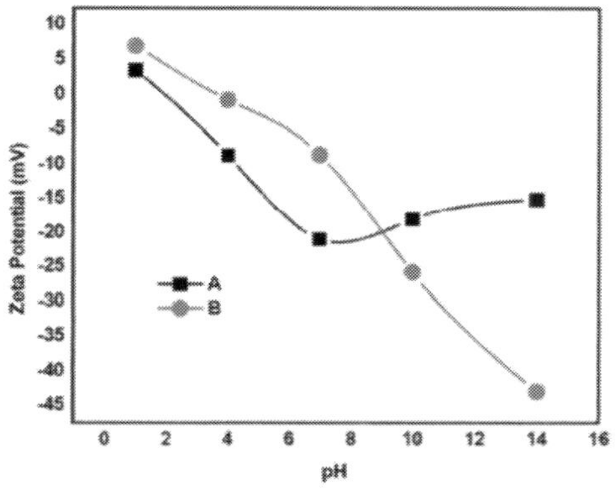

Fig. 5 ζ-Potential-pH curves of (A) fresh and (B) phenol CTAB loaded *Saccharomyces cerevisiae.*

3.7 SEM Analysis

It is evident from the Fig.6 that the *Saccharomyces cerevisiae* dead cells were of oval shaped aggregates with a smooth and glossy finish. They stay apart each other with almost no interaction. The change in surface morphology of the *Saccharomyces cerevisiae,* before and after biosorptive foam separation was observed at a magnification of 5 μm, using an accelerating voltage of 10 kV. All the samples were sputter coated with a gold ultrathin film prior to its mounting on a sample stub for SEM imaging. All the *Saccharomyces cerevisiae* dead cells were perfectly oval both before and after biosorptive foam separation. *Saccharomyces cerevisiae* cells were bit heftier indicating the presence of phenolic compounds after biosorptive foam separation.

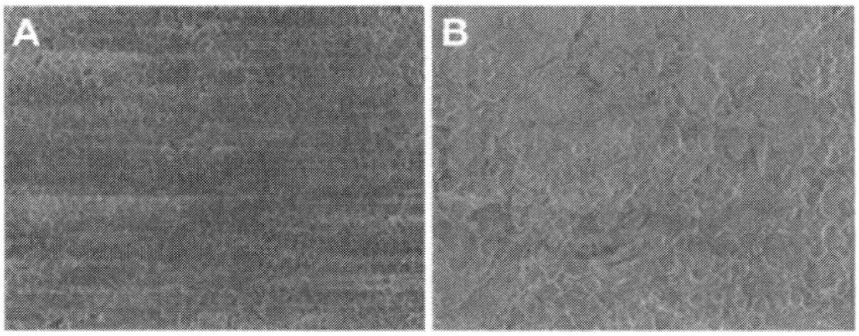

Fig.6 SEM images of *Saccharomyces cerevisiae***(A)** fresh biosorbent**(B)** after biosorptive foam separation.

4. CONCLUSION

The results of biosorptive foam separation of phenolic compounds by *Saccharomyces cerevisiae*, has been substantiated to be a most favorable technique. It is a two stage process where sorption onto dead cells of *Saccharomyces cerevisiae* constitutes the first stage and separation of phenol loaded biomass, and phenol lean solution without *Saccharomyces cerevisiae* cells constitutes the next. The conditions for maximum phenol removal are 100 mg/L of initial phenol concentration, sixty hours of contact time period, 0.1 g of cells, and a pH of 7, and the conditions for the second stage of biosorptive foam separation are a liquid pool height of 30 cm, cationic surfactant concentration of 0.2% (w/v), and an air flow rate of 5 L/m. Isotherm and kinetic studies revealed that the biosorption process followed the Fowler-Guggenheim isotherm model (R^2 = 0. 9703) and pseudo-second order kinetics with a kinetic constant of $5.5425X10^{-5}$ hr^{-1}. FTIR spectra for the fresh biosorbent, after biosorption, and after biosorptive foam separation were generated to indicate the interactions and bonding. The ζ-potential of the *Saccharomyces cerevisiae* cells was affected by the presence of phenol at different pH values showing isoelectric point at a pH of around 2. The results obtained from this study showed that *Saccharomyces cerevisiae* can be a good sorbent for the removal of phenol from an aqueous solution and CTAB can be a most appropriate surfactant. The recovery of phenol loaded biomass and final traces of phenol by floatation were found to be of the order of 99.75%.

Acknowledgements

The authors express their gratitude for the financial support extended by the Department of Science and Technology (DST-Government of India).Grant. No: SR/FT/ETA-0017/2012 and Technical Education Quality Improvement program (TEQIP II).

References

1. Abdelwahab O, Amin NK (2013), 'Biosorption of phenol from aqueous solutions by Luffacylindrica fibers: Kinetics, isotherm, and thermodynamic studies'. Egyptian Journal of Aquatic Research vol. 39, pp.215–223.
2. Aditi Banerjee, AlokeGhoshal K (2010); 'Phenol degradation by Bacillus cereus: Pathway and kinetic modeling', *Bioresource Technology*, 101, 5501-5507.
3. Aditi Banerjee, Aloke K. Ghoshal (2010), 'Isolation and characterization of hyper phenol tolerant Bacillus sp. from oil refinery and exploration sites', Journal of Hazardous Materials, vol.176, pp. 85–91
4. Aditi Banerjee, Aloke K. Ghoshal (2011), 'Phenol degradation performance by isolated *Bacillus cereus* immobilized in alginate', International Bio deterioration & Biodegradation, vol. 65, pp.1052-1060.
5. Afzal Shah,SuniyaShahzad,AzeemaMunir,Mallikarjuna N. Nadagouda,GulShahzada Khan,DilawarFarhan Shams,Dionysios D. Dionysiou, Usman Ali Rana(2016); 'Micelles as Soil and Water Decontamination Agents', *Journal of Chemical Reviews,* DOI: 10.1021/acs.chemrev.6b00132.

570638 Biosurfactant Fermented from Waste Cooking Oil and Its Application on Oil Sludge Celaning

The maximum biosurfactant concentration of 5.98 g L^{-1} was achieved from kitchen waste oil by *Pseudomonas aeruginosa*. Then biosurfactant was extracted from the broth and applied to wash oil-based drilling cuttings. Effect rules of six parameters on oil recovery were investigated separately, and the main influencing factors of concentration, pH, and temperature were screened out by Plackett-Burman design. And then the oil washing conditions were optimized by response surface method. The maximum recovery rate was predicted to be 70.48%, when rhamnolipid concentration was 325 mg L^{-1}, pH was 10.70, temperature was 63 °C, washing time was 35 min, stirring speed was 450 r min^{-1}, and liquid/solid ratio was 5:1. Finally, the optimal washing parameters and corresponding maximum recovery rate were verified through experiment, with the relative error less than 0.006%. The results indicating promising prospect of biosurfactant application in industrial oil recovery.

571053 Liquid-Liquid Extraction of Chlorophenols from Wastewater Using Hydrophobic Ionic Liquids

In this work, liquid-liquid extraction of 3-chlorophenol, 2, 5-dichlorophenol, 2, 4, 6-trichlorophenol and pentachlorophenol from wastewater was conducted using five hydrophobic ionic liquids (ILs). Moreover, the impact of key parameters like molecular structure, pH, temperature, phase ratio and initial concentration of feed on the removal of the chlorophenols (CPs) from their aqueous solution were studied. The results showed that the molecular structure of the IL has an important impact on the removal of CPs. The evaluation of different ILs with the same CPs revealed that piperidinium and pyridinium based ILs have greater extraction efficiency than sulfonium and imidazolium based ILs. It was found that the extraction efficiency decreased with the decrease of IL: wastewater mass ratio, with a maximum extraction at a ratio of 1:1. The results showed that temperature and initial concentrations in the range of 25-45°C and 4-1000 mg/L, respectively had no significant impact on the extraction for all studied systems.

ChemPren: A new technology for the conversion of mixed municipal waste to basic chemicals and fuels

Dr. Piyush Sabharwall, Dr. Lynn Wendt, Ms. Roxanne Pinsky and Dr. Anne M. Gaffney
Idaho National Laboratory
Idaho Falls, ID 83404

Dr. Gennaro J Maffia, Amanda E. Peterman and Xufang (Katie) Liu
Manhattan College
Riverdale, NY 10471

Currently, over 90% of used plastics are not directly recycled and end up in U.S. landfills or incinerators. As more composite plastic materials are developed, this problem will only get worse. Technologies to up convert waste plastics to petrochemicals that can be used in the production of new plastics and materials, have been developed in the U.S., Japan, and Korea [1-4]. Catalytic conversion of waste plastics to Benzene, Toluene and Xylenes (BTX) was initially developed by ARCO chemical in the early to mid-1990s [1,2]. The BTX mixture is a versatile feedstock and can be used for the production of polyethylene terephthalate (PET), polycarbonates, polyurethanes, nylons, SBR, ABS and other polymers which are in turn used to produce beverage bottles, clothing, carpeting, automotive components and a broad range of other products. Intellectual property for improvements in the process and catalyst to up convert waste plastics to BTX was recently signed over to INL [5]. This technology capitalizes on INL owned intellectual property and advances the technical readiness level from the current level of 4 to a level of 7 with the project culminating in demonstrating at least 1000 hours of pilot plant operation.

References:

1. Jones, A.C., Gaffney, A.M., Leyshon, D.W., Cozzone, G.E., Sofranko, J.A., Process for reclaiming a thermoset Polymer, U.S. Patent No. 5,192,809, March 9, 1993.
2. Gaffney, A.M., Jones, A., Process for treating polyether polyols, U.S. Patent No. 5,462,971, October 31, 1995.
3. Nishino, J., Itoh, M., Fujiyoshi, H., Uemichi, Y, Catalytic degradation of plastic waste into petrochemicals using Ga-ZSM-5, Fuel 87 (2008)3681-3686.
4. Jung, S., Cho, M., Kang, B., Kim, J., Pyrolysis of a fraction of waste polypropylene and polyethylene for the recovery of BTX aromatics using a fluidized bed reactor, Fuel Proc. Tech. 91 (2010) 277-284.
5. Gaffney, A.M., Alloyed zeolite catalyst component, Method for making and catalytic application thereof, U.S. Patent Application 0215043 A1, August 23, 2013.

559476 Anaerobic Digestion of Food Wastes into Energy and Fertilizer

Anaerobic digestion (AD) has been evaluated as one of the most cost-efficient and environmentally friendly biotechnologies for renewable energy production, in the form of heat and power, organic waste management, and recovery of valuable nutrient rich digestate [1]. The latter has the potential to be used as fertilizer for agricultural utilization, landscaping, and soil amelioration since it contains essential nutrients such as nitrogen (N), phosphorus (P), and potassium (K) [2]. Digestate is produced from anaerobic degradation of organic wastes, mainly derived from food waste, crops and sewage sludge from wastewater treatment processes. However, the use of digestate has become a crucial challenge due to the limitation of cost and quality for the production of fertilizers as well as the application location. This study examined the optimization of an anaerobic digestion waste-to-energy system for food waste management and conversion of digestate to fertilizers. Following which, how the use of digestate for the cultivation of vegetables was investigated in comparison with chemical fertilizers. The objectives of this study are: 1. to optimize the operational parameters of a distributed anaerobic digestion waste-to-energy system for food waste management. 2. to study the efficiency of various fertilizers extracted from digestate and chemical fertilizers by cultivating green vegetables and comparing the major indexes for the growth of vegetables.

571999 Subcritical Water Upgrading of Coal Refuse with Food Waste

The goal of this research was to improve the fuel characteristics of coal refuse through co-hydrothermal carbonization (co-HTC) with food waste. Coal refuse, also known as coal waste and gob, are by-products resulted from coal mining. Coal processing plants reject 30-60% materials from the mined coal during the cleaning procedure and it contains 40-70% energy compared to coal. However, toxic elements (sulfur, heavy metals, and ash) present in coal refuse possess serious environmental threat and that is why it is not possible to combust coal refuse without prior treatment. Co-HTC of coal refuse along with food waste can be an efficient option to utilize this huge amount of available fuel resource. Co-HTC is a thermochemical process, where two feedstocks are treated at high temperature (180-280 °C) and corresponding vapor pressure (5-60 bars) for a short period of time. Coal refuse from local coal mine and food wastes from Ohio University Dining Hall were co-hydrothermally treated at three different temperatures (180, 230, and 280 °C) for 30 min. The acidic medium created by the hydrothermal carbonization of food waste catalyzed the upgrading of coal refuse. The elemental carbon (%) increased to a maximum value of 49.4 % for co-HTC solids from 18.3 % for raw coal refuses. Additionally, the ash percentage decreased from 66.4 % to 27.4 % for co-HTC solids. Moreover, the van-Krevelen diagram showed that the co-HTC solids have fuel characteristics similar to bituminous and sub-bituminous coals. The total sulfur concentration also decreased to less than 1.4 % from 9.9 %. The raw coal refuse samples and treated samples were further analyzed using the standard test method (ASTM D2492-02) and an induced coupled plasma optical emission spectroscopy (ICP-OES) to determine sulfate sulfur and pyritic sulfur concentration, respectively. The organic sulfur was calculated by subtracting the summation of sulfate and pyritic sulfur from the total sulfur concentration. The study also evaluated the distribution of sulfur among the co-HTC products. Later, an economic analysis was performed to check the economic feasibility of large scale hydrochar production. The economic study revealed that the production cost of co-HTC solids to be $39 per ton in the basis of a 110 MWe power plant.

556642 Polar Solvents Facilitate Efficient Microwave Pretreatment of Woody Biomass

Microwave pretreatment of woody biomass may be the key to unlocking an energy efficient process for recalcitrant waste stream for the production of renewable chemicals and fuels. Sugars derived from non-food sources such as woody biomass have the potential to be converted into a wide range of chemicals. However, woody biomass requires pretreatment to open the structure to access cellulose. Steam explosion, organosolv and sulphite pretreatment are currently used to remove lignin from wood [1]. However, these methods require substantial energy to be effective [1]. Development of an alternative, more efficient pretreatment processes for the release of sugars and other useful fractions from renewable biomass sources is necessary for a sustainable future.

Ionic liquids have been extensively researched for woody biomass pretreatment as they have proven to yield exceptional results via ionic conduction mechanism. However, limited research has been reported on the intentional exploitation of polar solvents, which increase the microwave absorption capacity via dipolar polarisation mechanism, for the pretreatment of recalcitrant biomass. Microwave absorption can vary significantly depending upon the ratio of effective dielectric loss factor (") to the dielectric constant (), of which the ratio is commonly known as the loss tangent (tan).

In this research, we investigated the use of solvents with high tan value to increase the efficiency of microwave-assisted solvolysis of woody biomass through an increase in the heating rates, and hence reduction of the energy consumption. To the best of our knowledge formaldehyde pretreatment of wood has not yet been reported under microwave irradiation. We studied the microwave irradiation of *Eucalyptus Regnans* and *Pinus Radiata* sawdust (~0.5 mm) in the presence of formalin solution containing 1.8, 3.7 and 10 wt% formaldehyde and water as the control at 180°C for 10 minutes.

It was found that 3.7 wt% formaldehyde (FA) solution is more efficient than water only for the pretreatment of wood sawdust at 180 °C and 10 min microwave exposure. The presence of formaldehyde in the solution accelerated the heating rate by 50% in the microwave. Furthermore, >75% of the hemicellulose sugars were released under mild microwave treatment conditions for both the wood species. Whereas in water-only pretreatment 32% and 55% of the hemicellulose was released from *E. regnans* and *P. radiata*, respectively.

A proposed mechanism for formaldehyde-mediated deconstruction of hemicellulose was developed. The lignin in the remaining sawdust from the FA pretreatment was analysed by 2D NMR, which detected the presence of stabilised lignin monomers. Whereas, with water-only pretreatment this does not occur. This stabilised platform may open-up a more advanced continuous process which could allow near theoretical lignin monomer yields through the subsequent hydorgenolysis.

576948 Comparative Leading Pretreatments for Improving Lignin Waste Valorization to Biobased Products

Biorefineries offer a potential solution to the current energy and environmental challenges facing our generation. However, there are still many technical barriers in developing a sustainable biorefinery. Due to the intrinsic recalcitrance of lignocellulosic biomass (LCB), the inherent value of LCB is highly dependent on deconstructing and valorizing lignin polymer in a biorefinery concept, and not just exploiting the more uniform carbohydrates. Lignin is the largest renewable aromatic carbon source and the second most abundant terrestrial polymer after cellulose. Valorizing lignin solid waste to value-added products shows the potential to make a sustainability of biorefineries. Despite these advantages and the significant progress, lignin waste remains as an untapped resource and the application of lignin valorization is relatively unexplored due to the heterogeneity of diverse lignin substrates. We hereby compared the leading pretreatments including sulfuric acid, steam explosion, liquid hot water, AFEX, and alkaline pretreatment for tuning lignin chemistry and improving lignin waste valorization to biobased products. Lignin characterization results showed that the lignin chemistry and reactivity obviously depended on the pretreatment employed. After the delignification of pretreated solid, the lignin stream was used as carbon source for polyhydroxyalkanoates (PHA) production by engineered *Pseudomonas putida* KT2440. Results showed that multilevel fractionation and utilization process can achieve high-value conversion of lignin for PHA with high yield. Based on the further characterization of lignin form different fractionation streams, the molecular mechanism has been proposed to unveil the improved processibility of lignin from multilevel utilization process. Overall, improving lignin waste valorization to biobased products by employing leading pretreatments provided a unique set of features to deliver a sustainable biorefinery process.

571984 Waste to Resource: Synthesis of Electrolyte and Electrode Material in Vanadium Redox Flow Battery from Leachate of Industrial Waste

Fly ash, a typical waste driven out of boilers in combustion plants, contains a large range of metal oxides, and leads to environmental consequence without proper treatment. In this study, we focus on a fly ash generated from combustion of oil refinery waste, which includes substantial amount of metal oxides (mainly vanadium oxide and nickel oxide). Instead of landfill, chemical leaching was applied to extract and separate metallic components in the fly ash. Because of the high concentration of vanadium ion inside, the leachate can be used as the electrolyte in vanadium redox flow battery (VRFB). Unlike vanadium-related product synthesized from leachate, this application avoids solid-liquid separation and minimizes the wastewater generation. Meanwhile, as the secondary metallic composition, nickel ions become a low-cost resource for synthesis of nickel-dope nanomaterial that can be applied in VRFB as electrode materials to enhance V^{3+}/V^{2+} and VO_2^{+}/VO^{2+} reactions.

567686 Synthesis of $LiNi_{0.6}Co_{0.2}Mn_{0.2}O_2$ Cathode Material By a Carbonate Co-Precipitation from Spent Lithium-Ion Batteries

Synthesis of $LiNi_{0.6}Co_{0.2}Mn_{0.2}O_2$ cathode material by a carbonate co-precipitation from spent lithium-ion batteries

*Lei Xing (speaker)[1, 2], Shuying Sun[1,2], Sen Lin[1,2], Ping Li[1,2], Jianguo Yu[1,2]**

1. *National Engineering Research Center for Integrated Utilization of Salt Lake Resources, East China University of Science and Technology, Shanghai, China, 200237.*
2. *State Key laboratory of Chemical Engineering, East China University of Science and Technology, Shanghai, China, 200237.*

**Speaker and corresponding authors: E-mail: xinglei8406@126.com, jgyu@ecust.edu.cn*

Abstract:

The spent lithium ion batteries (LIBs) are classified as hazardous solid wastes because they contain toxic heavy metals and corrosive electrolytes. However, they contain various valuable metals such as copper, aluminum, magnesium, nickel, cobalt and lithium. Recycling of valuable metals from spent LIBs mitigate both resource shortage and environmental contamination problems [1]. So, the aim of this work is to present a promising approach for recycling high value-added metals from the cathode material of spent LIBs.

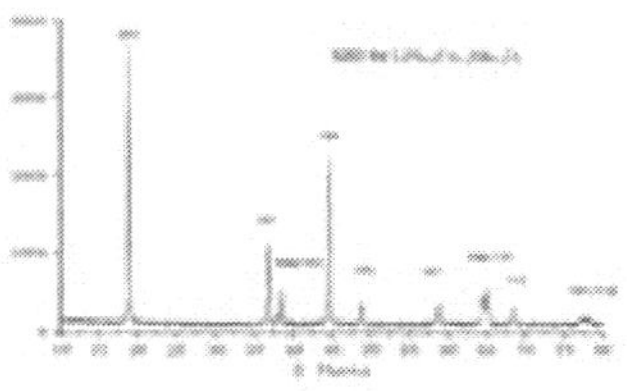

In this work, $LiNi_{0.6}Co_{0.2}Mn_{0.2}O_2$ cathode material is synthetized by a carbonate co-precipitation using the leaching liquid of spent LIBs as raw stocks. Firstly, the spent LIBs are manually dismantled to remove both the metal cases and plastic, and the cathodic material is separated from the Al foil by ultrasonic cleaning in the solvent 1-methyl-2-pyrrolidinone [2]. The valuable elements such as nickel, manganese and cobalt are obtained by leaching in sulfuric acid, and $NiSO_4$¡¤$6H_2O$, $CoSO_4$¡¤$7H_2O$ and $MnSO_4$¡¤H_2O are added into leaching solution to obtain the raw stocks for the $Ni_{0.6}Co_{0.2}Mn_{0.2}CO_3$ precursor through adjusting the molar radio of Ni: Co: Mn = 6: 2: 2. Secondly, the $Ni_{0.6}Co_{0.2}Mn_{0.2}CO_3$ precursor is synthesized by a carbonate co-precipitation method using the prepared raw stocks from leaching solution of spent LIBs. Thirdly, the $Ni_{0.6}Co_{0.2}Mn_{0.2}CO_3$ precursor is mixed with Li_2CO_3 at a Li/M (M=sum of Ni, Co and Mn) molar ratio 1.06, and the mixture is calcinated at 800-850¡æ under oxygen atmosphere to fabricate $LiNi_{0.6}Co_{0.2}Mn_{0.2}O_2$ cathode material. Finally, the advanced characterization methods (XRD, ICP-AES TG-DSC SEM-EDS) are used to demonstrate the synthetized $LiNi_{0.6}Co_{0.2}Mn_{0.2}O_2$ cathode material having a well-ordered layered structure and a low degree of cation mixing.

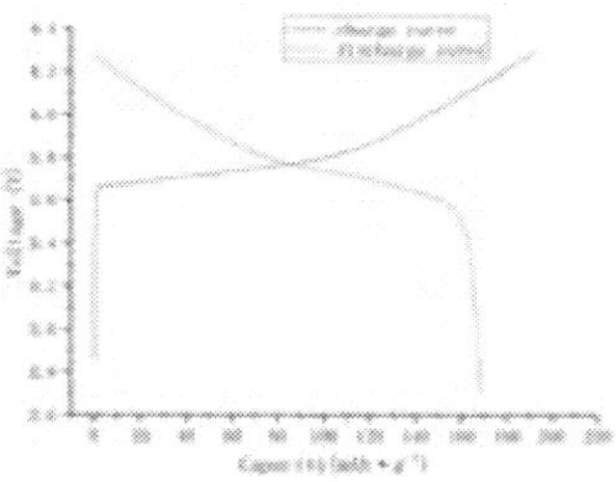

Furthermore, the electrochemical performance of the synthesized

$LiNi_{0.6}Co_{0.2}Mn_{0.2}O_2$ cathode material is tested in CR2025 coin cells. Electrochemical test methods include initial charge-discharge, continuous charge¨Cdischarge cycling, cyclic voltammetry and electrochemical impedance spectroscopy. It is found that the initial charge and discharge capacities of the $LiNi_{0.6}Co_{0.2}Mn_{0.2}O_2$ cathode material is 194.5mAh/g and 169.3mAh/g, and the initial coulombic efficiency is 87.05% when the initial charge-discharge curves are at a constant current density of 0.1C between 2.7V and 4.3V. The electrochemical performance of the prepared cathode material using the leaching liquid of spent LIBs is consistent with those of the cathode material synthetized from fresh materials. So, the proposed synthesis process has great potential for industrial-scale recycling from spent LIBs.

References:

[1] D.H.P. Kang, M. Chen, O.A.Ogunseitan, Potential environmental and human health impacts of rechargeable lithium batteries in electronic waste, Environ. Sci. Technol. 47 (2013) 5495-5503.

[2] L.H. He, S.Y. Sun, X.F. Song, J.G. Yu, Recovery of cathode materials and Al from spent lithium-ion batteries by ultrasonic cleaning , Waste Manag. 46 (2015) 523-528.

Keywords: lithium-ion batteries, cathode material, carbonate co-precipitation

557995 Sustainable Process for the Recycling of Polyethylene Phthalate

Polyethylene terephthalate (PET) is a thermoplastic polyester with excellent thermal and mechanical properties. PET is the third-most produced polymer (18% of world polymer production) and widely used in textile industry, X-ray films, food packaging particularly for water and soft-drink bottles.

Majority of plastic waste is going to land field. To recycle PET, chemical recycling has advantages over mechanical approaches in terms of better adaptability on heterogeneity of the solid wastes and no degradation of the product properties. In this paper, we will present a novel approach for chemical recycling of PET. This novel approach is featured by a simulated moving bed chromatography (SMBC) in the purification of Bis(2-Hydroxyethyl) terephthalate (BHET), which is the deploymerization product of PET. This novel approach can ensure a high purity of BHET that is suitable for reproducing chemical grade PET even with a low quality PET feedstock.

576903 Recycling Fe-P Slag As Energy Materials Via a Green Technics

Recycling waste is significant for the sustainable development of resource and environment. As one of the typical waste slags from phosphorus **industries, Fe-P generally has high melting point, plentiful iron and** phosphorus **elements and other impurities according to their raw materials and reaction technics. In order to recycle Fe-P slag as energy materials like $LiFePO_4$ and $FePO_4$, we have developed various routes including solid-state reaction and hydrothermal methods. The thermodynamics of different reaction processes are analyzed to obtain the possible routes with low energy consumption. Green technics with only water byproduct have been designed using the combined raw materials. $LiFePO_4$ and $FePO_4$ can be successfully obtained using $Fe_{1.5}P$ slag with ~1400°C melting point at no higher than 200°C. The purity and electrochemical performance of the as-synthesized samples are evaluated with various techniques. The conversion processes and key impact factors are discussed in detail. $FePO_4$ with good crystal structure and high purity can be obtained by adjusting the pH values of the product solution from the reaction of $Fe_{1.5}P$, H_3PO_4 and H_2O_2, which can meet the requirements for the production of $LiFePO_4$. On the other hand, $LiFePO_4$ can also be synthesized directly by using the raw materials of $Fe_{1.5}P$, H_3PO_4, H_2O_2 and LiOH. After being further coated with organic carbons using waste** sugarcane **slag, the as-obtained $LiFePO_4$ exhibits comparative electrochemical performance with that of current commercial samples. The discharge capacity can reach 150 mAh/g, and the capacity fade ratio is less than 1% after 100 cycles at 1.0C. The technics and equipment have also been proposed for mass production of $LiFePO_4$ and $FePO_4$ using Fe-P slag at low temperature. The green technics is interesting to recycle the waste with high melting point as function materials** through low-energy consumption way, **which is beneficial for the social sustainability with low environmental impact.**

576142 Overview of Waste Valorization Concepts and Principles

AIChE defines waste valorization as "the process of reusing, recycling or composting waste materials and converting them into more useful products including materials, chemicals, fuels or other sources of energy". "Waste-to-energy" aspects are becoming more prominent due to the rapid depletion of natural resources and increase in waste generation.

Waste valorization is the process of taking waste and changing it into constituent parts that can be utilized, and have value beyond the cost of the energy needed to process the transformation. The concept of waste valorization ties in with recycling and reuse technologies. Namely "enhancing the value", valorization is a process of changing residues into products at a much greater value. The results can include quality chemicals, materials, fuels, and energy, as well as many other products beneficial for a local economy.

It requires not only technical knowledge but also a global approach to accomplish this goal. Residues and products often need to be seen from a wider perspective that just one industry. Creating a closed system of valorization, industries may need to communicate with each other in the global economy pattern.

This presentation presents a technical overview on waste valorization addressing the principles and concepts involved.

571173 Conversion of Polyvinylchloride Containing Solid Wastes into Value-Added Products

Conversion of polyvinylchloride containing solid wastes into value-added products

Xianglan Bai, Joel Braden

Department of Mechanical Engineering, Iowa State University, Ames, IA, USA, 50011

Municipal Solid Waste (MSW) is a mixture of organic and synthetic materials. The organic material has the potential to be converted into valuable fuels through thermochemical conversion. On the other hand, main plastic wastes, such as PE and PP, can be an excellent source of chemicals or high-density liquid fuels. Previous studies by our group and others have shown that co-conversion of biomass and hydrocarbon-based plastics through catalytic or non-catalytic pyrolysis can be benefited from synergistic effects among the different materials. Nevertheless, MSW also contains significant quantities of halogenated plastics like polyvinyl chloride (PVC) that are difficult to recycle using traditional methods. When MSW is thermally converted (i.e., pyrolysis, gasification and incineration), PVC can not only produce corrosive hydrochloride gas, but also react with organic waste-derived products to form organohalogens and dioxins that create various environmental and health hazards. Our group has investigated a new approach to utilize PVC as an asset to promote MSW conversion to value-added products. We first co-converted cellulose and PVC in heated tetrahydrofuran (THF) solution. It was found that levoglucosan was the major monomer in liquid product. The highest monomeric sugar yield was 42% at 335 °C and 8 minutes with 2.5% cellulose weight equivalent PVC. The liquid products also contained up to 46% of soluble oligomeric sugars and small amounts of 5-HMF and furfural. Chlorinated organics were nearly absent in either liquid product or the pyrolyzates of the solid residue. Therefore, the results confirmed our hypothesis that hydrochloric acid released during thermal decomposition of PVC acts as an effective catalyst in THF to promote cellulose decomposition to soluble sugars. Encouraged by the result, we also converted a mixture of cardboard, PVC and PE in THF solution. As a result, not only PVC greatly facilitated rapid decomposition of cardboard to liquid product, but the presence of PE also helped to increase the conversion. Up to 68% of cardboard was converted into liquid product containing various organic compounds, such as furans, sugars and phenolics. The remaining solid residues could be easily separated into two fractions. The first fraction was fine char derived from cardboard and it had a HHV of 20.8 MJ/kg, compared to 11.7 MJ/kg with the cardboard. The second fraction of the residue was a waxy material. Upon pyrolyzing, the waxy residue produced a serious of aliphatic hydrocarbons identical to that are produced from pyrolysis of pure PE. More significantly, Cl atom was nearly theoretically converted to HCl and thus can be recovered from the solvent. Overall, we explored an environmentally friendly approach to convert mixed solid wastes into value added products.

574480 Development of High Performance Composites Using Recycled Materials

The purpose of the work is to develop and characterize high performance polymer matrix composites using recycled materials that otherwise are sent to landfills. Recycled polymer and recycled fibers are blended and molded to form a composite that possesses both high specific modulus and specific strength. Preliminary result of mechanical tests and thermomechanical analysis on the high performance composite material are presented.

576718 Sustainable Mass-Burn Refuse-Derived-Fuel-to-Energy System: A Municipal Solid Waste Management Case Study

Waste-to-energy (WtE) or energy-from-waste (EfW) is the process of generating energy in the form of electricity and/or heat from the primary treatment of waste. WtE is a form of energy recovery. Most WtE processes produce electricity and/or heat directly through combustion, or produce a combustible fuel commodity, such as methane, methanol, ethanol or synthetic fuels.

Gasification and pyrolysis processes have been known and used for centuries and for coal as early as the 18th century.... Development technologies for processing [residual solid mixed waste] has only become a focus of attention in recent years stimulated by the search for more efficient energy recovery.

Waste-to-energy can significantly reduce the amount of waste needing disposal, generate energy for a community, and also reduce municipal solid waste (MSW) transportation costs. It becomes more favorable for wastes that have high-energy content, low-moisture content, and low-ash content. These wastes include paper, plastics, textiles, rubber, leather, and wood. Therefore, MSW to a large extent is of biological origin (biogenic). Typically, more than half of the energy content in MSW is from biogenic material. Consequently, this energy is often recognized as renewable energy according to the waste input.

A case study of a WtE facility, operating at 200,000 tons of refuse annually, will be presented with pertinent information including but not limited to design and operation, annual electricity generation via turbine, life-cycle costs and overall profitability, total ash handling, emission control of PM, NO_x, SO_2, mercury, dioxins, and protection of health and environmental.

569146 Investigation of Reactions at the Gas/Solid Interface of Municipal Solid Waste Residues

Reactions at the gas-solid interfaces are present during the treatment/combustion of municipal solid waste (MSW) in waste-to-energy facilities, leaving a solid residue, ash, that has significant potential to valorize. Storage and transport of MSW ash after combustion in waste-to-energy facilities allows for reactions at the gas-solid interface such as corrosion and hydration. These reactions can potentially lead to issues such as production of undesirable gases, rise in temperature above regulation limits, and the consumption of recoverable metals and minerals from the ash. The issues of crucial concern in the waste management industry are elevated temperature in landfills, and the production of hydrogen from ash.

In order to minimize these issues in MSW management, it is necessary to understand the effect temperature, pressure, moisture content, and ash content on the rates of reaction and the products of the reactions. It is also important to study the characteristics of MSW ash and the its contribution in the reactions occurring in a landfill.

The first part of the talk focuses on understanding the impact of pressure, moisture, and temperature on MSW pyrolysis as a possible cause for elevated temperature in landfills (ETLFs). MSW provided from landfills is tested under simulated landfill conditions in controlled temperature and pressure environment. The effect on landfill gas concentrations (CH_4 and CO_2) in the reactors over a range of temperatures (49 to 177 °C) and moisture content (40 to 60 wt.%) in the presence of biological inhibitors are studied. Results from these tests identify a range of moisture content that impact the gas concentration during MSW decomposition. Gas analysis using microGC (gas chromatography) from these experiments identify the presence of hydrogen during MSW decomposition, which is also observed in elevated temperature landfills.

The second part of the talk focuses on characterization of MSW ash, and the reactions at the gas-solid interface of waste-to-energy ash after metal recovery. The ferrous and non-ferrous metals recovered from waste-to-energy ash are characterized using scanning electron microscopy (SEM), energy-dispersive X-ray spectroscopy (EDS), and X-ray diffraction (XRD). Hydrogen evolution resulting from the gas-solid interface reactions in MSW ash are recorded over time in controlled environment using gas analysis. The hydrogen production rate shows correlation with the loss of aluminum from MSW ash. Results also show higher hydrogen concentration in ash samples containing higher metal fractions, which points to ash hydration and carbonation as plausible reactions causing the elevated temperature behavior in landfills with co-disposal of ash.

577816 Microwave Enhanced Advanced Oxidation Treatment of Anaerobic Digested Effluent

Microwave enhanced advanced oxidation treatment (MWAOP) is a thermo-chemical treatment process which has been used in several studies for effective treatment of organic wastes (Saha, M. 2018). In MWAOP, hydrogen peroxide oxidant is converted to hydroxyl radicals in presence of microwave heating. Hydroxyl radical is a much stronger oxidant than hydrogen peroxide. MWAOP oxidizes organic waste to final products such as volatile fatty acids (VFA) and carbon dioxide (CO_2) (Liao et al. 2007). This process disintegrates solids and helps to release nutrients from organic waste (Saha, M. 2018). In this study, MWAOP has been applied to treat anaerobic digested effluent. Anaerobic digestion is extensively used for renewable energy production from organic wastes. The anaerobic digested effluent is rich in nutrients and can be used as a potential source for nutrient recovery. This study was focused on investigating the effectiveness of MWAOP to treat anaerobic digested effluent to produce more suitable substrate for the subsequent nutrient recovery. The digested effluent was collected from a dairy farm. All experiments were performed in a batch scale microwave system of 2450 MHz. The digested effluent was treated at two different temperatures (90°C and 110°C) using MWAOP. The results showed that settleability of MWAOP treated digested effluent was improved and soluble COD was increased after the treatment. Hence, MWAOP treated digested effluent might improve the subsequent nutrient recovery process.

References

[1] Liao, P. H., Lo, K. V., Chan, W. I., & Wong, W. T. (2007). Sludge reduction and volatile fatty acid recovery using microwave advanced oxidation process. Journal of Environmental Science and Health Part A, 42(5), 633-639.

[2] Saha, M. (2018). Microwave enhanced advanced oxidation treatment of organic slurries (Master's dissertation, University of British Columbia).

574960 Light Scattering from Soot Nanoparticle Agglomerates

Combustion-generated nanoparticles during material synthesis and energy generation (e.g. for transportation and electric power) as well as open fires form fractal-like structures by agglomeration and surface growth [1]. Yet, their optical properties are calculated typically by Rayleigh or Mie theories for spheres impeding characterization by optical diagnostics and selective sensing [2]. For example, fire (smoke) detectors are not selective enough to carbon black or soot light scattering and thus produce false alarms costing up to 1 billion £/y in United Kingdom alone [3].

Here, Soot agglomerate structure and optical properties are simulated by coupling Discrete Element Modeling (DEM) for surface growth and agglomeration with the Discrete Dipole Approximation (DDA) to account for multiple scattering [4]. The soot agglomerate morphology and light scattering estimated by DEM-DDA are compared to mass-mobility, microscopy and light scattering measurements in premixed ethylene flames. Using the Rayleigh-Debye-Gans theory and neglecting necking (aggregation) and polydispersity of constituent primary particles underestimates the differential scattering cross-sections for vertically-, C_v, and horizontally-polarized incident light, C_h, measured in premixed ethylene flames up to 50 %. In contrast, the DEM-derived C_v and C_h are in excellent agreement with the data. Thus, the DEM-DDA can be used to optimize optical diagnostics for soot or carbon black characterization and selective sensing by fire detectors.

References:

[1] Kelesidis, G. A., Goudeli, E., & Pratsinis, S. E. (2017). Carbon, 121, 527-535.

[2] Keller, A., Loepfe, M., Nebiker, P., Pleisch, R., & Burtscher, H. (2006). Fire Safety J., 41, 266-273

[3] Chagger, R., & Smith, D. (2014). http://www.bre.co.uk.

[4] Kelesidis, G. A., Kholghy, M. R., Zuercher, J., Robertz, J., Allemann, M., Duric, A., & Pratsinis, S. E. (2019). Powder Technol., doi.org/10.1016/j.powtec.2019.02.003.

553686 The Role of Organic Solvents in Cholestrol Cloud Condensation Nuclei Activation

Organic aerosol is ubiquitous and partially soluble particles can uptake water, form droplets and act as Cloud Condensation Nuclei (CCN). Cholesterol is a well-known organic aerosol. Cholesterol is insoluble in water (<0.002 gram in 100 ml of H_2O) but readily dissolves in organic solvents. In this study, we examine the ability of cholesterol generated in 7.2%, 10.4% and 18.9% (by volume) dilutions of 3 alcohols (ethanol, isopropanol and acetone) to act as CCN. The apparent particle hygroscopicity, κ, varies over two orders of magnitude, from ~ 0.001 to 0.1. We use statistical analysis of variance (ANOVA) to define significant physical and chemical factors that modify κ. Results show that aerosol formed in the presence of alcohols promote droplet formation. As the volume of water increases, κ decreases. However, the type of alcohol does not significantly modify the hygroscopicity ($p = 1.14\%$). Increases in alcohol concentration decrease droplet surface tension and change aerosol shape. Thus, the apparent κ is corrected with surface tension and shape factor data and estimated to be ~0.028 ± 0.02. The results imply that small amounts of azeotropic alcohol/water mixtures substantially modify the droplet forming potential of cholesterol and other insoluble organic particles.

579275 Role of Polarity Distribution of Organics in Aerosol Hygroscopicity

The polarity of organics is a key parameter that may determine the hygroscopic growth of aerosol, cloud droplet activation, and liquid-liquid phase separation (LLPS). However, direct measurements of OA polarity are limited. This study aims to determine the distribution of polarity in OA using the reversed-phase high-performance liquid chromatography (HPLC) and charged aerosol detection (CAD) for the first time. The new measurements showed that water-solubility C (v/v) of secondary organic aerosol (SOA) components produced by the ozonolysis of a-pinene spans nearly six-orders of magnitude. Model simulations of CCN activation show that despite the wide range of solubility, the majority of solutes in a-pinene SOA are dissolved at the point of activation, thus a-pinene SOA may be reasonably assumed as water-soluble especially in the presence of inorganic seeds. Implications of the polarity distribution on LLPS and the gap in sub-saturated and super-saturated aerosol hygroscopicity will be discussed.

568665 Estimation of Vapor Wall-Loss in Smog Chamber Experiments

Vapor wall-loss has been shown to be an important process in smog chamber studies, especially in yield experiments where organic aerosol (OA) concentrations can be underestimated due to semi-volatile vapor losses to chamber walls. Vapor wall-losses are also important, but largely unaccounted for, in active dilution experiments because an evaporating particle-phase component may never be able to fully reach equilibrium if the corresponding vapor-phase component is continually lost to the chamber walls. This study presents results that account for and estimate vapor wall-loss occurring during dilution experiments of secondary OA (SOA) generated from the ozonolysis of a-pinene and cyclohexene.

In this study, we generated SOA in a main chamber and then isothermally diluted the SOA close to ambient concentrations in a smaller dilution chamber, where the SOA evaporation over time was monitored. After correction for particle wall-losses in the dilution chamber, we generated areograms, particle mass fraction remaining as a function of time in the dilution chamber, by dividing the mass concentration at any given time by the initial mass concentration in the dilution chamber. We then used a dynamic mass transfer model coupled with an error minimization approach to estimate the SOA volatility distributions. However, the model was updated in this work to account for and estimate vapor wall-losses occurring in the dilution chamber. The model results indicated that the bulk vapor wall-loss constant for this chamber and the SOA species studied was ~0.3 h^{-1} and, when compared to model results that did not account for vapor wall-losses, significantly reduced the uncertainty in the volatility distributions and corresponding properties. This work signifies the importance of accounting for vapor wall-losses not only in yield experiments, but also in dilution experiments.

588410 Finding the right mass: Comparing measurements from commercially available differential mobility analyzer aerodynamic aerosol classifier & aerosol particle mass analyzer

There now exists an instrumentation trifecta for aerosol separation and classification by mobility diameter (Dm), aerodynamic diameter (as Daero or the relaxation time, t) and mass (mp) utilizing a differential mobility analyzer (DMA), aerodynamic aerosol classifier (AAC) or aerosol particle mass analyzer (APM), respectively. In principle, any combination of two measurements yields the third; i.e. relaxation time = mobility x mass. Here, we compare measurements of mp, effective density and mass-mobility scaling exponents (*D*fm, a surrogate for particle shape) utilizing different combinations of tandem measurements – DMA-APM, AAC-DMA and AAC- APM – utilizing ammonium sulfate, soot from a Santoro diffusion flame and water-soluble carbon black; notably, these particles represent a solid, near-spherical particle, a lacey aggregate and a near-spherical collapsed aggregate. Preliminary data suggests that deviations in mp and Dfm between tandem measurements are low (on the order of 5 to 10 %) and that the largest uncertainty is the resolution of particles bearing multiple charges. While the AAC by itself does not require charge to classify particles, mp (and hence Dfm) determination requires a secondary instrument (i.e. DMA or APM) that does separate based upon charge. Results show that the deconvolution of multiply charged particles with classical charging theory works well for AS but is especially problematic for lacey soot. Thus, measuring particle mass from the AAC may have resolution limitations. We will also present recommendations for data analysis and interpretation.

564605 Insight into the Mechanism of Alkali Resistance over Cunbti Catalyst for the Selective Catalytic Reduction of NO_x with NH_3

Insight into the mechanism of alkali resistance over CuNbTi catalyst for the selective catalytic reduction of NO_x with NH_3

Sujing Li*, Xiaoxiang Wang, Wei Li*

Key Laboratory of Biomass Chemical Engineering of Ministry of Education, Institute of Industrial Ecology and Environment, College of Chemical and Biological Engineering, Zhejiang University, Hangzhou 310027, China

Abstract: Alkaline metals in the flue gas usually deactivated the SCR catalyst seriously in the practical application. Compared with traditionally commercial VWTi catalyst, CuNbTi catalyst exhibited a higher alkali resistance. CuNbTi maintained 80% NO_x conversion while VWTi catalyst fully deactivated. Different alkali metals were individually employed to poison the catalysts and K showed the most toxic for VWTi and CuNbTi catalysts. For VWTi catalyst, the variation of structure and loss of active sites and reducibility resulted in the deactivation after 2%K_2O poisoning. By contrast, experimental and computational results indicated that the alkali resistance of CuNbTi was mainly due to the trapping effect, namely the interaction between Ti_2NbO_x support and K atoms. K atoms were preferentially bound to Nb-OH and Nb=O with a lower bonding energy of -2.33 eV - -2.83 eV to form $KNbO_3$ when Cu atoms were coordinated to Ti=O with a binding energy of -1.54 eV. This protected the active copper species from linking to alkali and the active sites were preserved. Moreover, Ti_2NbO_x support could weaken the impact of potassium on NH_3 adsorbing and activating on the surface of CuNbTi catalyst while the preserved copper species provided adsorption sites and redox ability for NH_3-SCR reaction. This mechanism would provide a general strategy for developing novel anti-alkali poisoning NH_3-SCR catalyst in the future. Meanwhile, both Eley-Rideal (E-R) and Langmuir-Hinshelwood (H-L) mechanisms with adsorbed NH_3 coordinated to the Lewis acid sites and bidentate nitrate as the dominating intermediate species existed during the NH_3-SCR reaction procedure over 2%K_2O-CuNbTi at 225 ^oC.

Key words: CuNbTi catalyst; VWTi catalyst; Alkali metal resistance

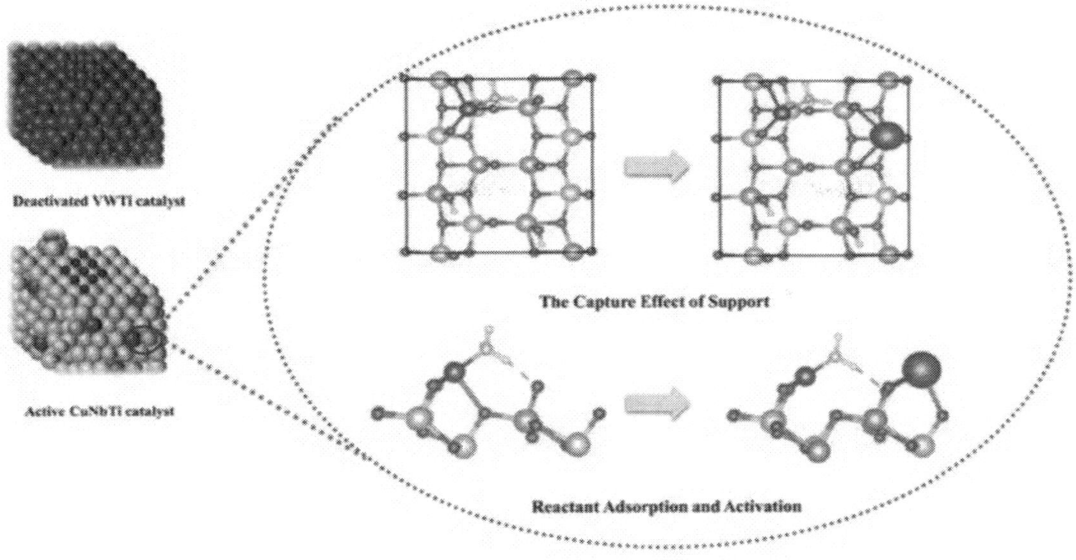

Fig.1. Schematic model of the alkali resistance mechanism over CuNbTi catalyst.

References

[1] X.X. Wang, Y. Shi, S. J. Li*, W. Li, Appl. Catal. B: Environ., 220 (2018) 234-250.

[2] X.X. Wang, Q.L. Cong, L. Chen, Y. Shi, Y. Shi, S.J. Li*, W. Li*, Appl. Catal. B: Environ., 246 (2019) 166-179.

567370 Simultaneous Product and Process Design for Tailor-Made Biofuels

By consuming roughly half of global oil produced [1], road transport contributes significantly to climate change. Internal combustion engines (ICEs) are currently the main propulsion system in road transport and are expected to stay important at least until 2050 [2]. Renewable fuels that are tailored for high-efficiency and low-emissions combustion pose a promising option for future ICEs. These fuels, which typically consist of multiple components, must be produced by economic and sustainable processes. In case the fuel components can be derived from the same feedstock in a single process, integrated design of process and product is needed to find the overall optimal process-product combination [3].

While many methods for process design and product design exist, few have dealt with simultaneous process and product design. Marvin et al. [4] propose a method for simultaneous fuel and reaction pathway design that is based on automatically constructed reaction networks. Following a similar approach, Dahmen and Marquardt [5] design tailor-made biofuels and their optimal production pathways using relatively detailed fuel property models that take into account non-ideal thermodynamics. The pathway models used by Marvin et al. [4] and Dahmen and Marquardt [5], however, neglect separations and their energy demands, thus hindering direct optimization of two key process performance measures, i.e., fuel production cost and global warming impact (GWI). By contrast, the optimization-based pathway screening method Process Network Flux Analysis (PNFA) proposed by Ulonska et al. [6] also evaluates downstream processing alternatives and their minimum energy demand based on reduced-order separation models [7]. PNFA requires a priori specification of a target product and therefore does not facilitate integrated product and process design; however, its more detailed mass- and energy-based pathway evaluation allows optimizing for fuel production cost and GWI.

We present a new method for integrated fuel and production process design that combines two of our aforementioned approaches, i.e., [5] and [6]. The new method is capable of performing simultaneous fuel and production process design with cost and GWI as objectives. To enhance computational tractability of the optimization problem, the new approach contains several model simplifications, e.g., an algebraic fuel distillation curve model based on the concept of the true boiling point curve [8]. These simplifications allow solving the resulting mixed-integer nonlinear program (MINLP) with the global deterministic solver BARON [9]. We demonstrate the new method with a case study on future tailor-made spark-ignition engine fuels consisting of up to seven potential bio-derived components. An analysis of the tradeoff between fuel production cost and GWI reveals how the composition and the process configurations of the rationally formulated fuels change along the Pareto-optimal curve from the minimal cost design to the minimal GWI design.

References:

[1] Statista. Distribution of oil demand worldwide in 2016 by sector. https://www.statista.com/statistics/307194/top-oil-consuming-sectors-worldwide/ (accessed March 2019).

[2] ERTRAC Working Group on Energy & Environment. Future light and heavy duty ICE powertrain technologies., 2016, https://www.ertrac.org/uploads/documentsearch/ id42/2016-06-09_Future%20ICE_Powertrain_Technologies_final.pdf (accessed March 2019).

[3] Hechinger M, Voll A, Marquardt W. Towards an integrated design of biofuels and their production pathways. Comput. Chem. Eng. 2010;34(12):1909–1918.

[4] Marvin WA, Rangarajan S, Daoutidis P. Automated generation and optimal selection of biofuel-gasoline blends and their synthesis routes. Energy Fuels. 2013;27(6):3585–3594.

[5] Dahmen M, Marquardt W. model-based formulation of biofuel blends by simultaneous product and pathway design. Energy Fuels. 2017;31(4):4096–4121.

[6] Ulonska K, Skiborowski M, Mitsos A, Viell J. Early-stage evaluation of biorefinery processing pathways using process network flux analysis. AIChE J. 2016;62(9):3096–3108.

[7] Bausa J, Watzdorf Rv, Marquardt W. Shortcut methods for nonideal multicomponent distillation: I. Simple columns. AIChE J. 1998;44(10):2181–2198.

[8] Reiter AM, Wallek T, Pfennig A, Zeymer M. Surrogate generation and evaluation for diesel fuel. Energy Fuels. 2015;29(7):4181–4192.

[9] Tawarmalani M, Sahinidis NV. A polyhedral branch-and-cut approach to global optimization. Math. Program. 2005;103(2):225–249.

573407 Improving the Location Robustness of Spatially Distributed Variables for Biomass Supply Chains Under Supply Uncertainty

Due to climate change and other issues surrounding fossil fuels, finding alternative fuels and feedstocks is of vital importance. By expanding the role of the bioeconomy, biomass can be utilized as a substitute for both uses. Supply chain optimisation (SCO) is a method that can been used to help the biomass industry gain a foothold and expand while remaining economically viable. One issues is while supply chain network optimization offers a dependable method of searching a large solution space to find results that are feasible and maximized (or minimized) for the objective function, it does have a weakness in that small differences in initial conditions can cascade and result in non-negligible variations in the form of the optimized result. Frequently the actual value of the objective is similar for small differences in initial conditions, but the organization of the network that is recommended by the optimization can be measurably different. Spatially distributed variables can be noticeably different for different starting conditions.

Spatially distributed variables are variables in the biomass supply chain optimisation which may be distributed over the real map space, such as locations to build biochemical plants, locations for feedstock storage, or locations to grow biomass. These variables are decided in the process of optimising the supply chain for the objective variable – whether to place a plant in a specific location, or how much biomass to grow in each location. Each potential organisation is a solution which will have different values of the objective variable. These spatially distributed variables can be sensitive to changes in input values, as the supply chain optimises for the new starting conditions. The new optimal solution recommends placing refineries, storage facilities and growth sites in different locations depending on the starting conditions. Ideally these variables would be locationally robust.

Location robustness is the idea that there should be some consistency in the final design of the optimised supply chain for minor variation in the inputs. Input variation is bound to happen, and ideally these supply chain designs should be able to perform to a high level regardless of the variation in input. For some applications this is not a significant concern; for high investment situations, this may be problematic. A final supply chain design can be very expensive and time-consuming to implement, and a design which is optimal for the given input but is sensitive to variations in that input would be less useful than designs which can perform across a range of conditions. When a company is deciding where to locate refineries, where to grow crops or where to store products, the investment demands that a location be viable regardless if the input prices or yield rates vary slightly.

One major source of input variation for these supply chain systems is from uncertainty in the supply of biomass, such as year-to-year variation in biomass yield, or in land quality drift due to environment changes. It is also one that is particularly difficult to control – weather is still a factor in the productivity of farming. Biomass yield is a key input for biomass supply chains, which can have cascading impacts on the final biomass supply chain design. Understanding how uncertainty from this input in particular can be mitigated is key to designing supply chain frameworks and supply chains which are locationally robust.

This work seeks to develop a biomass supply chain optimization framework that considers location robustness as part of the framework. The work first seeks to identify the causes of this locational instability within some supply chain frameworks to understand how this arises. It then develops techniques that may be more locationally consistent due to the construction of the framework.

Biomass supply chain models often use the mean biomass yield of cellularised tracts of land to calculate yield for that cell. However, there can be large variety in the biomass yield within those tracts, losing useful information to aggregation and simplification. A biomass SCO framework has been developed which can incorporate some of this information about the quality of land available by piecewise linearly approximation of the biomass yield distribution, and incorporating this information into the optimisation.

Using linear approximations of the biomass yield distributions allows the SCO model to make more accurate decisions about quantity and location of biomass growth operations, thereby affecting all downstream decisions as well. It is postulated that providing the optimisation with the option to focus on developing high yield land, that high quality land will be used more consistently. This would have the dual effects of being more reasonable from a development perspective – better to develop high quality land – and of potentially increasing the locational robustness. The good land will be used preferentially, reducing the number of new or different areas which may be developed for biomass growth under supply uncertainty.

A case study of the United Kingdom for potential biomass industry viability will be examined using the framework to illustrate the impact of retaining this biomass yield information in the optimisation. Initial results indicate that using the more accurate biomass yield estimates reduced the land overall usage. The framework results in a better use of the available land to meet available demands, reducing land use. Further, it allows for improved biomass output, which in turn increased the quantity of bioproducts which can be produced. All of this leads to an increase in the overall profit. The next step is to implement supply uncertainty and to examine the locational robustness of the framework with respect to these variations. This framework could be used with other optimisations with spatially distributed resources, where more detailed information could improve results, such as solar insolation or wind resource availability.

571188 Process Intensification in By-Product Recovery from Biobased Effluents

Low-value by-product processing is an economic challenge common to many biobased processes. For recovery, the downstream processing must be very efficient, Value-adding conversion via chemical reaction may be beneficial. Both goals can be met with reactive separations. Chemical conversion of target constituents may simplify separation, and the market value of by-products may be increased by conversion.

Beside the main product fibres the pulping industry produces a black liquor containing lignin, hemicellulose, hydroxycarboxylic acids and carboxylic acids with a high water content. After thickening, this effluent is mainly used for thermal energy production via incineration. The volatile carboxylic acids are removed with the water during thickening. Separation of acids from the condensate is recommended for environmental reasons. State of the art recovery of carboxylic acids from aqueous effluents is reactive extraction followed by distillation for solvent regeneration. However, distillative solvent regeneration produces an azeotropic mixture of carboxylic acids and water causing energy intensive product isolation. Therefore, reactive distillation was investigated to intensify solvent regeneration and product isolation. Methanol was added as esterifying agent to the laden solvent containing formic acid and acetic acid, converting the carboxylic acids into their corresponding low boiling methyl esters. Esterification improves the separation of the acids and the reactive extractant and simplifies separation of the distillate products due to lower boiling points of the methyl esters compared to the acids. A homogeneous (4-dodecylbenzenesulfonic acid) and a heterogeneous catalyst (Amberlite IR120) were tested to speed up the reaction rates in reactive distillation. The admixture of 5 w% of 4-dodecylbenzenesulfonic acid and a proton equivalent amount of Amberlite IR120 in batch reactive distillation experiments provided complete conversion of formic acid and 88 % yield of methyl acetate in the distillate. High solvent regeneration efficiency at low temperature and ambient distillation pressure is achieved. The heterogeneous catalyst enables simple separation from the solvent in the reboiler. In homogeneous catalysis catalyst separation step may be challenging. To circumvent an additional separation step the influence of the homogeneous catalyst in the solvent phase on extraction efficiency of acid extraction was investigated. Loss of solvent load capacity is of minor impact. Due to the preference of the reactive extractant to strong acids, the catalyst 4-dodecylbenzenesulfonic acid is kept in the solvent phase during extraction without losses. During distillation the catalyst-extractant aggregate is cleaved at operation temperature in the reboiler and therefore available for catalysis of esterification, enabling combination of extraction and reactive distillation without catalyst separation.

568081 Jet Fuels Production from Waste Cooking Oil Via Catalytic Transfer Hydrogenation

The progressive depletion of fossil fuels and the environmental impacts associated with such exploitation has fostered research on alternative and renewable fuel sources. Currently, the transportation sector, especially the aviation industry, relies on liquid hydrocarbon-based fuels which requires kerosene-range hydrocarbons with strict quality standards to power gas-turbine engines. These hydrocarbons can be obtained from renewable sources, such as triglycerides. Waste cooking oil (WCO) is a suitable renewable fuel and is not an edible source like virgin vegetable oil. It can represent a sustainable alternative feedstock as it is widely produced globally. Moreover, its use in jet fuels production converts waste into a valuable product resulting in a circular economy.

In this work, a novel process for jet fuels production from WCO (catalytic transfer hydrogenation (CTH)) is thoroughly investigated. CTH utilizes hydrogen-donating compounds and a suitable catalyst to release H_2 *in situ* which saturates and reduces triglyceride molecules. CTH is potentially beneficial relative to conventional hydrogenation because it avoids the handling of large quantities of H_2 gas at high operating pressures. Jet fuels production from WCO was investigated experimentally in a continuous-flow lab-scale fixed-bed reactor, using 2-propanol as hydrogen donor and a granular activated carbon catalyst. The experiments were carried out at low pressure (2 bar) and temperature (300-400°C). The reaction products were fully characterized using GC-MS. An optimum yield of 72 wt% liquid hydrocarbon products was obtained at 380°C.

Using the experimental results obtained in the lab, a detailed techno-economic analysis (TEA) and life cycle assessment (LCA) were performed to assess the profitability and environmental impacts of CTH. Both processes were simulated via Aspen Plus® process simulation software to determine material and energy balances. For the LCA, a "well-to-pump" system boundary is used, incorporating all material and energy flows during resource extraction up to the delivered product (i.e. jet fuel). A functional unit of 1,000 ton-d^{-1} WCO feedstock was employed for the life cycle inventory (LCI). Characterization factors were obtained from Argonne's GREET model and SimaPro's Ecoinvent database to determine three (3) key environmental impacts: (1) total energy use (including fossil and petroleum), (2) greenhouse gas (GHG) emissions, and (3) energy return-on-investment (EROI). For the TEA employing a 25-year life, cost information was collected to determine capital cost, direct cost, indirect cost, fixed operating cost, variable operating cost, and energy cost. Modeling results for this holistic system is benchmarked against a conventional hydrogenation process to fully capture the novel process' overall sustainability.

The results of the analysis highlight that although both processes achieved comparable jet fuel/diesel product yields, CTH would require significantly lower capital and energetic costs. However, the cost of the 2-propanol feedstock might hinder the process profitability under current market scenarios. CTH's "well-to-pump" total energy consumption is 25% lower than conventional hydrogenation, for all products and co-products. CTH also reflects a 35% GHG reduction compared to conventional hydrogenation. Both processes were highly-sensitive to the cost of 2-propanol. While the market price for 2-propanol is significantly variable, this study shows that CTH could potentially be a promising, greener and safer process for jet fuels production from WCO. This work demonstrated that with reduced feedstock supply costs, the tradeoff between environmental impacts and profitability of CTH-based jet fuels is achieved.

562866 The Conceptual Framework and Modeling Analysis of Compact Energy Systems

The conceptual frame work for compact energy system is presented to make energy from a biomass like wood/switchgrass, as an alternative to the construction of large scale biorefineries. The compact energy system consists of four individual components: biomass preparation, biomass converter, fuel reformer, and powered-engine. The individual unit processes within each component and the possible types of compact energy systems with different biomass converter technologies like fermentation, pyrolysis, and gasification are presented. The size, weight, and energy efficiency of upgrading biomass to energy using compact energy system with various gasification technologies have estimated. For a compact energy system with the hydrogen fuel cell as a powered-engine component and that processes 10 kg of dry biomass per day, the net energy (kWh) of compact energy system is found to be -7.5, -30, 18.7, 13.1, and 11.7 with the super-critical, microwave assisted, catalytic, steam, and conventional gasification technologies as biomass converter technologies, respectively. The low yields of super-critical gasification and a low efficacy of converting electric energy to heat via electromagnetic waves with the microwave assisted gasification are results in such negative net energy with the respective compact energy systems. Finally, the challenges and opportunities with the development of compact energy systems built around gasification are discussed.

575367 Synthesis and Evaluation of Dipicolinic Acid (DPA) – Based Interwell Tracers for Reservoir Surveillance

Our efforts to address the known deficiencies of modern-day chemical tracing in hot and high salinity reservoirs has led to the adaptation of developed technology such that it touches upon multiple facets of tracer management, with particular emphasis on materials development and their subsequent detection. Fundamentally, this technology comprises classes of optically detectable and chromatographically distinguishable molecular tracers that serve as selective chromophore ligands for trivalent lanthanide metals capable of sensitizing their luminescence upon ligand-metal ion complexation, thus enabling high-sensitivity detection. Lanthanide-based chelation chemistry already has a known basis in applications involving the detection of biological entities; in particular, dipicolinic acid (DPA), a biomarker of bacterial spores, exhibits high-specificity binding to sensors containing Tb^{3+}/Eu^{3+} metal ions. In continuing to exploit the optical properties of such systems, we herein report the synthesis of DPA-derived molecular inter-well water tracers, placing particular emphasis on backbone functionalization of native DPA to generate "barcodes" that can enable their simultaneous full-field deployment. Preliminary coreflood results benchmarking DPA against conventional fluorobenzoic acid (FBA)-based and ionic (KCl) tracers indicate that DPA exhibits similar breakthrough times, is non-retentive and also non-partitioning. We have also demonstrated that DPA and its variants are amenable to chromatographic separation and detection by high performance liquid chromatography (HPLC), providing ultra-trace limits of detection (LOD) using a method complementary to optical detection modalities. Studies regarding the physicochemical properties of the lanthanide ion complexes generated are being conducted concurrently alongside evaluation of synthesized tracer candidates, the integrity of which are also being evaluated via long-term thermal stability testing. We hope that the research conducted herein will ultimately lead to the deployment of new classes of chemical tracers capable of facilitating increased waterflood efficiency, particularly in mature reservoirs where optimization of oil recovery techniques requires high levels of detail.

573670 Design Development of a Novel Sour Water Stripper

With an increasing focus on energy performance improvements and strict environmental laws, there is a need to recycle water resources and reuse them in the industry. Refineries consumes huge amounts of water which offer an opportunity to reprocess it in an efficient manner. This study focusses on the design and simulation of sour water stripping (SWS) unit in the refinery setup for the reuse of sour water. The stripper removes the H_2S and NH_3 content from the sour water making it recyclable for various processes. In this study, process simulation has been performed to analyze the energy saving and cost reducing prospects in the sour water stripping unit. First, a base case design has been simulated using a reference sour water stripping unit data. The developed model is validated against the plant data in order to ensure the model validity. Various alternative designs have been developed to seek process improvements either by improving the energy efficiency or by reducing the sour water stripping unit cost. All the process designs were compared in terms of energy requirement and economics while maintaining the stripped water purity of 10 ppm H_2S and 100 ppm NH_3 content.

A novel design has been proposed in this study that can significantly reduce the energy requirement compared to the conventional designs. The proposed design employs mechanical work using a compressor instead of hot and cold utilities. In this design, the stripped water stream leaving at the bottom of the column is flashed to a reduced pressure, whereby, some amount of vapors are generated. The produced vapors leave at the top of the flash drum and are fed to the two-stage compression train. A compression ratio of three is assumed in each compressor stage with an inter-cooler. The compressed vapor stream at 2.1 bar and 240.7 °C then heats up the incoming feed stream and is fed to the bottom of the column at 121.7 °C.

The results of this study provide a meaningful analysis in terms of energy and economic considerations to improve the performance of the sour water units in the refineries. The results show that the proposed vapor recompression design can reduce the hot and cold utilities requirement by 90 % and 22 % respectively compared to the base case design. Despite the fact that vapor compression design requires higher initial capital investment because of expensive compressors, the economic analysis shows that over the 25 years plant life, the operating energy savings are substantial which makes vapor compression design more economical. The results reveal that the vapor compression design requires 53 % less operational energy compared to the base case design which translates into the annual operating cost savings of more than 60 %. This results in total annual cost of M$ 3.09 for the vapor compression design compared to the M$ 6.49 for the base case design.

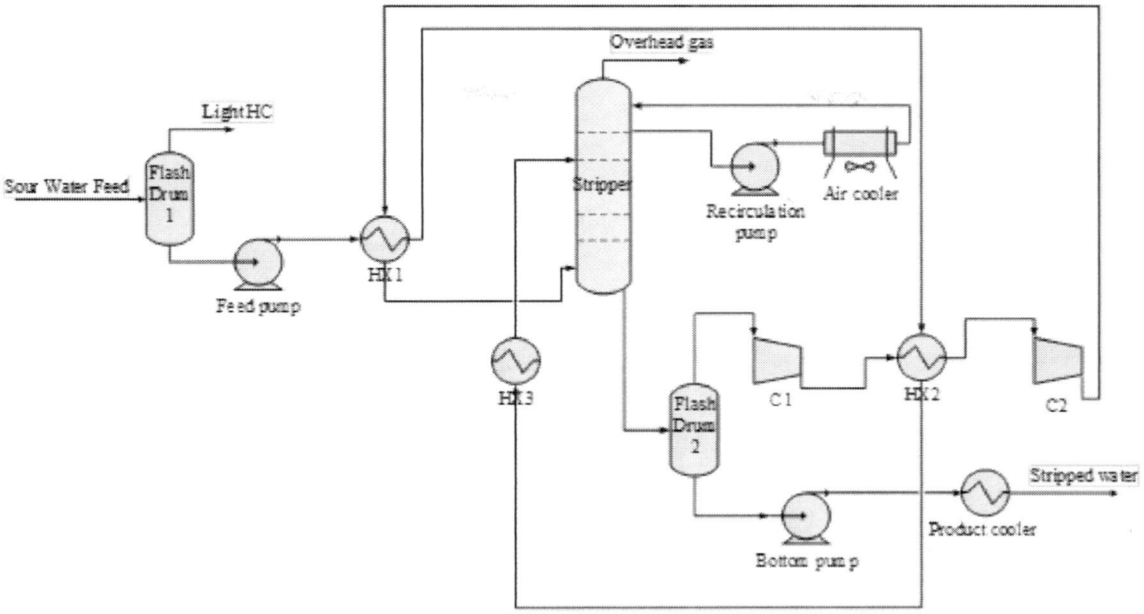

Figure 1: Process flow scheme for the proposed vapor compression design

576034 Two-Step Sub/Supercritical Water and Ethanol Processes for Non- Catalytic Biodiesel Production

Two-step Sub/supercritical Water and Ethanol Processes for Non- catalytic Biodiesel Production

Aso A. Hassan, Hayder Al-Hameedi, Dr. Joseph Smith

Chemical and Biochemical, Engineering/ Missouri University of Science & Technology, USA

ABSTRACT

The catalyst-free two-step process has been developed for biodiesel production using low-grade feedstocks. The first step consists of triglycerides hydrolysis under subcritical water conditions to generates and increase a free fatty acid (FFA) contains for ethyl ester production. In its subcritical state, water can be used as both a solvent and reactant for the hydrolysis of triglycerides. The hydrolyzed products mixture is separated by decantation into the oil phase of FFA (upper layer) and a water phase that contains glycerol (lower layer). In the second step, the hydrolyzed products of free fatty acids were successfully esterified to their ethyl ester in supercritical ethanol conditions without any catalyst. Under the sub- and supercritical conditions of water and ethanol the hydrolysis and the esterification reactions proceed quickly, with the conversion of greater than 98% after 10-20 min. This two-step process for biodiesel production offers several advantages such as milder reaction conditions and pollution reduction due to water using instead of organic solvents. Also, the glycerol is removed after the hydrolysis reaction so that the backward reaction between the glycerol and the ethyl ester eliminated and lead to the biodiesel yield and quality improvement. Finally, the water-glycerol layer (lower layer) could use directly in supercritical water gasification or dehydration processes for hydrogen or acrolein production, respectively. The aim of this study is making a comparison between our previous one-step process and the two-step reactions process (see Figures 1 and 2) to find the best pathway for designing and building an integrated reactor. Indeed, the two-step process is more applicable for the low-grade feedstocks with a high amount of FFA and water.

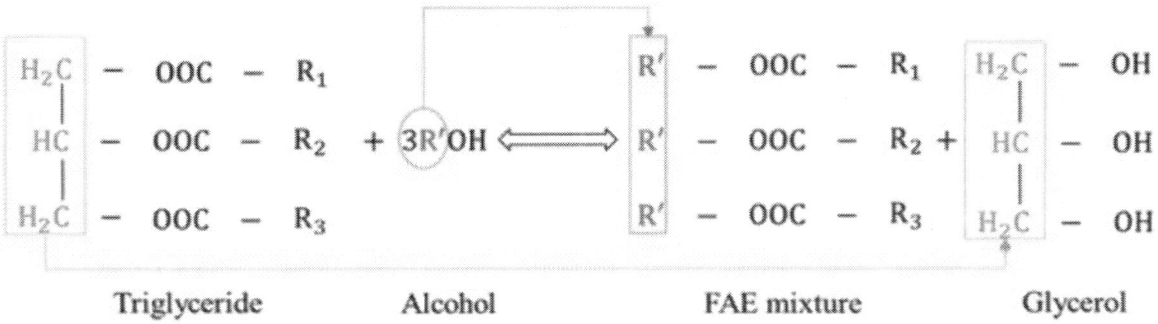

Figure 1: The Transesterification reaction (One-step)

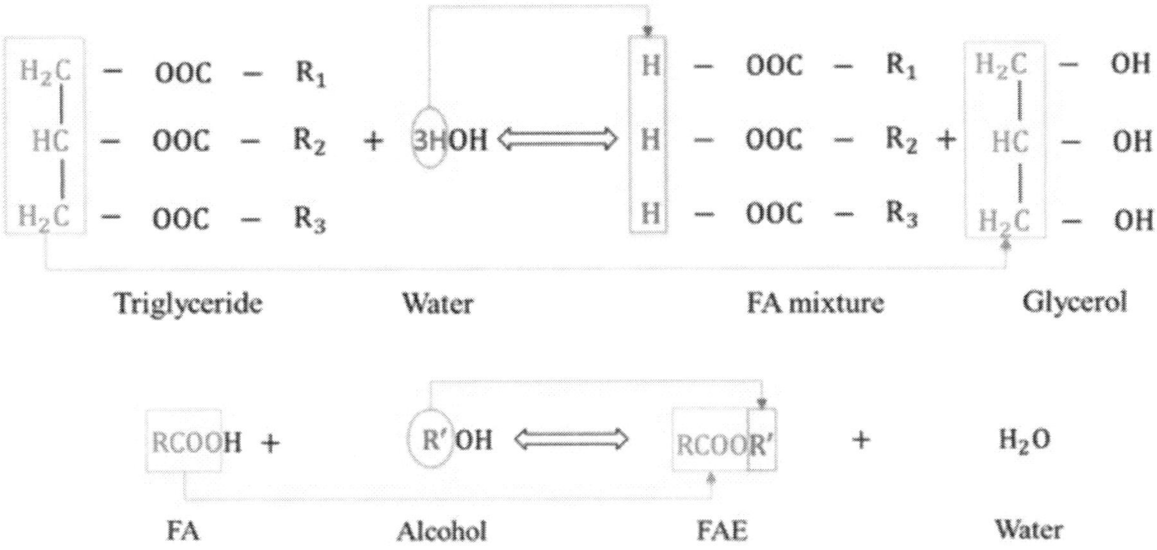

Figure 2: The hydrolysis and esterification reactions (Two-step)

Keyword: Biodiesel, supercritical ethanol, subcritical water, hydrolysis.

577001 Advances in Ultrasound- and Cavitation-Assisted Catalyzed Conversions of Biomass to Biofuels

The continuing effort to develop novel, eco-friendly, renewable and cost-effective bio-based processes to solve our energy scarcity and sustainability problems, is of great technological, societal, cultural, economic and political importance. However, the conversion of lignocellulosic biomass for biofuels and bio-refinery applications is limited by the cost of pretreatment to separate or access the biomass's three main components, cellulose, hemicelluloses, and lignin. Ultrasound and cavitation-assisted technologies are green techniques that have the potential to enhance the separation and hydrolysis of lignocellulosic materials for use in biofuel production and bio-refineries, through physical and/or chemical mechanisms. We studied the enzymatic hydrolysis of cellulose using a multi-frequency ultrasonic reactor, and optimized glucose yield at 50°C and 5.2 pH. The best cellulose hydrolysis was observed at 611 kHz and 104 W. The effects of ultrasound and enzyme on the surface and structure of the cellulose were studied using BET, SEM, XRD and SHIM. We also demonstrated through laboratory experimentation coupled with optimization and mass transfer/reaction modeling that high-frequency ultrasound could achieve very high conversions of soybean oil to biodiesel at relatively low times and energy inputs via base-catalyzed transesterification using methanol, and both KOH and calcium methoxide catalysts. We also demonstrated that sulfated zirconia was highly efficient for esterification of free fatty acids (FFAs) in oils. This talk will discuss the state-of-the-art advances on the use of ultrasound and cavitation for the intensification of biomass conversion and synthesis of biofuels.

References:

- Adewuyi, Y.G. Intensification of Enzymatic Hydrolysis of Cellulose Using High-Frequency Ultrasound. Chapter 8, Page 166-187. In: Intensification of Bio-Based Processes, A. Gorak and A. Stankiewicz (editors), Green Chemistry Series No. 5, Royal Society of Chemistry (RSC), 2018, 166-187.
- Adewuyi, G., and Deshmane, V.G., Intensification of Enzymatic Hydrolysis of Cellulose Using High Frequency Ultrasound. An Investigation of the Effects of Process Parameters on Glucose Yield, Energy & Fuels, 2015, 20: 4998–5006.
- Deshmane, V.G., and Adewuyi, Y.G. Mesoporous Nanocrystalline Sulfated Zirconia Synthesis and Its Application for FFA Esterification in Oils. Applied Catalysis A: General, 2013, 462-463: 196-
- Deshmane, V.G., and Adewuyi, Y.G. Synthesis and Kinetics of Biodiesel Formation via Calcium Methoxide Base-Catalyzed Transesterification Reaction in the Absence and Presence of Ultrasound, Fuel, 2013, 107: 474-482.
- Mahamuni, N.N., and Adewuyi, Y.G., Application of Taguchi Method to Investigate the Effects of Process Parameters on the Transesterification of Soybean Oil Using High Frequency Ultrasound, Energy & Fuels, 2010, 24: 2120-
- Mahamuni, N.N., and Adewuyi, Y.G., Fourier Transform Infrared Spectroscopy (FTIR) Method to Monitor Soy Biodiesel and Soybean Oil in Transesterification Reactions, Petrodiesel-Biodiesel and Blend Adulteration with Soy Oil, Energy & Fuels, 2009, 23: 3773-
- Mahamuni, N.N., and Adewuyi, Y.G., Optimization of the Synthesis of Biodiesel via Ultrasound-Enhanced Base-Catalyzed Transesterification of Soybean Oil Using a Multifrequency Ultrasonic Reactor, Energy & Fuels, 2009, 23: 757-

577971 Biogasification of Cellulosic Ethanol Stillage in an Anaerobic Fluidized Bed Bioreactor

Global concerns about climate change and other environmental impacts of fossil fuels have accelerated efforts to commercialize and introduce biofuels into the transportation fuel supply chain. The Energy Independence and Security Act (EISA) passed by US Congress and Renewable Fuel Standards (RFS) administered by USEPA, mandates that biofuels should account for 36 billion gallons of the fuel mix used by the transportation sector by year 2022. Ethanol from corn-starch having been capped at 15 billion gallons, major fraction of the remaining mandate would be required to be sourced from cellulosic feedstocks like agricultural wastes, forestry residues and dedicated energy crops. Ethanol from such feedstocks are referred to as 'cellulosic ethanol'. To be classified as a cellulosic biofuel, in addition to the feedstock, the fuel must also generate 60% less lifecycle greenhouse gas emissions compared to the baseline emissions for gasoline or diesel in the year 2005.

Several initiatives are underway to demonstrate the production of cellulosic ethanol. One such operation is the Stan Mayfield Biorefinery in Perry, FL located within the premises of Georgia-Pacific's Foley Cellulose Plant. This biorefinery was funded by the State of Florida and operated by the University of Florida until 2018. In this biorefinery cellulosic ethanol was produced from sugarcane bagasse using the following process steps. Bagasse was first soaked in dilute phosphoric acid and steam exploded, followed by enzymatic saccharification with cellulase enzymes and then fermentation using recombinant *E. coli* engineered to ferment both five and six carbon sugars to ethanol. Ethanol was recovered by distilling the whole fermentation broth. The waste stream from the distillation step is referred to as 'stillage'. Large volumes of stillage are generated in an ethanol plant. Stillage generation will be 10 to 14 times the volume of ethanol of produced and is a highly polluting waste stream containing high content of organic carbon and other nutrients like nitrogen and phosphorus. Stillage from corn-starch ethanol plants is usually dewatered and dried to produce distillers dried grain with solubles (DDGS) which is a high-value animal feed. However, the nutrition value of cellulosic ethanol stillage is poor so the waste stream will have to be treated prior to disposal. Anaerobic digestion (AD) may be employed to treat stillage. AD is a biochemical process that mineralizes organic carbon compounds to biogas (a mixture of methane and carbon dioxide) through the syntrophic action of several groups of anaerobic microorganisms. Biogas can be used as a fuel for process heat or converted to electricity. In a cellulosic ethanol facility, on-site utilization of biogas can lead to further reductions in lifecycle greenhouse gas emissions and help to meet EISA mandates. AD is a mature technology and many installations are operating world-wide for treatment of municipal, agricultural and industrial wastes. The AD process can be implemented in many types of bioreactor designs. Choice of design is dependent on waste characteristics and can influence capital and operating costs as well as the footprint of the facility. There are only a couple of studies reported in the literature on the anaerobic treatment of cellulosic ethanol stillage.

In this research, stillage generated from the Stan Mayfield Biorefinery was treated in a semi-continuously fed laboratory-scale fluidized bed reactor. Among several designs that are available for anaerobic digestion, it has been shown that a fluidized bed reactor could be operated at high organic loading rates with very short hydraulic retention times, thus requiring a small footprint. Hence, a fluidized bed design was chosen for the study here. The fluidization medium used was activated carbon granules. The bioreactor was a 7 liter total volume (5.5 L working volume) glass vessel with a conical bottom. Heating was provided by a heating tape wound on the outside of the vessel. The reactor was operated at 55 ± 2 °C and temperature was measured by a thermocouple and controlled by a CR10 data logger. Long term experiments were conducted to investigate effect of dilution rate on reactor performance, and pH control and nutrient requirements for reactor operation. Reactor performance was measured in terms of biogas production rate, methane composition of biogas, pH, total volatile fatty acid concentration, and organic carbon reduction (as soluble chemical oxygen demand or sCOD). The methane yield on the basis of both volume and COD of feed was compared to the true yield obtained by operating the bioreactor in a batch mode.

Stillage characteristics were as follows: pH = 5.15; TS = 6.89 ± 0.33 %(w/w) wet weight; VS= 5.36 ± 0.33 %(w/w) wet weight; sCOD = 36 g/L ± 2.65 g/L; Ethanol = 0.75 g/L. The stillage also contained levulinic acid, hydroxymethyl furfurals and furfurals at concentrations of 1.6, 0.09 and 0.05 g/L, respectively. True methane potential of stillage was 342 ± 5 ml methane at STP/ g sCOD or 12.31 L methane at STP/ L of stillage. Currently experiments are underway to determine bioreactor performance at various dilution rates. The dilution rate was increased from 0.027 d^{-1} to 0.0545 d^{-1} and again to 0.082 d^{-1}. At each dilution rate the reactor was operated until steady state was reached (as determined by consecutive daily measurements of methane composition, biogas production rate, and total volatile fatty acid concentration). As dilution rate is increased the methane yield dropped from 12 L methane at STP/ L of stillage at 0.027 d^{-1} to 11 L methane at STP/ L of stillage at 0.082 d^{-1}. Total volatile fatty increased from 4 -7 mM to 50 – 60 mM. Until now the reactor has been operated for over 100 days without the need for any nutrient addition or pH control. Based on the alkalinity, it is expected that the dilution rate could be further increased without acidifying the system.

A techno-economic analysis of the integrating a process for anaerobic digestion of stillage to a cellulosic ethanol plant was conducted using ASPEN Plus flowsheeting software. The cellulosic ethanol plant simulations were calibrated using data from Stan Mayfield Biorefinery. Laboratory scale anaerobic digestion data was scaled up to simulate AD in ASPEN Plus. An energy analysis showed that biogas produced by anaerobic digestion if used on-site can displace about 40% of the fossil energy consumed for generating steam for distillation.

The optimal dilution rate and the performance of the fluidized bed reactor under these optimal conditions will be presented. A technoeconomic analysis of integrating an anaerobic fluidized bed bioreactor to treat stillage will also be presented. The impact of introducing anaerobic digestion on the overall economics and energy inputs of a cellulosic ethanol facility will be presented.

571717 Biomass Utilisation Network Life-Cycle Optimisation for the Satisfaction of Energy, Fuels and Chemicals in the European Union

The utilisation of biomass resources is one of the most promising renewable alternatives to satisfy several of our current needs. In particular, several crops and vegetal residues have demonstrated their capacity to produce more sustainable alternatives to conventional demands such as electric power, fuels and chemicals through a wide range of conversion technologies.

The numerous potential biomass production and processing strategies, however, ultimately complicate the clear identification of optimal biomass utilisation strategies, each strategy presenting a particular economic and environmental performance to be evaluated and compared with other strategies. Furthermore, the identification of optimal biomass utilisation strategies is also hindered by the local availability limitations of particular biomass resources or by specific energy, fuel and product demands.

Under this scenario, we present an optimisation approach based on a biomass-conversion network combined with local biomass availability limitations and demand constraints. In particular, the considered network involves several cultivated crops and residual biomass types to be used in more than 200 biomass-conversion technologies including both mature and promising pre-treatments, chemical, bio-chemical and thermal conversion processes. In addition to economic factors, the proposed optimisation approach also accounts for the life cycle environmental performance of the network, taking into account several environmental criteria to assess and the full sustainability of the optimal biomass utilisation strategies identified.

This approach is here applied to the European Union case study, aiming to identify the best biomass resources and conversion routes to satisfy the demands of electricity, fuels for the transport sector and ethylene using locally produced biomass resources.

The identified optimal utilisation strategies clearly indicate the potential of energy crops combined with carbon capture bio-energy production technologies as well as the combined use of different biomass resources for the production of bio-ethanol blends to satisfy fuel requirements, while the ethylene demand is fully satisfied with conventional fossil production routes. Furthermore, the results obtained determine the insufficiency of locally produced biomass resources to cover the combined demand of electricity, transport fuels and ethylene in the European Union.

Overall, the presented approach demonstrates its capacity to assist decision and policy makers in the development of fully sustainable biomass utilisation strategies for a particular set of regional limitations.

572621 Comparative Economic and Environmental Assessments of Methanol Production from CO_2: Transport Vs. Energy

Methanol from captured CO_2 has been promoted as an alternative green fuel due to its low carbon footprint. In this study, the techno-economic and environmental impacts of the electricity and transport sectors were analysed simultaneously at endpoint levels e.g. human health, ecosystems quality and resource depletion. Different scenarios were investigated with different uses of the carbon-free energy source (electricity from wind in this case). Six countries were selected according to their differences in the energy mix grids. The obtained results show that decarbonising the energy grid with CCS offers greater environmental savings with no significant increase in the economic assessment compared to producing methanol from captured CO_2 over their full life cycle assessment. Our results highlight the importance of performing a transport-electricity nexus analysis when assessing the proposed alternative fuels along with the demand for detailed LCA in policy making for both sectors. Life cycle thinking should be applied to the sectors with the highest CO_2 emissions such as transport and energy to fully understand the implications for each scenario and to prevent emissions shifting from the transport sector to the energy sector.

579588 Characterizing Aerosol Liquid Water at Two Contrasting Locations in the Eastern and Western United States

Aerosol liquid water (ALW) is an important component of the atmosphere as it is widely present in ample concentrations, and it could thus promote the formation of particles in the aqueous phase. This study characterizes ALW in two cities of contrasting metrological conditions and source emissions influences in the United States. A climatology of ALW was derived based upon ten years of aerosol composition measurements at an urban site in Maryland and a remote site in Colorado, in the eastern and western United States, respectively. Annual organic aerosol concentrations and inorganic speciation were provided by the Maryland Department of the Environment in the east and a combination of filter collection from Colorado State University and IMPROVE were used in the west. ALW associated with inorganic components (W_{inorg}) was simulated using the ISORROPIAv2.1 aerosol thermodynamic model, while the liquid water associated with organics (W_{org}) was calculated according to Koehler theory. The total ALW was quantified as the sum of W_{inorg} and W_{org} content. ALW was correlated to sulfate concentrations in the eastern United States ($R^2 = 0.45$), however, it was tightly correlated to nitrate concentrations in the west ($R^2 = 0.60$). Despite average ALW concentrations of 6 $\mu g\ m^{-3}$ observed in the east throughout the year, concentrations in the west were dependent on the season. Summertime ALW concentrations were higher (average = 1.4 $\mu g\ m^{-3}$) compared to lower wintertime concentrations (average = 0.4 $\mu g\ m^{-3}$) in the west, mainly due to the abundance of W_{org}. In the cold seasons, the contribution of W_{org} was 10% and 20% of the total ALW in the west and east, respectively. On the other hand, the contribution of W_{org} in the warm seasons was 40% in the east and 70% in the west, possibly due to biogenic emissions and wildfires in the two locations, respectively. This work demonstrates the importance of accounting for organic-associated water. Effects of these organic concentrations on pH will further be examined. Additionally, an analysis of ALW will be presented from other locations in the United States to examine the spatial distribution of ALW concentrations.

560130 Chemical Fingerprinting of Particulate Organic Compounds in Fresh Smoke Produced from Burning Individual Western U.S. Wildland Fuels

Wildfires in the western U.S. are becoming larger and more frequent due to climate change and historical fire suppression. Smoke particles emitted from wildfires are composed of thousands of organic compounds that can vary based upon the fuel burned. The purpose of this study is to identify particulate chemical tracers in order relate smoke composition to a specific fuel or fuel type. Chemical fingerprints were developed from molecular speciation of smoke particles collected during the FIREX campaign at the USDA Fire Sciences Laboratory (Missoula, MT). Western U.S. wildland fuels were burned individually and the emitted smoke particles were collected onto filters. Filters were analyzed using thermal desorption, online derivatization, and 2-dimensional gas chromatography with electron ionization (70 eV) or vacuum ultra-violet light (10.5 eV) high-resolution time-of-flight mass spectrometry. Emission factors of individual organic compounds were normalized to the measured total organic carbon emission factor. Compounds were considered part of the fingerprint for a specific fuel when their ratios exceeded two standard deviations relative to other fuels. These chemical fingerprints were then compared to smoke collected during two western U.S. wildland fires: low-intensity prescribed burning at a managed forest (Blodgett Forest, Georgetown, CA, Nov. 2017) and high-intensity wildfires in northern California (Santa Rosa County, CA, Oct. 2017). Fuels that were identified by their fingerprints, such as manzanita and coniferous rotten log, matched the plant inventory of the forest plots/region. These results indicate that the chemical fingerprints derived from detailed molecular speciation can be used to help determine the specific fuels that burned during wildland fires.

565679 Development of a Calibration Chamber to Evaluate the Performance of Low-Cost Particulate Matter Sensors

Design and Evaluation of Cost-Effective Chamber for Calibrating Low-cost Particulate Matter Sensors

Abstract

Low-cost particulate matter (PM) air quality sensors are becoming widely available and are being increasingly deployed in ambient and home/workplace environments to provide highly resolved spatiotemporal PM concentrations. However, PM measurements from these sensors are often of questionable quality, and the sensors need to be characterized individually for the conditions of use. In this study, we designed and evaluated a cost-effective calibration chamber capable of continuously providing a uniform PM concentration to multiple low-cost PM sensors. The chamber development and assessment were based on a Computational Fluid Dynamics (CFD) model and a rigorous experimental protocol. We also used this new chamber to calibrate 242 Plantower PMS 3003 PM sensors from two manufacturing batches (I and II) by exposing them to two aerosol types: ammonium nitrate and alumina oxide. Our CFD models and experiments demonstrated that the chamber is capable of providing uniform PM concentration to 8 PM sensors at once within 6% error and with excellent reliability (intraclass correlation coefficient > 0.771). The study identified two malfunctioning sensors and showed that the remaining sensors have high linearity compared to a DustTrak reference monitor that was calibrated for each aerosol type (R^2>0.978). Finally, the results revealed statistically significant differences between the responses of batch I and II sensors to the same aerosol (P-value<0.001) and the batch I sensors to the two different aerosol types (P-value<0.001). This chamber design and evaluation protocol can provide a useful tool for those interested in laboratory characterization of low-cost air quality sensors in a systematic and rigorous manner.

Drs. K. Kelly and P.-E. Gaillardon have an interest in the company Tetrad: Sensor Network Solutions, LLC, which commercializes solutions for environmental monitoring.

573534 Comparing Cyclist Air Pollution Exposures in Major Cities in the United States

Air pollution exposure during outdoor activity is associated with decreased maximum oxygen consumption, increased heart rate, decreased maximum activity time, decreased maximum heart rate, and decreased performance. While the current scientific evidence suggests that the benefits of increased activity outweigh the consequences of air pollution exposure, at least in areas like the United States, it also suggests the importance of taking measures to minimize exposure by carefully considering the time and location of the activity. In this project, we estimate the air pollution exposure to cyclists in major cities throughout the United States. We use publically shared cycling route data to estimate the routes cyclists take in each city. We use satellite-derived land-use regression estimates for NOx and particulate matter concentrations for each census block-group developed by the Center for Air, Climate, and Energy Solutions (Kim et al. 2018). Based on initial results, cyclists in cities with lower cycling-commuter rates tend to experience lower air pollutant exposures, likely due to decreased amounts of cycling in the dense urban core. For example, Los Angeles, CA and Chicago, IL, which have relatively high cycling commuter rates (1.8% and 1.2%, respectively), have relatively high average NOx exposure concentrations (21.714 ppb and 20.617 ppb, respectively). By contrast, cyclists in Detroit, MI and San Antonio, TX, both cities with low cycling commuter rates (0.6% and 0.2%, respectively), had much lower average NOx exposure concentrations (12.153 ppb and 7.129 ppb, respectively). Ultimately, the findings will help active commuters on a personal level recognize the effect that their route and mode of transportation have on their exposure to air pollution. On a larger scale, these findings may encourage infrastructure development which mitigates some of the air pollution exposure of active commuters.

570737 Comparisons and Contrasts of Jet Engine Emissions with Other PM2.5 Combustion Sources

Jet engine aircraft exhaust contains combustion byproducts and particulate matter in the form of non-volatile particulate matter (nvPM), black carbon (BC) is used synonymously for nvPM throughout this paper. Aircraft cruise emissions are the only direct source of anthropogenic BC particles at altitudes above the tropopause.[1] Black carbon aerosols are strong solar radiation absorbers and have long atmospheric lifetimes.[2] Therefore, BC results in positive radiative forcing and is believed to be the second largest contributor to climate change.[3] Additionally, upper troposphere and lower stratosphere BC particles contribute to climate forcing indirectly by acting as ice nucleation sites and cloud activators.[4-6] With regards to human health, a link between cardiopulmonary diseases and carbonaceous black particulate matter has recently been suggested.[7] As concern for human health risks and environmental impacts caused by aviation BC emissions increases, emission reduction strategies will need to be implemented. An ambitious carbon, solid and gaseous, emission reduction goal of 50% reduction by 2050 as compared to 2000-2005 levels have already been defined by the International Air Transport Association and Advisory Council for Aviation Research and Innovation in Europe.[8] Meeting these goals will require significant engineering advancements requiring a long implementation period. However, in the near term, alternative jet fuels with reduced aromatic content are an attractive solution for reducing BC emission.[9-12] Alternative aviation fuels containing synthetic blend components with near zero aromatic content (synthetic paraffinic kerosenes, SPKs) such as those synthesized via the Fischer-Tropsch (FT-SPK) process and hydrotreated esters and fatty acids (HEFA-SPK) overall contain highly reduced aromatic content compared to conventional fuel and thus significantly reduce aircraft engine BC emissions.[9-12]

Currently there is not a direct regulation on BC emissions from jet engines. Rather, BC emissions during the landing and take-off (LTO) cycle are limited by the International Civil Aviation Organization (ICAO) through regulations on smoke number (SN).[13] The smoke number regulation introduced in 1981 was put in place with the purpose of reducing plume visibility and no engines have failed this regulation since 1990.[14] With increasing concern on both human health and environmental impacts caused by jet engine BC emissions the EPA is expected to place regulations on such emissions.[15] The ICAO's Committee on Aviation Environmental Protection is currently developing a regulatory standard for BC emissions.

To better understand regulatory needs and health impacts and related regulatory needs, further physical-chemical data on particulate emissions from jet engines is needed. To-date there is a large deficit of detailed characterization compared to emissions from other sources such as diesel engines, despite field campaigns in recent years. Accordingly, we have performed electron-based microscopy and spectroscopy upon samples collected from various jet engines in service. Fuel compositions have also been varied to test effects on EI_{BC} with further interest in changes to the nature of the emitted particulate. Analysis results from NASA campaigns, APEX, AAFEX I & II, and the recent ND-MAX conducted in Germany will be presented.

References

1. Peck, J.; Oluwayemisi, O; Wong, H; Miake-Lye, R. An algorithm to estimate cruise black carbon emissions for use in developing a cruise emissions inventory. *J. Air Waste Manage. Assoc.* **2013,** *63*, 367-375.
2. Lee, D. S.; Fahey, D. W.; Forster, P. M.; Newton, P. J.; Wit, R. C. N.; Lim, L. L.; Owen, B.; Sausen, R. Aviation and global climate change in the 21st century. *Atmos. Environ.* **2009**, *43*, 3520-3537.
3. Bond, T.; Doherty, S.; Fahey, D.; Forster, P.; Berntsen, T.; DeAngelo, B.; Flanner, M.; Ghan, S.; Karcher, B.; Koch, D.; Kinne, S.; Kondo, Y.; Quinn, P.; Sarofim, M.; Schultz, M.; Schulz, M.; Venkataraman, C.; Zhang, H.; Zhang, S.; Bellouin, N.; Guttikunda, S.; Hopke, P.; Jacobson, M.;

Kaiser, J.; Klimont, Z.; Lohmann, U.; Schwarz, J.; Shindell, D.; Storelvmo, T.; Warren, S.; Zender, C. Bounding the role of black carbon in the climate system: A scientific assessment. *J. Geophys. Res.: Atmos.* **2013,** *118,* 5380-5552.

4. Haywood, J. M.; Shine, K. P. The Effect of Anthropogenic Sulfate and Soot Aerosol on the Clear-Sky Planetary Radiation Budget. *Geophys. Res. Lett.* **1995**, 22, 603-606
5. Karcher, B.; Peter, T.; Biermann, U. M. Schumann, U. The Initial Composition of Jet Condensation Trails. *J. Atmos. Sci.***1996**, *53*, 3066-3083.
6. Heymslield, A. J.; Lawson, R. P.; Sachse, G. W. Growth of Ice Crystals in Precipitating Contrail. *Geophys. Res. Lett.* **1998**, *25*, 1335-1338.
7. Pope, C. A.; Dockery, D. W. Health effects of fine particulate air pollution: Lines that connect. *J. Air Waste Manage. Assoc.* **2006**, *56*, 709-742.
8. Realising Europe's Vision for Aviation: Strategic Research & Innovation Agenda (Executive Summary); Advisory Council for Aviation Research and Innovation in Europe (ACARE): Brussels, Belgium, 2012.
9. Moore, R.; Shook, M.; Beyersdorf, A.; Corr, C.; Herndon, S.; Knighton, W.; Miake-Lye, R.; Winstead, S.; Yu, Z.; Ziemba, L.; Anderson, B. Influence of Jet Fuel Composition on Aircraft Engine Emissions: A synthesis of aerosol emissions data from the NASA APEX, AAFEX, and ACCESS missions. *Energy Fuels* **2015,** *29,* 2591-2600.
10. Timko, M. T.; Herndon, S. C.; Blanco, E. R.; Wood, E. C.; Yu, Z.; Miake-Lye, R. C.; Knighton, W. B.; Shafer, L.; DeWitt, M. J.; Corporan, E. Combustion products of petroleum jet fuel, a Fischer–Tropsch synthetic fuel, and a biomass fatty acid methyl ester fuel for a gas turbine engine. *Combust. Sci. Technol.* **2011**, *183*, 1039-1068.
11. Corporan, E.; Dewitt, M. J.; Belovich, V.; Pawlik, R.; Lynch, A. C.; Gord J. R.; Meyer, T. R. Emissions characteristics of a turbine engine and research combustor burning a Fischer–Tropsch jet fuel. *Energy Fuels* **2007**, *21*, 2615–2626.
12. Cain, J.; DeWitt, M.J.; Blunck, D.; Corporan, E.; Striebich, R.; Anneken, D.; Klingshirn, C.; Roquemore, W.; Vander Wal, R. Characterization of gaseous and particulate emissions from a turboshaft engine burning conventional, alternative, and surrogate fuels. *Energy Fuels* **2013**, *27,* 2290-2302.
13. ICAO. *ICAO Annex 16: Environmental Protection, Volume II – Aircraft Engine Emissions;* International Civil Aviation Organization: Montreal, Canada, 2008.
14. EASA ICAO Engine Emissions Databank [online]. Available at: http://easa.europa.eu/document-library/icao-aircraft-engine-emissions [Accessed 4 Jan 2016].
15. Wayson, R. L.; Fleming, G. G.; Lovinelli, R. Methodology to estimate particulate matter emissions from certified commercial aircraft engines. *J. Air Waste Manage. Assoc.* **2009**, *59*, 91-100.

588339 A visible colorimetric sensor based on a two-fold preconcentration/detection system for the determination of trihalomethanes in treated drinking water

Water is an integral part of our society, and it is of the utmost importance that this water is clean and safe for potable use. Drinking water sources can become contaminated with halogenated organic compounds (HOCs) via improper disposal of refrigeration or air conditioning units, industrial usage, and even through the disinfection of our drinking water (chlorination). One class of these HOCs regulated by the U.S. Environmental Protection Agency (USEPA) is trihalomethanes (THMs). THMs have been found to cause adverse health effects and can even be carcinogenic, they are limited to 80 ppb in treated water by the USEPA. Currently THM concentrations are detected in treated water using expensive analytical equipment, which have high sensitivity, but lack any portability or ease-of-use. Here we show a syndiotactic polypropylene (sPP) electrospun fiber mat sensor capable of visible colorimetric detection of trihalomethanes (THMs) at environmentally relevant levels (ppb-scale), without a separate preconcentration step. The developed detection method includes a two-fold preconcentration technique coupled with a colorimetric detection reaction, integrated into a single sensor device. Here, the utilization of the colorimetric Fujiwara reaction serves as the detection reaction. This long-studied reaction has a response that is normally limited to a beaker scale visible detection range on the order of 80 ppm THMs, when a separate preconcentration step is not present. The two-fold integrated preconcentration technique consists of a thermodynamic method based on the utilization of the THM: water equilibrium in the vapor phase of the system as well as a physical method using a sPP fiber mat capable of concentrating THMs in a contaminated aqueous sample. The developed device has successfully lowered the portable single-step visible THM detection concentration by an order of magnitude (80 ppm→80 ppb). This sensor was also successfully applied for the determination of the total THM (t-THM) concentration in spiked and actual treated water samples, and these results correlated well with the GC/MS results of the corresponding sample, showing its real world applicability.

568835 Fairness-Guided Design of Water Distribution Systems for Agricultural Lands

Allocating resources and associated wealth (utility) among multiple stakeholders is a fundamental problem in social planning. Allocation is often guided by maximizing the total utility (the sum of players utilities). This solution, also known as the social welfare approach, is intuitive but has significant defficiencies. In particular, this approach might identify optimal allocations that are nonunique (different allocations give the same maximum social welfare) and that improperly capture system scales [1]. A wide variety of alternative utility allocation schemes have been explored in the literature. Rawls [2] proposed an allocation scheme that seeks to address the scaling issue of the social welfare by maximizing the utility of the smallest stakeholder. Unfortunately, this approach ignores large stakeholders and can yield non-unique solutions. Nash [3, 4] proposed an allocation approach that maximizes the product of the stakeholder utilities. This approach is equivalent to maximizing the sum of the logarithms of the utilities and thus captures stakeholder scales more naturally. Moreover, this approach provides a unique solution. Recently, it has been established that the Nash allocation approach maximizes entropy (a measure of diversity) and that it can be generalized by using a measure called proportional fairness. An important application of fair allocation schemes is infrastructure systems, which are systems that must be shared with a large number of stakeholders that cover a large spectrum of scales. For instance, electricity markets allocate supply and generation by maximizing the social welfare [5]. Proportional fairness schemes are used to allicate bandwidth in telecommunication networks and air traffic flow management [6]. Similarly, the Rawlsian allocation scheme has been used in networking, telecommunication, routing, and load balancing problems [7].

Fairness measures are also widely used to quantify income inequality. Recently, it has been establishes that fairness can also be measured using descriptive statistics, such as entropy [8, 1]. This conceptually makes sense, since the distribution of resources among a set of stakeholders can be interpreted as a task that seeks to shape a distribution of outcomes (a fundamental problem in statistics and stochastic programming) [9, 10]. Allocation of resources in agricultural systems is essential for maintaining a sufficiently diversified infrastructure that enables long-term sustainability [11]. The optimal water allocation problem has been studied by Li and Guo [12]. Here, they proposed a conflict resolution (multi-objective) optimization model that factors in economic, social, and ecological functions. A social welfare scheme is used as economic objective. Similarly, a conflict resolution model including the minimization of fresh water consumption and the minimization of the total annual cost in agricultural systems was proposed in Arredondo-Ramırez et al. [13]. Here, the authors observed that, when irrigation water is limited, crops water requirements cannot be met and this creates an economic conflicts between the stakeholders. Allocation model for optimizing regional water resources among various crops was reported in [14]. The problem considers insufficiency of water supply and the objective function is to maximize the total benefit of all the crops productions (using a social welfare approach). A stochastic programming method has been reported in the literature to conduct crop planning and water resource allocation under uncertainty. In that work, the formulation includes the maximization of the agricultural system benefit given limited water resources [15]. Recent work has emphasized on the need for developing design tools meet and prioritize water demand needs under scarcity events [16]. Here, one can use optimization techniques to design sophisticated water distribution networks that make optimal decisions on use, reuse, and recycling of water by accounting for the water exchange among crops, storage tanks, and treatment units. Existing work has designed such infrastructures by maximizing the overall utility of all crops [17]. In summary, the vast majority of work on resource allocation in agricultural systems has focused on maximizing the social welfare [18].

This work presents an optimization formulation for the design of distribution networks that allocate water and associated utility to multiple stakeholders in a fair manner. The developed model accounts for the water requirements, yields and sales of each crop, as well as for the costs related to the water exchange, storage and distribution. We use our framework to highlight deficiencies of the social welfare approach as well as to illustrate the benefits of using alternative fairness measures in utility allocation. We demonstrate that such measures become particularly critical under extreme events (e.g., with scarse resources).

References:

[1] A. M. Sampat and V. M. Zavala. Fairness measures for decision-making and conflict resolution. 2018. [Under Review].

[2] J. Rawls. A theory of justice, harvard. Press, Cambridge, 1971.

[3] J. F. Nash Jr. The bargaining problem. Econometrica: Journal of the Econometric Society, pages 155–162, 1950.

[4] A. E. Roth. An impossibility result concerningn-person bargaining games. International Journal of Game Theory, 8(3):129–132, 1979.

[5] V. M. Zavala, K. Kim, M. Anitescu, and J. Birge. A stochastic electricity market clearing formulation with consistent pricing properties. Operations Research, 65(3):557–576, 2017.

[6] M. Zukerman, L. Tan, H. Wang, and I. Ouveysi. Efficiency-fairness tradeoff in telecommunications networks. IEEE Communications Letters, 9(7):643–645, 2005.

[7] D. Bertsimas, V. F. Farias, and N. Trichakis. The price of fairness. Operations research, 59(1):17–31, 2011.

[8] V. Venkatasubramanian. How Much Inequality Is Fair?: Mathematical Principles of a Moral, Optimal, and Stable Capitalist Society. Columbia University Press, 2017.

[9] Alexander W Dowling, Gerardo Ruiz-Mercado, and Victor M Zavala. A framework for multistakeholder decision-making and conflict resolution. Computers & Chemical Engineering, 90:136–150, 2016.

[10] Jian Hu and Sanjay Mehrotra. Robust and stochastically weighted multiobjective optimization models and reformulations. Operations research, 60(4):936–953, 2012.

[11] OECD. OECD compendium of agri-environmental indicators (summary). 2013. https://doi.org/10.1787/9789264186217-sum-en.

[12] M. Li and P. Guo. A multi-objective optimal allocation model for irrigation water resources under multiple uncertainties. Applied Mathematical Modelling, 38(19-20):4897–4911, 2014.

[13] K. Arredondo-Ram´ırez, E. Rubio-Castro, F. N´apoles-Rivera, J. M. Ponce-Ortega, M. Serna-Gonzalez, and M. M. El-Halwagi. Optimal design of agricultural water systems with multiperiod collection, storage, and distribution. Agricultural Water Management, 152:161–172, 2015.

[14] Z. Shangguan, M. Shao, R. Horton, T. Lei, L. Qin, and J. Ma. A model for regional optimal allocation of irrigation water resources under deficit irrigation and its applications. Agricultural Water Management, 52(2):139–154, 2002.

[15] G. Niu, Y. P. Li, G. H. Huang, J. Liu, and Y. R. Fan. Crop planning and water resource allocation for sustainable development of an irrigation region in china under multiple uncertainties. Agricultural Water Management, 166:53–69, 2016.

[16] I. J. Lorite, L. Mateos, F. Orgaz, and E. Fereres. Assessing deficit irrigation strategies at the level of an irrigation district. Agricultural Water Management, 91(1-3):51–60, 2007.

[17] E. Rubio-Castro, J. M. Ponce-Ortega, M. E. Cervantes-Gaxiola, O. M. Hern´andez-Calder´on, J. R. Ortiz del Castillo, J. Mil´an-Carrillo, J. F. Hern´andez-Martınez, and J. A. Meza-Contreras. Optimal design of integrated agricultural water networks. Computers & Chemical Engineering, 84:63–82, 2016.

[18] S. Vedula, P. P Mujumdar, and G. Chandra Sekhar. Conjunctive use modeling for multicrop irrigation. Agricultural Water Management, 73(3):193–221, 2005.

575070 Experimental Study of the Potential of Nano-Fibrous Membranes in Water Treatment Systems for Rural Communities

Water is a crucial component for expecting life on earth. Having covered 70% of earth's total surface, lack of drinking water is an escalating problem worldwide. Apart from the decreasing ground water level, available ground water in most of the places is contaminated by various ions above safe limits. Also, the increasing environmental pollution degrade the quality of fresh water available to human use. Rising population, industrialization, agriculture sector and urbanization have played a major role in depleting underground water level and polluting surface water resources. Desalination appears to be a promising practice to meet the fresh water demand. Effective techniques of water treatment and desalination need major focus to combat the issue.

Membrane Distillation is one such technology having high potential to produce fresh water and it has gained popularity due to the unique advantages it offers. Membrane distillation as a process for supplying fresh water, using microporous hydrophobic membranes, was introduced by Bodell for the first time. In 1967, Findley reported the basic theory and findings of the process.

The following experimental study concentrate on treatment of waste water using membrane distillation process. In membrane distillation systems, a warm feed solution comes in contact with microporous hydrophobic membrane, mass transfer occur due to vapor pressure gradient, and the vapor pass through the membrane pores to the permeate side. The vapors are collected and condensed outside the membrane cell. The hydrophobic membranes are fabricated using electrospinning process, under application of high electric field, which forms the fiber due to charge repulsion. The fiber mat properties i.e. pore size, thickness, porosity, roughness are controlled by electrospinning parameters and polymer solution characteristics. The fabricated membranes are characterized and tested for performance in membrane distillation systems. The effect of process parameters like feed temperature, flow rate, concentration on system performance is observed. The performance is defined in terms of permeate flux, permeate conductivity and rejection rate. The percentage reduction in permeate flux was also investigated at regular intervals of experimental runs. The process is run for longer durations to observe scaling of membrane. This study shows that membrane distillation is a viable technique for producing fresh water from wastewater/saline water. The process can be studied further for high performance membranes and alternative energy sources to obtain low cost treatment systems.

582275 Effects of Cadmium and Mercury on the Germination of Lolium Perenne

The plants germination represents one of the main factors for the development of technologies of phytoremediation of heavy metals. In this research the effects of cadmium and mercury on germination time and emergence as markers of toxicity of the Lolium perenne plant were determined. The seeds were exposed to increasing concentrations in the range of 0 to 25 mg/L of cadmium (Cd^{2+}) and mercury (Hg^{2+}) in Petri dishes, by independent experiments carried out in triplicate for 14 days. The germination rate of seeds exposed to 25 mg/L Hg^{2+} at day 14 did not show significant differences with respect to the control group (P-value = 0.6030). A similar behavior was observed at 25 mg/L Cd^{2+} (P-value = 0.4140). Concentrations higher than 5 mg/L Hg^{2+} decreased root length by 38.7% (P-value = 0.0005), same effect was observed at 25 mg/L Cd^{2+} (P-value = 0.0000). The heavy metals Cd and Hg in their inorganic form present toxic effects on the development of stem and roots of Lolium perenne, however, suitable conditions of germination and development were observed within a wide range of concentrations of heavy metals in which the plant can potentially act as a phytoremediator.

582274 Analysis of the leachates of the monoliths from cementhazardous waste ash mixture

Fly ash may come from different industries, whether from thermoelectric plants or incineration plants. Fly ash may have a high toxic character derived from heavy metals, chloride salts and organochlorine compounds. Generally, the content of heavy metals in fly ash is higher due to the vaporization of metals during the combustion process and in parallel with the process of adsorption of metals on the surface of the ash. They may contain traces of polycyclic organic compounds, PCBs, dioxins, furans, among others. They have high contents of silicon and aluminum oxides (crystal formers), sodium and potassium oxides (fluxing agents), calcium, magnesium, zinc and lead oxides (stabilizers). The objective of this work was to evaluate the capacity of Portland cement to encapsulate ashes of hazardous waste by means of analysis of leaching of metals by simulation of acid rain conditions. The percentage of replacement in cement monoliths for hazardous waste was 10% and for treated hazardous waste it was 3% and 5%, all at a water-cement material ratio of 0.5. The already processed monoliths are stored in a curing room that is at a temperature of 23 °C and a relative humidity of 95%. The leaching tests were carried out at the ages of 1, 3, 7, 14 and 28 days. A monolith was taken from each type of ash and the target for the test. The SPLP fluid will be prepared by water and a mixture of H2SO4/HNO3 in a 60/40 weight-to-weight ratio adjusting the pH to 4.2±0.05. The monolith is introduced into a bottle and 250 ml of SPLP fluid is added, and left to stand for 24 hours, and then the fluid is filtered into extraction bottles preserved with nitric acid for metal reading by ICP-OES. It was found that the only parameters that exceed international standards are calcium and sodium, but these are not considered within international standards as dangerous heavy metals. Likewise, the heavy metals detected in the sample are below the limit of detection of the equipment and the regulations. Therefore, the confinement that is given from the cement-ash matrix is considered good and this type of stabilization processes can be carried out.

580828 An Integrated CO_2 Capture and Conversion Device based on Metal-Organic Framework -and Cu Nanostructured Electrocatalyst

There is no doubt that CO_2 emissions and its proportion in the atmosphere are rising sharply due to growing world population and as a result of fossil fuel production and consumption outcomes. This is a significant motivation for the scientific community to focus on CO_2 capture and utilization strategies to make carbon-based chemical feedstocks. One such process, the electrochemical CO_2 reduction reaction, enables the conversion of CO_2 into valuable hydrocarbons and alcohols. However, efficient electrocatalysts need to be employed for CO_2 reduction reaction to become competitive with traditional fossil fuel-based technologies.

On the other hand, emerging research into organic and inorganic hybrids and Metal-Organic-Frameworks "MOF" have shown promise as a new class of porous materials with micropores or mesopores. Because recent MOF research has shown high capacity and selectivity towards CO_2 capture even at ambient temperature, it appears promising as a possible material to serve as a catalyst or concentrator for CO_2 conversion. In this study, we are developing an integrated device, wherein MOF based porous materials serve as CO_2 concentrator and subsequently a Cu based nanostructured catalyst electrochemically converts CO_2 into valuable C_2+ based products. Carbon fiber sheet was used as a platform to conduct the current and support the MOF and Cu catalyst for CO_2 capture and electrochemical reduction reaction, respectively. One side of the carbon fiber is coated with Zinc-based MOF with 500 m^2/g Langmuir surface area, and an adsorption capacity of 4 mmol/g CO_2 at 20°C. The MOF particles were synthesized in-house and followed with a set of characterizations including, TGA, XRD, FTIR, and gas sorption. The MOF layer served as a CO_2 concentrator and a screening stage, where it allows higher percentage of CO_2 molecules to pass through the fiber to the other side. The other side of the fiber was coated with Cu_2O particles to catalyze the electrochemical reduction reaction. Spray coating technique was used to spray coat MOF and Cu_2O onto the carbon fiber. The spray coating process took place after individually mixing the Cu_2O and MOF particles in Methanol/Nafion solution to maintain a certain level of rigidness and stability on the substrate surface. Binary gas feed N_2 50%/CO_2 50% was used for the electrochemical reduction experiments in a flow-cell configuration. The initial results showed high C_2+ products selectivity, conversion efficiency, and catalyst stability.

557750 Triboelectric Particulate Matter Purifier Using Electrospun Ethyl Cellulose

Triboelectric Particulate Matter Purifier Using Electrospun Ethyl Cellulose

Min Hee Cho, Jeong Gil Seo*

Department of Energy Science and Technology, Myongji University, 116, Myongji-ro, Cheoin-gu, Yongin-si, Gyeonggi-do 17058, Republic of Korea

*E-mail: jgseo@mju.ac.kr

As a result of ever-increasing demand for transportation, industrial production, and power plants, air pollution has become one of the most serious environmental problems. One of the major pollutants, particulate matter (PM), has attracted much attention because of its catastrophic effect on human health such as cardiovascular and respiratory diseases. PMs are categorized by their aerodynamic diameter; PM10 and PM2.5 are defined as a particulate matter with an aerodynamic diameter less than 10 ¥im and 2.5 ¥im, respectively. Very recently, South Korea has suffered from PMs especially during winter and spring seasons. Although the main source of PMs is still controversial, it is urgent to protect human health from the highly concentrated PMs. There are typical ways to remove PMs such as electrical precipitation and filtration. However, in the case of the electrical precipitation, it needs a very high voltage requiring a lot of energy consumptions and O_3 which is very hazardous to a human being is generated during the operation. In the case of filtration, it usually uses a High-Efficiency Particulate Absorber (HEPA) filter. However, it also requires high maintenance and administrative cost for exchanging and discarding. Furthermore, in a high concentration of PMs, life time of the filter is drastically decreased. Therefore, the reusable, economical and high-efficiency technology that minimizes the energy consumption and maintains the high PM particles removal efficiency is needed. In this study, a new method that utilized the electrostatic effect induced by triboelectrification was developed in order to make better PM purifier more than former ways. Triboelectrification is a phenomenon that material surface becomes electrically charged as a result of contact or collision between different materials. This phenomenon can be applied in the PM purifying process. Our previous experiments demonstrated the triboelectrification effect in PM capture process using the electrification between PTFE and aluminum (Figure 1) and the electrification between PTFE and nylon (Figure 2). Generally, PTFE has a very high electronegativity so its surface can receive the electrons when it contacts aluminum or nylon. Consequently, PTFE becomes negatively charged and the opposite material becomes positively charged. So when polluted air flow into these materials, PM particles can be captured on their surface. This mechanism is based on the electrostatic filtration induced by triboelectrification. And there is also another mechanism of capturing PM particles. That is mechanical filtration such as interception, inertial impaction, diffusion and gravitational settling because of its structural characteristics. In other words, electrostatic filtration and mechanical filtration are accompanied and make a synergy effect in the PMs capture process (Figure 1 and 2). On the basis of previous studies, ethyl cellulose was selected as the replaced material to be positively charged instead of nylon or aluminum. Ethyl cellulose is one of the cellulose derivatives called cellulose ether. Cellulose is the most abundant natural polymer and it is easily accessible, biodegradable and economical cost. According to the triboelectric series, ethyl cellulose has a big tendency to be positively charged much more than cellulose and even bigger than nylon. The main components of Triboelectric PM Purifier (TPP) is a triboelectric layer (TL) consisting of ethyl cellulose and PTFE spheres. In order to induce electrification effectively on the surface of ethyl cellulose and PTFE sphere, the structure of ethyl cellulose should be non-woven structure so it was conducted by using electrospinning. Ethyl cellulose non-woven was placed in the top and bottom of the triboelectric layer and PTFE spheres got into between top and bottom space. Under some of the mechanical movement, PTFE spheres contact and collide with the ethyl cellulose surface. As a result, PTFE becomes negatively charged and in turn ethyl cellulose becomes positively charged. In this state of the TL, PM particles can be effectively captured by the synergy of mechanical filtration and electrostatic filtration induced by triboelectrification. In order to test its capacity to remove PM particles in the indoor environment, the indoor environment was simulated by an acrylic box (W: 800mm, D: 550mm, H: 500mm, thickness: 10 mm). The mass concentrations (¥ig m^{-3}) of PM1, PM2.5, and PM10 were observed in real time by the measurement instrument (Y09-PM handheld air quality monitor, Sujing Group automation instrument equip CO, Ltd). And Rocker 3D digital shaker (IKA) was used as a mechanical movement source (Figure 3). PMs were generated by incense burning as it produces PMs much more than cigarettes and also generates gas products such as CO, CO_2, NO_2 and SO_2 and organic compounds such as benzene, toluene, xylene, aldehydes and polycyclic aromatic hydrocarbons. The performance of the TPP was tested by the various ways to demonstrate its capacity to remove PMs. In order to verify the effect of triboelectrification in the PMs capture process, the comparison test between the TL and other TL without PTFE spheres was conducted. In the case without PTFE spheres, it needed more time to reduce PMs. And from this result, the synergy of mechanical filtration and electrostatic filtration induced by triboelectrification was verified. And for the purpose of estimating its ability, comparison with a commercial air purifier using HEPA filter was conducted and the cycle test (> 30 cycles) of TPP was also conducted to investigate its durability. In addition, the reusability test was conducted by just washing the TL with water. After washing, it was dried in the room temperature and then the PMs removal test was conducted again. Through the screening of surface condition of ethyl cellulose and PTFE by using FE-SEM, the surface difference before and after the capturing of PMs and the surface after washing were observed in detail. The captured PMs could be found easily on its surface and particle size distributions were from 1 nm to several micrometers. From the images of washed TL components, it showed that the captured PMs were removed well by washing with water. This work proves that the TPP is the reusable, economical, and high-efficiency technology in the air purification field.

Keywords: triboelectrification, ethyl cellulose, particulate matter

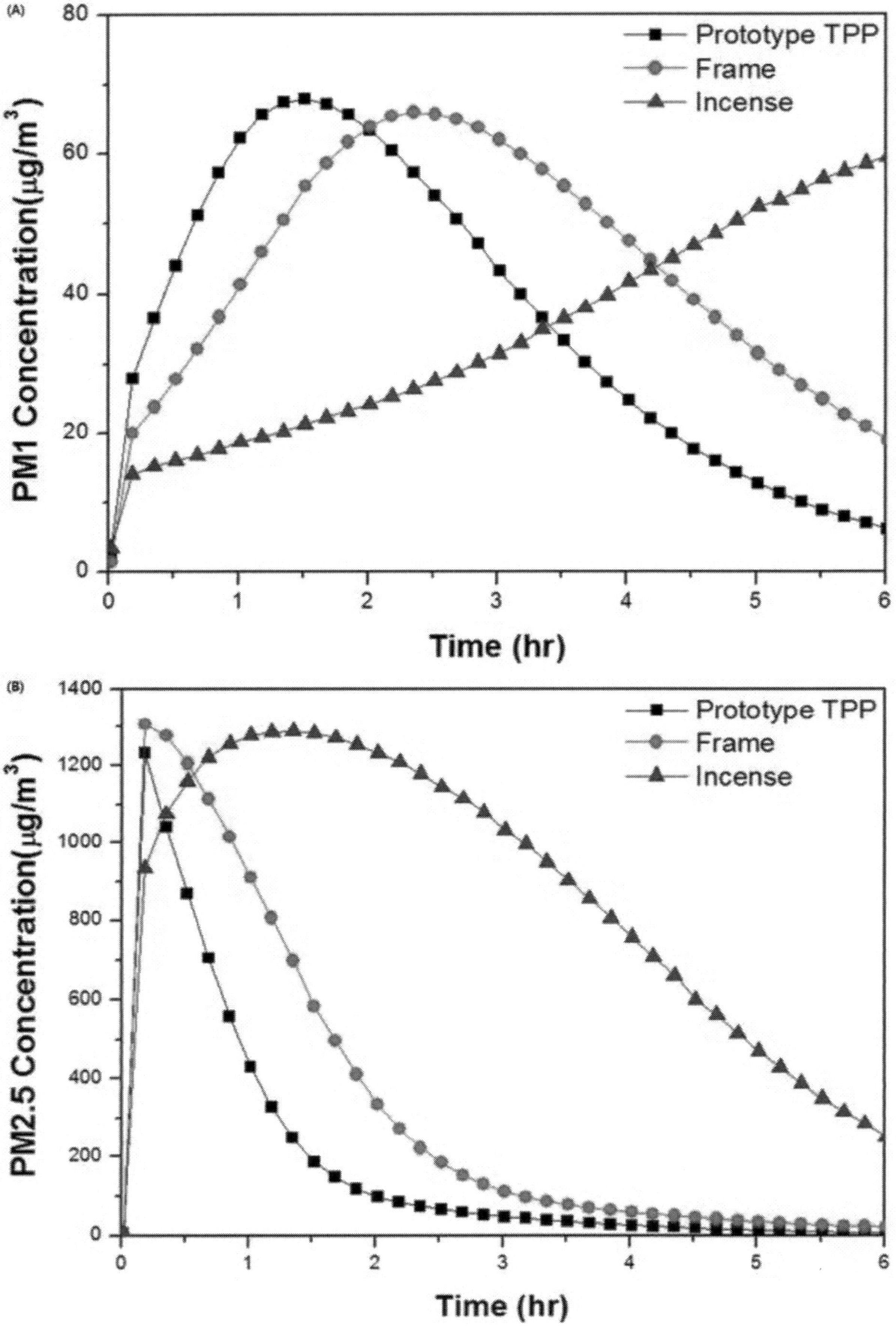
(A)
PM1 Concentration(μg/m³)
Time (hr)
Prototype TPP
Frame
Incense
(B)
PM2.5 Concentration(μg/m³)
Time (hr)
Prototype TPP
Frame
Incense

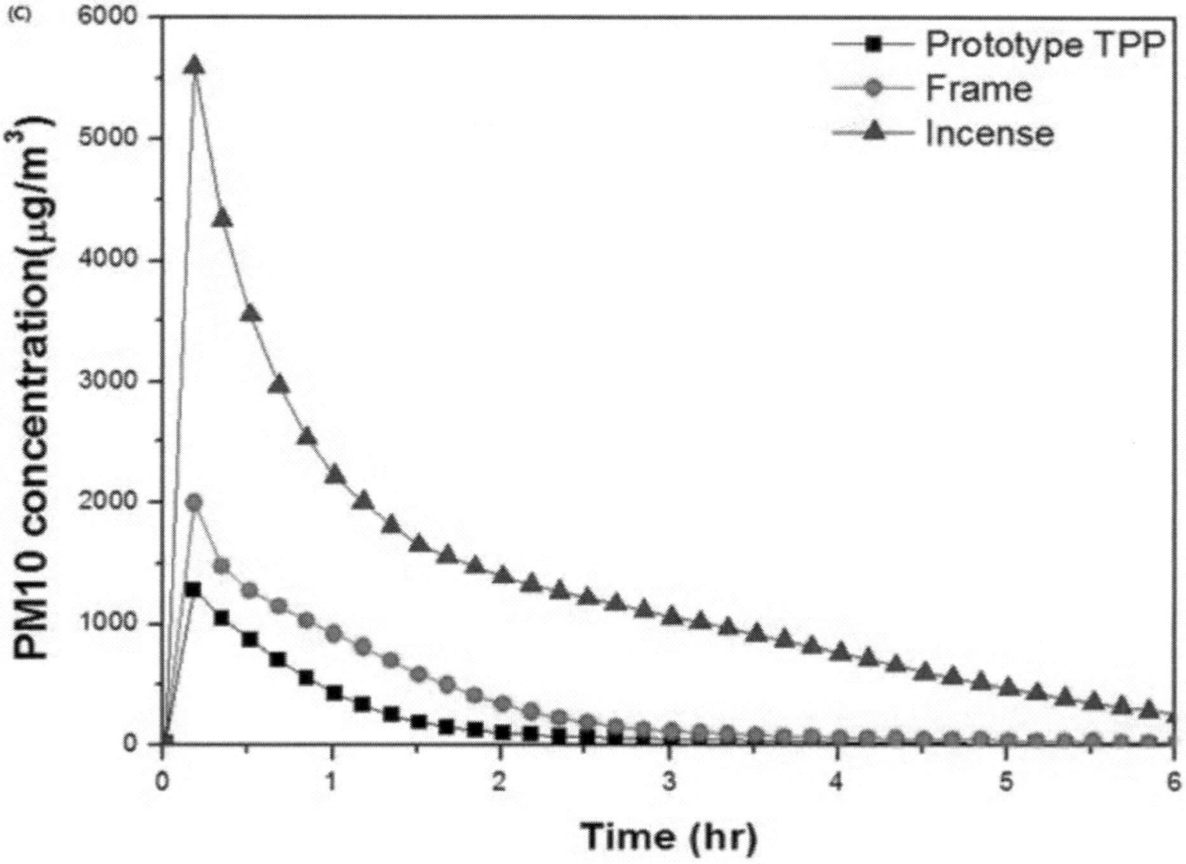

Figure 1. PM removal of prototype TPP consisting of PTFE sphere and aluminum plate. In the graph, the frame means that PTFE and aluminum are removed from the device.
(A) PM1 concentration graph. (B) PM2.5 concentration graph. (C) PM10 concentration graph.

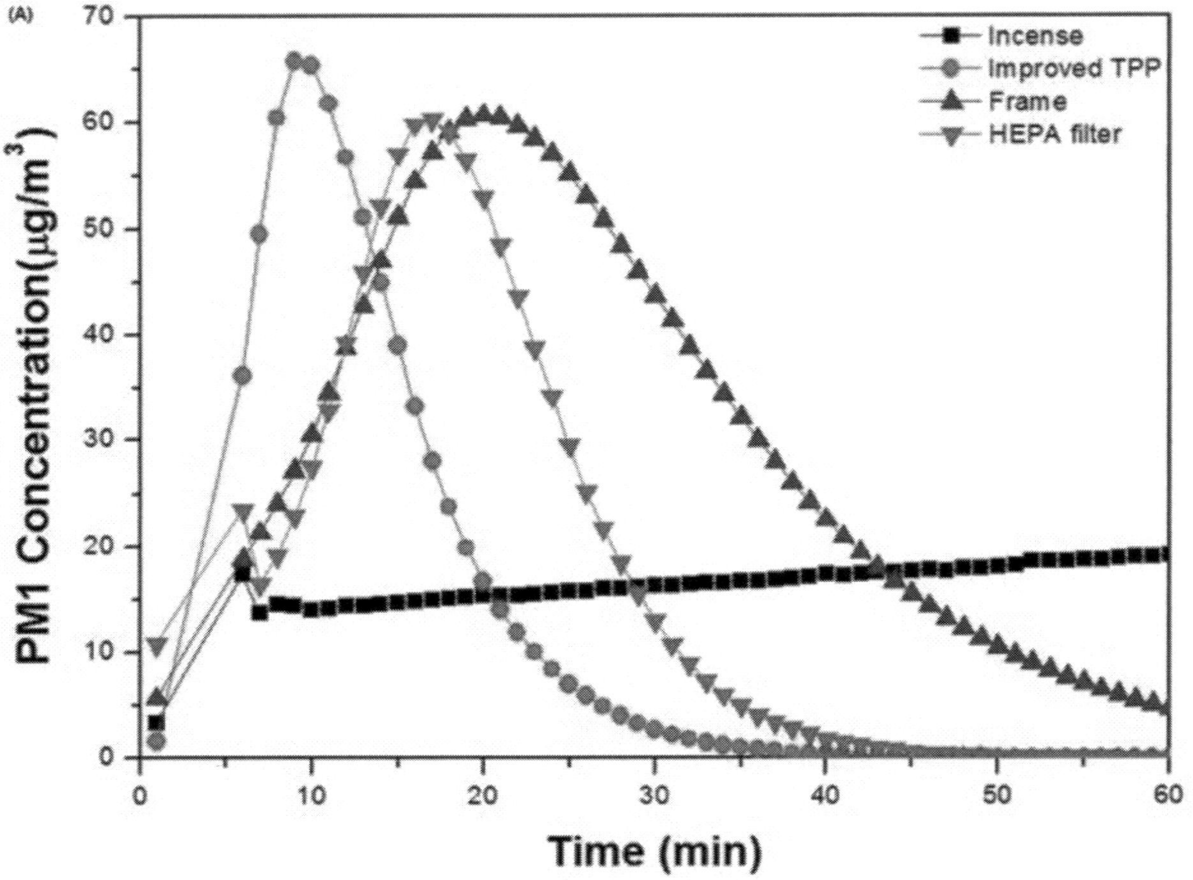
(A)
PM1 Concentration(μg/m3)
Time (min)
Incense
Improved TPP
Frame
HEPA filter
0
10
20
30
40
50
60
70

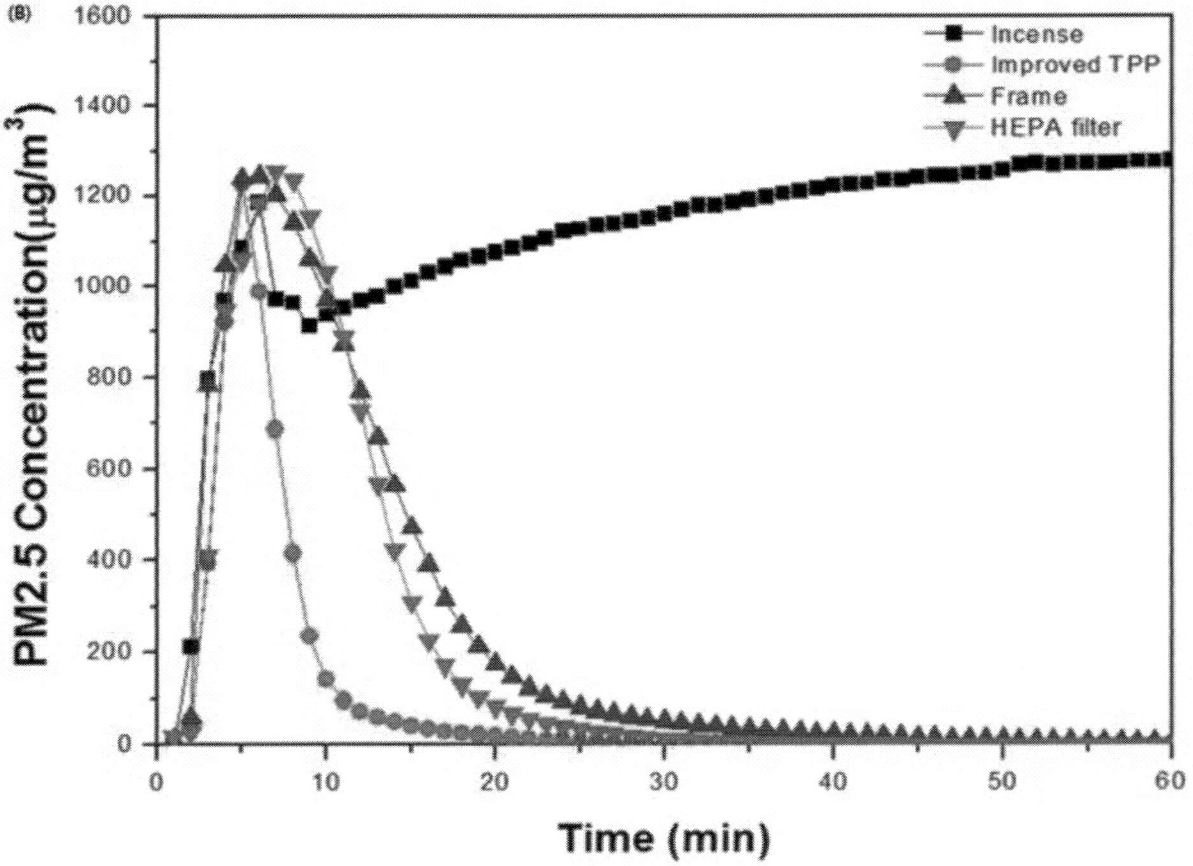
(B)
PM2.5 Concentration(μg/m3)
Time (min)
Incense
Improved TPP
Frame
HEPA filter
0
200
400
600
800
1000
1200
1400
1600
10
20
30
40
50
60

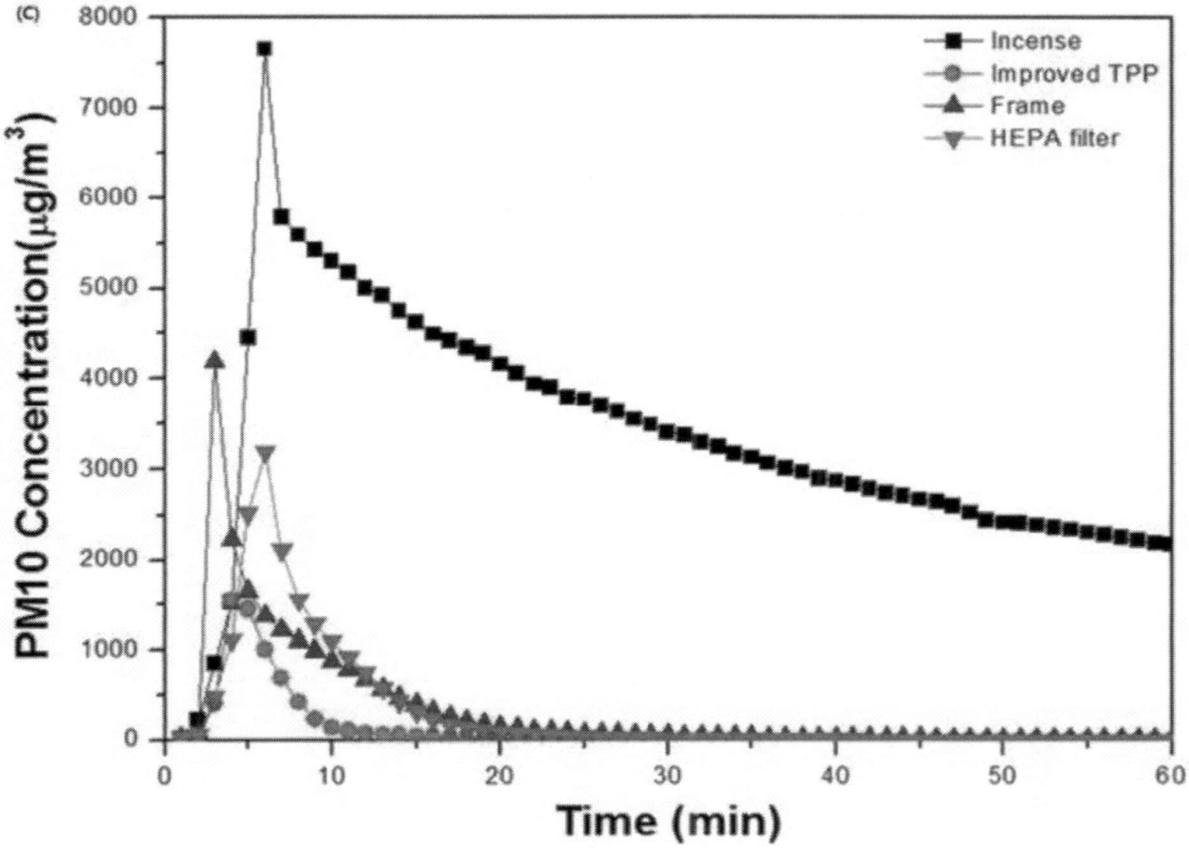

Figure 2. PM removal of improved TPP consisting of PTFE sphere and Nylon mesh. In the graph, the frame means that PTFE and nylon are removed from the device. It was compared with a commercial purifier using HEPA filter.
(A) PM1 concentration graph. (B) PM2.5 concentration graph. (C) PM10 concentration graph.

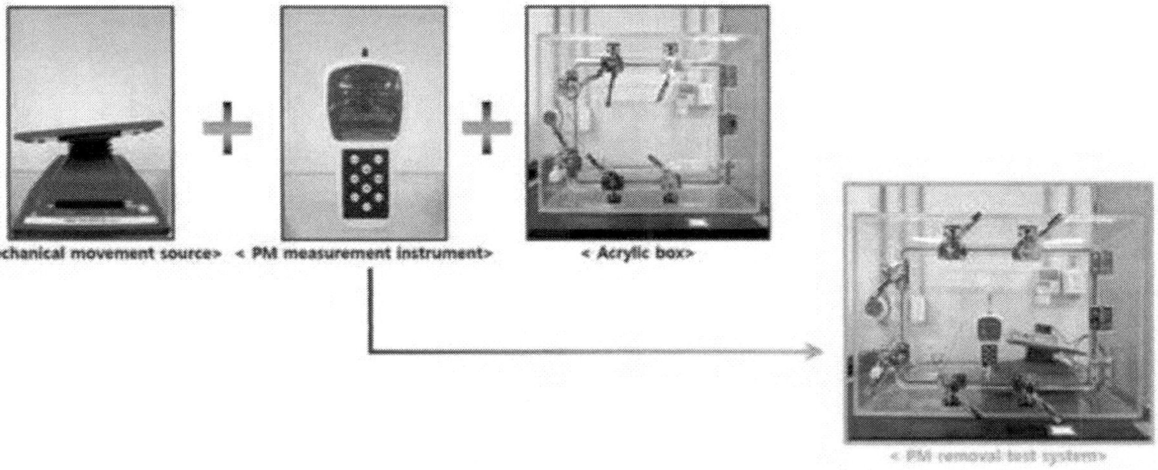

Figure 3. Simulated system for PM removal test.

The application of a Box-Behnken design in the study of microcystin production

Gisella Lamas-Samanamud[a,b], *Fabiano Naves*[c], *Carla Loures*[d], *Tony Reeves*[e], *Heather Shipley*[b]
[a] *Department of Chemical & Materials Engineering, University of Kentucky - Paducah Campus*
[b] *Department of Civil and Environmental Engineering, University of Texas at San Antonio*
c Department of Chemical Engineering, Universidade Federal de Sao Joao del Rei
d Department of Mechanical Engineering, Centro Federal de Educacao Tecnologica de Rio de Janeiro
e Department of Internal Medicine, Wake Forest Baptist Medical Center

Abstract

Microcystin is a cyanotoxin produced by harmful algal blooms such as *Microcystis aeruginosa* PCC7806. Warmer temperatures favor algal bloom and algal growth. The biggest concern is if more incidents of algal bloom would also produce microcystin. The purpose of this study was to determine whether microcystin is favored by other environmental parameters such as: volume, concentration, aeration, pH and temperature or the combination of some or all of them. In this sense, a Box-Behnken design was used as a statistical tool. The optimization of this design uses fewer experiments which is a great asset for the duration of 30 days of the growth of this bacterial strain. Thus, more possibilities are tested during the same period of time as opposed to test every possible combination. Results revealed that microcystin is affected by temperature and the combination of temperature and pH. To have a higher concentration of the toxin, the water body needs to have been exposed to cooler temperatures (20°C) and neutral pH (7). This does not mean that a water system has to be at cool temperatures or neutral pH, but it means that microcystin production is enhanced by stressful conditions such as abrupt changes in temperature and pH from high to low in a short period.

Keywords: Box-Behnken, microcystin, environmental factors, cyanobacteria

1. Introduction

Microcystin is a cyanotoxin produced by *Microcystis aeruginosa*. It depletes oxygen from water bodies and imposes a severe environmental concern and economic damages where they are found. This toxin causes serious damages to the liver and thus, it affects aquatic plants, fish, pets and even human health (Dittmann et al., 2013; Paerl & Huisman, 2008; Sun et al., 2013). One of the largest events associated to algal bloom of *M. aeruginosa* happened in 2014, when microcystin concentration increased to 200 µg/mL (Steffen et al., 2014). This was the highest record for the presence of this toxin in Lake Erie.

M. aeruginosa growth is also favored by warmer temperatures. During summer time the water temperature can reach up to 23°C. Therefore, water contamination and warming conditions could potentially increase the frequency and the scale of events such as the one in Lake Erie in 2014 (Paerl & Huisman, 2008; Steffen et al., 2014). Lately, cyanobacteria can be found in several water bodies. Media is constantly presenting cases of animals that were affected by the toxin. The presence of this type of algae is constantly monitored especially during summertime.

The purpose of this study was to determine whether microcystin is favored by other environmental parameters such as: volume, concentration, aeration, pH and temperature or the combination of some or all of them. In this sense, a Box-Behnken design was used as a statistical tool.

2. Methodology

2.1 Microbial growth

Microcystis aeruginosa were grown in BG11 media for 30 days and they were inoculated from a culture of an OD>2. Microcystin was detected using ELISA kit. The parameters studied in the Box-Behnken design were temperature (20, 25 and 30 °C), pH (7, 9, 11) and aeration (0, 50 and 100 rpm). Microbial growth was monitored on EPOCH (96-well plate) on a daily basis. Microcystin assay was done in duplicate. The pH of the solution was monitored on a daily basis and corrected with 1 M and 0.1 M solutions of both NaOH and HCL.

2.2 The Box-Behnken design

The response value was microcystin and the Box-Behnken surface response can be seen in Tables 1 and 2. This design generates fewer experiments when compared to the Central Composite Design (CCD), and responses are obtained from quadratic models. Table 1 corresponds to the factors and levels considered in this study with their respective codified and non-codified levels.

Table 1 – codified and non-codified levels for the factors used in the Box-Behnken design

Factors		Codified levels		
		-1	0	1
		Non-codified		
A	Aeration (rpm)	0	50	100
B	Temperature (°C)	20	25	30
C	pH	7	9	11

This design of experiment provides a good assay by reducing the number of experiments performed and no need for replicas except on the center point. The order of the experiments described in Table 2 is designed as such in order to randomize any errors or bias that may be inherent with the execution of experiments.

3. Results and Discussion

A few preliminary experiments revealed that pH significantly affects the production of microcystin. At lower pHs, contaminants increased microcystin content to 2 ug/L. In the lab, the maximum concentration had been 0.5 ug/L. These results combined revealed that *M. aeruginosa* forms algal bloom and the pH of the environment raises to pH 11. If no biofilm is formed, water system remains at neutral pH. If pH is reduced, it increases the chance of microcystin production and/or release. It is hard to tell whether microcystin is induced when pH varies but it is certain to say that under stressful conditions, cells release this toxin to the environment. Microcystin values were as high as 1.71 µg/L, which is higher than 3x more than the average value for *M. aeruginosa* cells and almost as twice as dangerous than the maximum limit for microcystin in the environment.

To determine whether this is harmful, or it affects environmental conditions and water parameters, the experiment also considered COD values to verify any correlation between cell content and chemical oxygen demand. Microcystin increases with lower pH, but COD values increase with pH. This indicates that cell content and bacterial growth is favored by higher pH. This can suggest that the increase in microcystin content is due to death of the cells and lysis of bacterial cell wall which in turn releases more microcystin in the environment.

Considering that changes in pH play a role in favoring microcystin release, this study focused on a few environmental factors that could affect microcystin production: pH, temperature and aeration. Therefore, in order to apply a statistical tool that determines which parameter influences the microcystin content the most, a box-Behnken was designed with microcystin as a response variable. Therefore, the new range of pH used was 7, 9 and 11. pH 11 is the pH in which there is algal bloom. Any pH that is reduced back to neutral pH will increase microcystin content. pH 9 is the maximum pH to prevent algal bloom and consequently, harmful algal bloom. The other two parameters were aeration: zero, 50 and 100 rpm and temperature: 20, 25 and 30 °C as shown in Table 1.

Table 2 shows all results for the levels and factors using microcystin as variable response.

Table 2 – Box-Behnken design and microcystin results

order	aeration (rpm)	Temperature (°C)	pH	Microcystin (μg/L)
4	100	30	9	0.14
14	50	25	9	0.13
3	0	30	9	-0.18
6	100	25	7	0.07
12	50	30	11	-0.21
2	100	20	9	0.8
9	50	20	7	1.27
1	0	20	9	1.04
10	50	30	7	-0.04
15	50	25	9	0.09
5	0	25	7	-0.07
13	50	25	9	-0.12
8	100	25	11	-0.22
7	0	25	11	0.03
11	50	20	11	0.17

All values below zero were considered zero in the following analysis. This is because ELISA kit has a limitation in terms of concentration range of its standards which vary from 0 – 2 ppb (μg/L).

All statistical calculations and significance are shown below. An ANOVA (Analysis of Variance) was used to test this Box-Behnken model. Temperature was highly significantly different (p-value <0.01) and pH was significantly different (p-value < 0.05). The 2-way interaction showed that temperature x pH were significantly different (p-value <0.0).

Table 3- Response Surface Regression: Microcystin versus Temperature; pH (ANOVA)

Source	DF	Adj SS	Adj MS	F-Value	P-Value
Model	9	2.32108	0.25790	8.05	0.017
Linear	3	1.39535	0.46512	14.52	0.007
Aeration	1	0.00045	0.00045	0.01	0.910
Temperature	1	1.23245	1.23245	38.46	0.002
pH	1	0.16245	0.16245	5.07	0.074
Square	3	0.58463	0.19488	6.08	0.040
Aeration*Aeration	1	0.00693	0.00693	0.22	0.661
Temperature*Temperature	1	0.52850	0.52850	16.49	0.010
pH*pH	1	0.03103	0.03103	0.97	0.370
2-Way Interaction	3	0.34110	0.11370	3.55	0.103
Aeration*Temperature	1	0.03610	0.03610	1.13	0.337
Aeration*pH	1	0.00250	0.00250	0.08	0.791
Temperature*pH	1	0.30250	0.30250	9.44	0.028
Error	5	0.16022	0.03204		
Lack-of-Fit	3	0.15135	0.05045	11.38	0.082
Pure Error	2	0.00887	0.00443		
Total	14	2.48129			

The model predicted was close to the actual model (1.6 % difference). The regression equation in uncoded units is the following:

$$\begin{aligned} \text{Microcystin} = {} & 0.073 - 0.0075\ \text{Aeration} - 0.3925\ \text{Temperature} - 0.1425\ \text{pH} \\ & + 0.0433\ \text{Aeration*Aeration} + 0.3783\ \text{Temperature*Temperature} - 0.0917\ \text{pH*pH} \\ & + 0.0950\ \text{Aeration*Temperature} - 0.0250\ \text{Aeration*pH} + 0.2750\ \text{Temperature*pH} \end{aligned}$$

The following graphs and figures show statistical results based on the variable response.

The Pareto chart (Figure 1) reveals that the most significant parameter is temperature in which the highest value of was relevant to this factor. This has already been discussed in literature. It was expected to be a significant parameter for algal bloom and microcystin production. The novelty in this study is that the interaction between temperature and pH is also significant. The following figure (Figure 2) shows main effects.

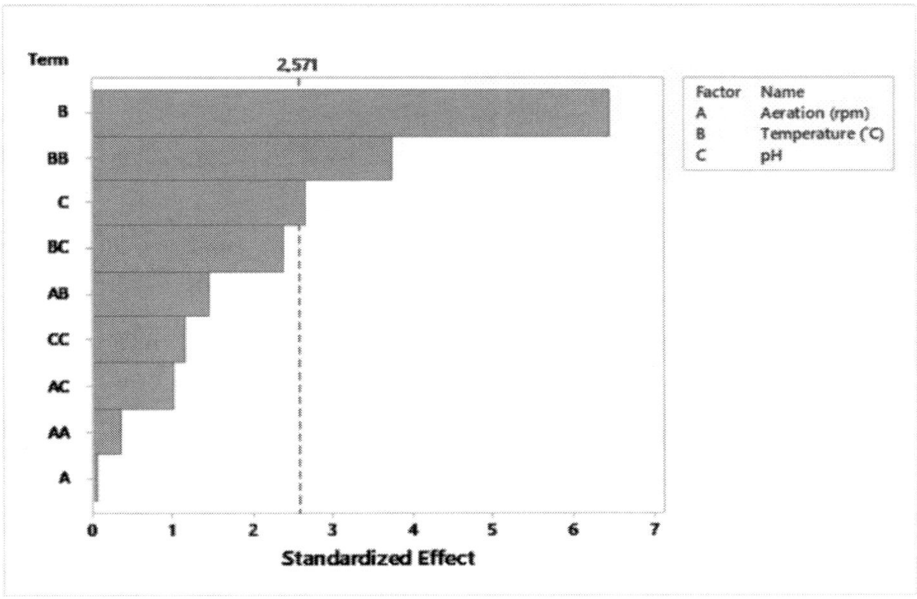

Figure 1 – Pareto Chart of the Standardized Effects

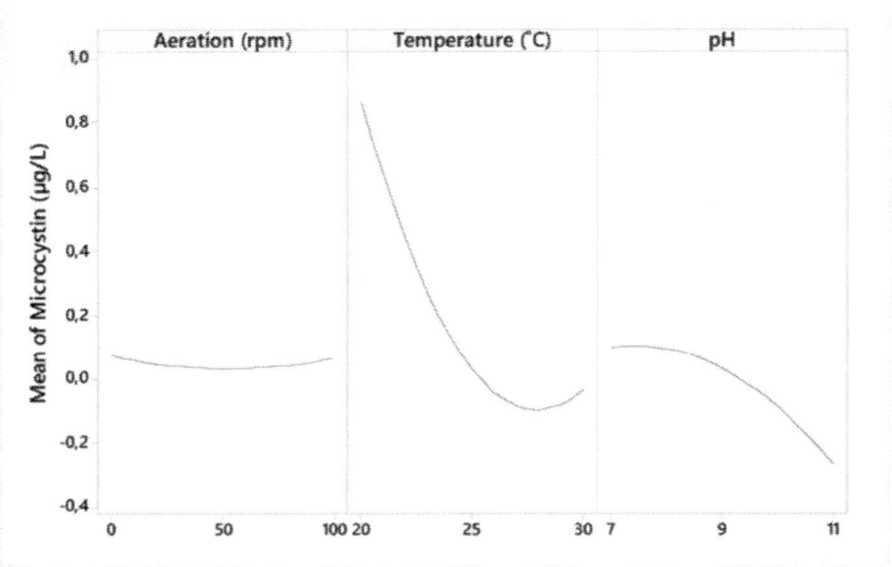

Figure 2 – Main Effects Plot for Microcystin (coded levels)

Aeration did not show any significant results, however both temperature and pH lead to an increase in microcystin content depending on the level. Temperatures at 20 °C and neutral to lower pH influence toxin production the most.

The following figure (Figure 3) shows the Interaction chart for microcystin growth.

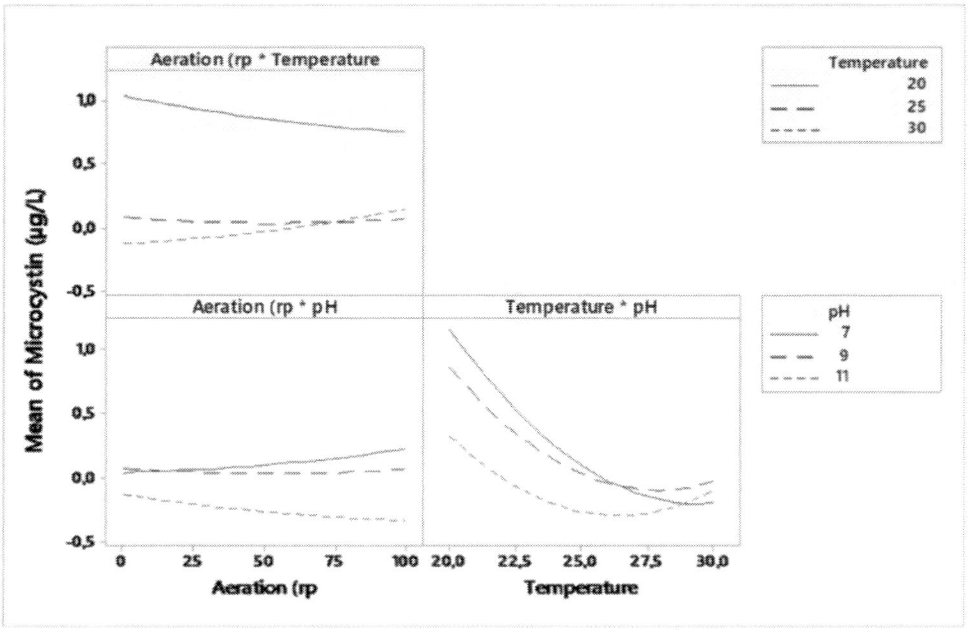

Figure 3 – Interaction chart for microcystin growth

Similar to the main effect plots, toxin production is increased when temperature is the lowest (20°C) and pH is also the lowest (pH=7). This confirms that after algal bloom, if the pH from the water system goes back to a neutral pH, the microcystin production will be increased. If water temperature is maintained above 25°C, this effect will be minimal. However, if water temperature is below 25°C, the microcystin level will increase dramatically.

This means that any episodes of cold front reaching hot water temperatures will influence in the release of microcystin content. This does not mean that cooler temperatures will promote toxin production, it is the abrupt change in temperature from hot to cold that promotes microcystin release, as show in Figure 4.

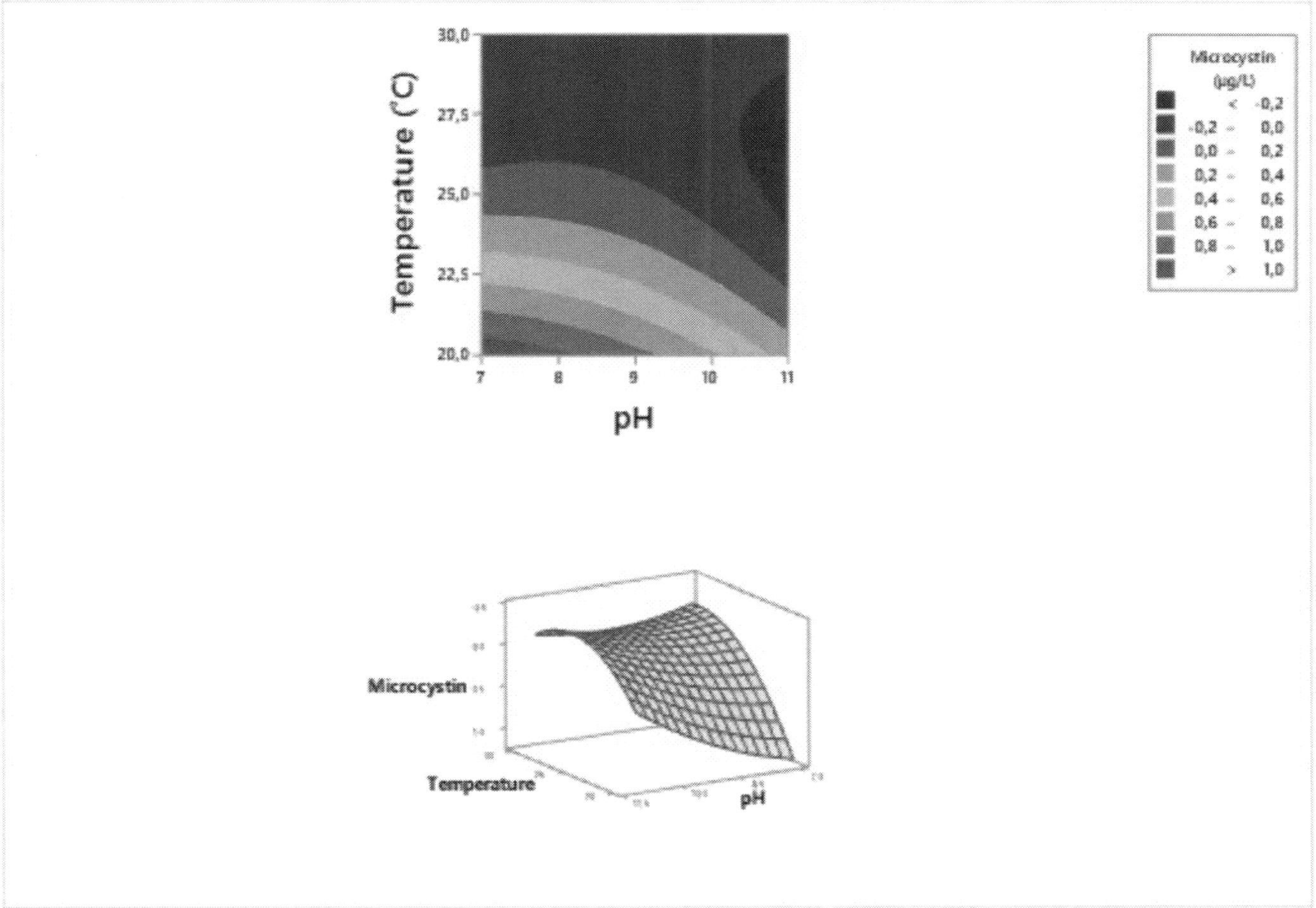

Figure 4. Contour plot graphic and surface plot

With a neutral pH, the temperature influences directly on the production of microcystin. However, in more basic pH (values close to 11), the variation on temperature becomes less significant on the microcystin production. The higher the temperature, the higher the concentration of microcystin in the environment. This suggests that if a stressful condition produced microcystin, it will only worsen the situation to that particular water body if temperatures would keep increasing. This confirms the importance of temperature in microbial growth and also in the microcystin release.

4. Conclusions

Algal blooms are present in several water bodies on different locations. Because some cyanobacteria produce toxins, there is the concern of finding ways to prevent algal blooms and contamination. This study showed that warmer temperatures influence in algal blooms but colder temperatures and pH change were responsible for toxin production. The conditions used in this study are based on drastic changes of temperature and pH since the natural course of microbial growth is an increase in temperature and pH. These stressful conditions influenced the toxin production. If algal bloom cannot be prevented, stressful conditions could be prevented or at least serve as an alert to the population, livestock and local authorities that toxin could have been released in the water body. Recreational activities during summer when temperatures are higher bringing a higher concern in regards to the exposition of this contaminated water. The statistical tool used in this study could produce significant results in fewer steps. A model was produced which could be used in the monitoring of water bodies.

5. Acknowledgements

The authors would like to thank Dr. Vikram Kapoor for the use of his laboratory and all the students involved in the sampling process: Andrea Mertins. Armando Montante, Brian Foor, Duc Phan, Fernando Lopez, Muhammad Bilal, Nick Carwile, and Tanvir Pasha. This study was supported by National Science Foundation (1438622) and CONNECT Program.

References

Bridgeman, T. B., Chaffin, J. D., & Filbrun, J. E. (2013). A novel method for tracking western lake erie microcystis blooms, 2002–2011. *Journal of Great Lakes Research, 39*(1), 83-89. doi:10.1016/j.jglr.2012.11.004

Dittmann, E., Fewer, D. P., & Neilan, B. A. (2013). Cyanobacterial toxins: biosynthetic routes and evolutionary roots. *FEMS microbiology reviews, 37*(1), 23-43.

Paerl, H. W., & Huisman, J. (2008). Blooms like it hot. *SCIENCE-NEW YORK THEN WASHINGTON-, 320*(5872), 57.

Steffen, M. M., Belisle, B. S., Watson, S. B., Boyer, G. L., & Wilhelm, S. W. (2014). Status, causes and controls of cyanobacterial blooms in Lake Erie. *Journal of Great Lakes Research, 40*(2), 215-225.

Sun, F., Pei, H.-Y., Hu, W.-R., Li, X.-Q., Ma, C.-X., & Pei, R.-T. (2013). The cell damage of Microcystis aeruginosa in PACl coagulation and floc storage processes. *Separation and Purification Technology, 115*, 123-128.

561654 Surface Change of Lean-NO_x Trap Adsorbent through CuO Co-Impregnation and Its NO_x Storage-Desorption Performance

In major industrial countries, fuel economy regulations have curbed carbon dioxide (CO_2) emissions from vehicles, especially automobiles. In diesel vehicles, the lean-burn engine technology, which is the most popular solution for improving fuel economy, has been used. This technology operates at high air/fuel ratios to make fuel economy efficient while reducing the CO_2 emissions. However, using the lean-burn engine for automobile not only increases NO_x emissions but also reduces the NO_x removal efficiency of conventional catalytic converter. Therefore, the lean-NO_x trap (LNT) has been used to further reduce NO_x emissions from automobiles. In order to develop the vehicles having both low energy and low emission, LNT should be improved in terms of the adsorption performance for NO_x gas at low temperature (the emission condition of a lean-burn engine). In this study, we developed LNT adsorbent in which CuO was co-impregnated with BaO in γ-Al_2O_3 and evaluated its NO_x storage-desorption performance. The characterization of the adsorbents was performed using X-ray diffraction (XRD) and BET surface area was calculated by nitrogen adsorption analysis results. The adsorption capacity of the adsorbents was measured by breakthrough tests using a quartz flow reactor with a NO_x analyzer. The desorption temperature of adsorbed NO_x on the adsorbents was figured out by temperature programmed desorption (TPD) followed by breakthrough tests. Also, the surface change of adsorbents during the NO_x storage-desorption was observed by *in-situ* FT-IR.

561855 Measurement of Hansen Solubility Parameter on Pollen Particle Surface and Application to Pollen Removal in Air

Pollen from various vegetation such as cedar and cypress can negatively affect human health. For example, it shows allergic reactions to the human, such as rhinitis and fever. Allergies caused by pollen are spread all over the world and especially in Japan, one in four people has developed allergic symptoms. So, various attempts have been made to mitigate the effects of pollen on human health. Studies of the effects of removing pollen from air using air purifiers, pollen filters, and nasal filters have also been performed. The possibility of preventing exposure to pollen by repelling the pollen using electrostatic barriers has also been studied. Studies of methods of alleviating allergic symptoms using antihistamines, immunotherapy, and other methods have also been performed. I gave most attention to remove pollen from air by efficiently collecting the pollen with spray mist. In order to efficiently collect pollen, it is necessary to design a solvent having high affinity with pollen. To achieve that requires, we have to understand the physical properties of the pollen surfaces. So, I focused on Hansen Solubility Parameter (HSP), which is used for evaluating the physical properties of the fine particle surface.

HSP is physical property value of each substances. HSP is based on the idea that "two substances having similar interactions between molecules are high affinity". And HSP is consisted of 3 components: dispersion forces term, polar forces term and hydrogen-bonding forces term. The dispersion force term is the general van der Waals interactions between just about everything. Polar force term is the familiar "positive attracts negative" electrical attractions arising from dipole moments. hydrogen-bonding forces term is due to hydrogen bonding. HSP is measured experimentally by evaluating the affinity with organic solvents and using Hansen solbility sphere method. Since HSP is possible to grasp the surface physical properties of the material quantitatively, it has been used for many fields, such as petroum, pharmacy and food.

In this experiment, I measured HSPs for pollen particle, cedar and cypress which are major allelgen in Japan, using Hansen solbility sphere method. For HSP measurement on a fine particle surface, the affinity evaluation by the sedimentation method and dynamic light scattering method (DLS) are commonly used. However, HSP measurement of low-density and non-spherical particles like pollen by the above-mentioned measurement methods is difficult. So, I measured HSPs of each pollen by the capillary penetration method which is I developped.

Finally, considering these HSPs, I designed mixed solvents highly compatible with pollen and presented guidelines for future research and development.

563422 Design of a Hierarchical Fe-ZSM-5@CeO_2 Catalyst and the Enhanced Performances for the Selective Catalytic Reduction of NO with NH_3

Selective catalytic reduction of NOx by NH_3 (NH_3-SCR) is an effective technique for the abatement of NOx. Development of SCR catalyst is the core of the technique. Herein, a hierarchical Fe-ZSM-5@CeO_2 catalyst with CeO_2 nanoparticles decorating is originally fabricated via dopamine polymerization for NH_3-SCR. Fe-ZSM-5@CeO_2 possess a NOx conversion over 90% and N_2 selectivity over 98% in the temperature range of 250-400 °C, showing an advanced catalytic activity compared with Fe-ZSM-5/CeO_2. The synergistic effect of Fe-ZSM-5 and CeO_2 over Fe-ZSM-5@CeO_2 facilitates the formation of redox cycle between Ce^{3+}/Ce^{4+} redox-couple and high rate active oxygen species (O_β and O_γ), leading to the enhanced NH_3-SCR performance. The feature of hierarchical structure with CeO_2 outside provides a pathway to convert NO to NO_2, facilitating the fast SCR. Meanwhile, the formed NO_2 can activate Fe^{2+}/Fe^{3+} redox-couple, offering an extra redox cycle which can further contribute to fast SCR. The outer CeO_2 layer can also improve the water resistance of Fe-ZSM-5 as the catalytic activity of Fe-ZSM-5@CeO_2 was impervious when exposed to 10% H_2O. To be concluded, the hierarchical Fe-ZSM-5@CeO_2 is proved to be a promising candidate for NH_3-SCR due to its uniform distribution of active sites and the synergistic effect of Fe-ZSM-5 and CeO_2 species.

- Abstract Text

563443 Evaluation on the Removal Performance of Dichloromethane and Toluene from Waste Gases Using an Airlift Packing Reactor

Biological removal of dichloromethane (DCM) from pharmaceutical industry is limited by its recalcitrance. In this study, an airlift packing reactor (ALPR), which combined the suspended and fixed-film microbial growth system, was set up to remove DCM and co-existed toluene. The removal performance of the ALPR for DCM was greater than traditional airlift reactor (ALR). The maximum elimination capacity (EC_{max}) of the ALPR for DCM reached 108 g m^{-3} h^{-1} with removal efficiency (RE) of 41%, increased by 145% if compared to the ALR. The EC_{max} for toluene was 172 g m^{-3} h^{-1} with RE of 70%, decreased by 25% than the ALR, which was mainly due to the higher liquid-phase biomass in the ALR. The results of high-throughput sequencing showed that the microbial composition on the packings of the ALPR had a large difference from its liquid-phase or the liquid-phase of the ALR. *Gemmobacter*, *Rhizomicrobium*, *Chitinophaga*, *Vampirovibrio*, and *Fodinicurvata* were genera with great abundance fixed on the packings and *Rhizomicrobium, Chitinophaga*, *Vampirovibrio*, and *Fodinicurvata* are first to be reported in VOCs biological removal. This study indicated that the ALPR can augment the microbial community and effectively improve the removal of recalcitrant VOCs.

564765 Hexavalent Chromium Removal from Treated Water Via Electrospun Composite Nanofibers

Uniform electrospun polyacrylonitrile / iron (III) nitrate nanofibers with small diameters were successfully prepared by systematically optimizing the electrospinning parameters including polymer concentration, salt loading, applied voltage, tip to collector distance, and flow rate. These parameters are shown to affect the fiber morphology and the addition of iron (III) nitrate salt was found to be able to reduce the fiber size and required electrospinning voltage. The improved thermal properties of the composite fibers were investigated and shown by thermal gravimetric analysis (TGA) and differential scanning calorimetry (DSC). The composite nanofibers were further carbonized to carbon nanofibers to show their potential for hexavalent chromium removal from water. The removal percentage has been found to show a pH dependence that is related to the iron loading. This material has shown promising results towards an effective water purification method.

567114 CO_2 Utilization By Calcium Carbonate, Magnesium Carbonate and Sodium Bicarbonate Production Using Metal Cations in Reject Brine Solutions

Since the Industrial Revolution, the usage of fossil fuels such as petroleum and coals has resulted in a large amount of carbon dioxide emitted into the atmosphere. Normally, the average temperature of the Earth is kept constant by the greenhouse effect. However, when the concentration of carbon dioxide in the atmosphere is too high, the thermal equilibrium between solar radiation and Earth radiation energy is disrupted and the temperature of the Earth's surface rises. Nature and ecology have an ability to cope with temperature changes up to a certain degree. However, climate change results in serious environmental problems such as changes in the ecology and coast lines, floods and many others. These happen when normal greenhouse effects adapt to the changing climate. To prevent such events, many developed nations are developing technologies that can prevent carbon dioxide emissions into the atmosphere. While there are other types of greenhouse gases that have the potential to trigger serious greenhouse effects or climate change, due to the massive amounts emitted, carbon dioxide is thought to be the main cause of climate change and global warming. In order to prevent carbon dioxide produced from industrial facilities using fossil fuel from being emitted into the atmosphere, various technologies, commonly called Carbon Capture and Storage (CCS) technologies, have been invented. In the well-known wet absorption process, carbon dioxide in flue gas is separated using a wet absorbent such as alkanolamines or metal hydroxides in an absorber. When this absorbent is saturated with carbon dioxide, it is transported to a desorber where captured carbon dioxide is separated by the heat of hot steam. Separated carbon dioxide is then compressed and transported to a storage site, resulting in the isolation of carbon dioxide from the atmosphere. However, there are some drawbacks associated with wet absorption. First, a large amount of energy is required to operate the whole process. In particular, hot steam is required in the desorption process to separate carbon dioxide from the absorbent solution. The energy used in this process is called the regeneration energy, which represents 70% to 80 % of the total CCS process operation energy. The process cost is high, requiring part of the energy production in the facilities to be provided to the CCS process. Second, many nations lack the storage sites for this process. In Europe, underground storage of carbon dioxide is not feasible since there are no massive underground energy sources such as petroleum or natural gases. Also, the main component of the crust in this continent are limes that react with carbon dioxide to form calcium bicarbonate, which is soluble to water and thus is potentially weakened upon reaction with CO_2. For some nations located in the circum-Pacific orogenic zone, the high level of geological activities makes difficult to guarantee the stability of stored carbon dioxide. Hence, a new way of treating carbon dioxide is needed, preferably by converting carbon dioxide into useful materials. This concept is commonly referred to as Carbon Capture and Utilization (CCU). CCU can be divided into two processes: organic and inorganic utilization processes. In organic utilization, carbon dioxide is converted to organic substances such as diesel, plastics and other organic chemicals. However, plenty of energy is required to trigger the reactions involved, since carbon dioxide is very stable. Although catalysts may be used, their cost is too high to make it feasible for commercialization. In inorganic utilization, inorganic substances such as metal carbonates are produced. When the wet absorption process is applied to the inorganic utilization process, a massive amount of carbon dioxide capture and conversion can be achieved, due to the absorption characteristics of absorbents and carbon dioxide. When carbon dioxide is absorbed in alkanolamine-type absorbent solutions, it undergoes the following reactions for primary and secondary amines:

$RR'NH + CO_2 \leftrightarrow RR'NH^+COO^-$(zwitterion) ------------------- (1)

$RR'NH^+COO^- + RR'NH \rightarrow RR'NCOO^- + RR'NH_2^+$--------------- (2)

$2RR'NH + CO_2 \leftrightarrow RR'NCOO^-$(carbamate) + $RR'NH_2^+$ -------- (3) <overall reaction>

For ternary amines:

$CO_2 + R_3N + H_2O \leftrightarrow HCO_3^-$(bicarbonate) $+ R_3NH^+$------------------- (4)

As shown in the chemical equations above, carbon dioxide exists in the form of reactive ionic CO_2 (*ri*-CO_2), which reacts with metal cations. Therefore, this allows for large scale CO_2 capture and the formation of metal carbonate. Moreover, this process requires no heat to separate the captured carbon dioxide from the absorbent solutions, since most of the CO_2 is removed when it precipitates in the form of metal carbonate salts. These metal carbonate salts can be used for various industrial applications such as cement, paper and others. The inorganic utilization process has received plenty of attention. However, most of the studies have used metal cations from natural sources such as limestone; or from some construction materials such as waste concrete. Upon consideration of the economic feasibility, it seems inappropriate to use natural resources to treat waste gas. On the other hand, waste concrete contains plenty of aggregates that can be reused after separation. Hence, the pretreatment process is complicated and has to be applied before the metal cation sources in the inorganic utilization process can be used. In this study, seawater-based industrial wastewater was utilized as the metal cation source. In the refined salt production process, concentrated wastewater is produced as a byproduct. When released to the near shore without any treatment, this wastewater may negatively impact the near shore ecology. Hence, when metal cations contained in this wastewater are utilized, wastewater treatment and metal cation capture along with carbon dioxide conversion can be achieved simultaneously. In this study, we considered both metal cation separation and chemical conversion of carbon dioxide. In the former, calcium and magnesium cations contained in the wastewater were separated by using sodium hydroxide that can be obtained from the electrolysis of seawater. Calcium and magnesium components are separated in the form of calcium hydroxide and magnesium hydroxide, respectively. The major cationic component of the remaining solution after the Ca/Mg separation are sodium ions.

After cation separation, carbon dioxide is captured using an absorbent, i.e., monoethanolamine (MEA) solution. When this absorbent is saturated, calcium and magnesium hydroxides are added, resulting in the formation of calcium and magnesium carbonates, respectively. To form magnesium carbonate, the pH of the solution is controlled using sodium hydroxide to enhance the reaction. After Ca/Mg separation, sodium bicarbonate was produced from the remaining solution by applying the steric hindrance effect of alkanolamines.

567226 Inverse Vulcanized Polysulfide-Based Foam with Hierarchical Pore Structure for Effective and Efficient Heavy Metal Sequestration

Introduction of inverse vulcanization (IV) led to the development of new class of polysulfides (PS) with interesting properties for various applications. IV involves ring opening reaction of elemental sulfur (i.e. S-S chain scission at T > 159°C) with organic comonomer. Due to the high sulfur content, PS synthesized via IV show potential as mercury (II) sequestrants. However, almost no reported PS has competitive mercury adsorption performance with capacity only ranging 1-30 mg g^{-1}. This could be attributed to the hydrophobicity of the materials since PS were synthesized with hydrocarbon- or aromatic-rich comonomers. The low wettability prohibits intimate adsorbent/aqueous feed contact thereby disfavoring Hg^{2+} capture.

In this work, hydrophilic PS was synthesized by copolymerizing elemental sulfur with ethylene glycol dimethacrylate (EGDMA). Different S-EGDMA PS were prepared according to different S:EGDMA molar ratio and were thoroughly analyzed to propose their structure and characterized in terms of morphology and other physico-chemical properties. Their ability to sequester mercury (II) were evaluated via adsorption isotherm, kinetics, and thermodynamics.

The S-EGDMA PS was configured into a macroporous foam. PS were incorporated in poly(vinyl alcohol) (PVA) matrix via blending, cryo-dessication, and chemical cross-linking. This ensured the high quality and convenient loading control of PS in the foam. The composite foams were thoroughly characterized and evaluated in terms of mercury (II) adsorption capacity, kinetics, and thermodynamics. Mechanical durability vis-à-vis reusability was also evaluated. This study features significant advantages not only to the competitive mercury (II) adsorption performance of the materials but also to the cost-effectiveness of the material preparation.

This research was supported by the National Research Foundation of Korea (NRF) funded by the Ministry of Science and ICT (No. 2018R1D1A1B07047503 and No. 2017R1A2B2002109) and by the Ministry of Education (No. 22A20130012051(BK21Plus)).

567240 Development of Adsorbent Based on Layered Double Hydroxides Structure with High Mg/Al Ratio for Heavy Metal Removal

As industrialization progresses, various environmental problems arise due to the influx of wastewater from various routes into rivers. Especially, the heavy metal ions introduced into the water cause a lot of diseases in the human body, and various methods for reducing the heavy metal ions in wastewater have been studied. Among several methods, the adsorption technology has various advantages such as easy application and low cost, but its performance is lower than the existing heavy metal removal technologies such as ion exchange and membrane separation. Therefore, it is necessary to develop effective adsorbent for efficient adsorption processes. Among various adsorbents, layered double hydroxide has attracted lots of attention as an adsorbent of heavy metals because it is capable of substituting various anions in the interlayer region and giving various applications. In this study, we focused on the fact that $Mg(OH)_2$ on the surface of layered double hydroxides exhibits high affinity to heavy metals such as copper and chromium, and developed an adsorbent with increased $Mg(OH)_2$ ratio while maintaining the existing structure. For layered double hydroxides different ratios of Mg/Al, the adsorption capacity of heavy metals including copper and chromium was measured and the effect of Mg/Al ratio of layered double hydroxides on heavy metal adsorption capacity was investigated by XRD analysis.

567472 Characteristics of Heat Transfer in Carbon Dioxide Methanation Process

In recent years, there has been great interest in developing carbon capture, utilization and storage (CCUS) technology to mitigate the carbon dioxide emissions and address global warming issue. In a similar vein, the development of the utilization technology for intermittent electrical energy from renewable energy sources, such as solar photovoltaic, wind, wave, geothermal, etc., is considered as a promising research area, and water electrolysis technology for hydrogen production using renewable energy sources is one of viable options. In terms of energy storage capacity and discharge time, the synthesis of hydrogen from water electrolysis and carbon dioxide from carbon capture and storage process into methane, i.e. methanation, is a more favorable process and is carried out by the Sabatier reaction. During the reaction, the carrier of chemical energy is converted from hydrogen with low energy density to methane with high energy density. The CO_2 methanation process can be carried out with high GHSV and steam can be produced with high temperatures using heat exchange system. In power-to-gas process, heat dissipation gradients inside reactors and temperature control are of major interest since the methanation reaction is a relatively high exothermic reaction ($\Delta H° = -165$ kJ/mol) and a renewable electricity could not supplied constantly. Various reactor types have been developed and the fixed bed reactors are most used for the CO_2 methanation. However, the main disadvantages of fixed bed reactor is the hot spots and the poor flexibility with load changes due to fluctuation of renewable energy sources. A fluidized bed reactor is known as a good chemical reactor due to its excellent mass and heat transfer properties. Moreover, a fluidized bed reactor has advantages in easy temperature control and good contacting between reactants compared to the fixed bed reactor. The aim of this study is to investigate the reactor and heat exchange system suitable for the demonstration of the CO2 methanation process and discuss the design factors of the heat exchanger according to the selected reaction system.

567769 Development of a Biofilm-Based Technology for the Treatment of High Saline Process Wastewater

Global competition and environmental legislation imposing strict emission limits lead to an increasing demand on the efficiency of wastewater treatment in the chemical industry. Identifying the adequate method for such treatment, biological technologies are increasingly preferred over physical or chemical processes due to their capability to remove pollutants in a more environmentally friendly and cost-effective way. In the field of biological wastewater treatment, fixed-biofilm technologies provide several advantages over suspended-growth systems. In particular, attached biomass becomes more specialized and adapted to not readily biodegradable components. This is attractive in view of the challenging complexity and diversity characteristic to industrial wastewater. Many industrial sectors, for example the petroleum and leather industry, are likely to generate wastewater rich in both organic matter and salt. High salinity is a microbial stress condition and thus constitutes a considerable difficulty for the development of a reliable biofilm-based wastewater treatment process.

In the course of our investigations, a fixed-film process for the treatment of high saline wastewater is developed at laboratory- and pilot-plant scale. With special attention to the difficulties accompanied by high salinity, our research aims at deriving a promising reactor design and evaluating possible process improvements regarding their impact and practical implementability. In addition, the aim is to reduce uncertainties about key input parameters to ensure steady-state operation at industrial scale. The exhaustive oxidation of carbon content is chosen as the main treatment objective, representing a widely-spread and industrially relevant example. To further contribute to the generation of insight transferable to a broad field of application, the conceptual approach implies to use municipal activated sludge as the starting point of the process development.

The poster presentation will give an introduction into the overall concept to meet the challenges of our research with special attention to the requirements resulting from the high salinity conditions. Furthermore, the experimental set-up will be presented and a choice of results from the initial acclimation phase will be shown.

569930 Synthesis of Fluorinated Hydrocarbon Anion Exchange Resins for the Removal of Perfluorinated Chemicals

Perfluoroalkyl compounds (PFCs) are anthropogenic industrial compounds used in many applications, including surface coatings, surfactants, and flame retardants. They are persistent and very stable in air, water, and soil environments. PFCs have been detected in water and soil samples all over the globe. They have been reported as bioaccumulative, with detectable concentrations in humans and in both aquatic and terrestrial animals. For reasons of their stability and bioaccumulative nature, efforts have been made to restrict their production and use worldwide. We will present the synthesis and characterization of a novel fluorinated anion exchange sorbent for removal of perfluoroalkyl compounds including PFOA and PFOS from drinking water. Our idea improves upon adsorption technologies such as activated carbon and anion exchange resins, which are not specific for perfluoroalkyl compounds. While it has been demonstrated that these adsorption materials are effective in removing perfluoroalky compounds (PFCs), they are prone to reduced effectiveness in the presence of co-existing organic matter in water. Organic compounds can outcompete PFCs for binding sites reducing effectiveness. We will present details of the synthesis of a mixed-mode fluorinated polymer-anion exchange resin. The surface chemistry, morphology and surface area of the resins along with adsorption equilibrium and kinetic studies in both the batch and column modes will be described.

569971 Hybrid Technology for Soil Remediation: Foam Injection and Bioremediation

Remediation of soil contaminated with hydrocarbons is a concern in the world, especially in Mexico; there is an increase in the sinisters caused by fuel theft and bad handling of operation systems. The polluted regions comprise agricultural, recreational, ecological and habitational soil, thus giving a solution to this problem is priority.

We are proposing a hybrid technology to apply in fields where the incidence has occurred in a short term and that for its activity, it is fundamental the recovery of the natural state of the soil. The system comprises foam injection and soil flushing to remove the major amount of free hydrocarbon adsorbed to the soil. The foam technology can help drag the hydrocarbons in the soil with a very low amount of chemical and better control than water based solutions. The formulation of the foam includes a mixture of synthetic and biosurfactant and cactus leave mucilage that increases the viscosity of the water matrix inducing stability to the foams. By microbiologic protocols it has been demonstrated that cactus leave mucilage does not affect the connate microbiota and does not develop harmful microorganisms to the system.

The development of hidrocarbonoclastic bacteria has been observed in contaminated soil without any previous treatment and after flushing the soil with the foaming formulation. Bacteria can develop after flushing with the foam, leading to the possibility of promoting a remediation train comprising on a short tem soil flushing process followed by bio-remediation. Soil restauration is still being studied at laboratory and field scale.

569974 Operational Experience of 0.5 Mwth Pressurized Chemical Looping Combustion System for Inherent CO_2 Separation

Chemical looping combustion (CLC) is a promising technology for fossil fuel combustion with inherent CO_2 capture and sequestration that is able to mitigate greenhouse gases (GHGs) emission. The CLC process employs metal oxide as an oxygen carrier circulating between fuel reactor, where the fuel is combusted to CO_2 and H_2O, and air reactor, where the oxygen carrier returned to its oxidized form by aeration. In Korea, KIER(Korean Institute of Energy Research) has been doing research on chemical looping combustion technology since 2000 and KIER and KEPCO RI(Korea Electric Power Research Institute) started a collaborative research from 2006 to develop chemical looping combustion process and oxygen carriers, respectively. We installed a 200 kWth chemical looping combustor and demonstrated long-term operation more than 100 hours using natural gas as fuel and Ni-based oxygen carrier in 2011. After that, we started new research project to develop a 0.5 MWth pressurized chemical looping combustor from 2015. In this project, our target is development of pressurized chemical looping combustor up to 5 bar-abs and achievement of high CO_2 selectivity (> 98%) using natural gas and syngas as fuels. Thesedays, Continuous solid circulation test at high temperature and high pressure conditions and batch type reduction-oxidation tests were performed to check feasibility of a 0.5 MWth chemical looping combustion system. Pressure drop profiles were maintained stable during continuous solid circulation up to 16 hours. Therefore, we could conclude that the solid circulation between an air reactor and a fuel reactor could be smooth and stable. The measured fuel conversion and CO_2 selectivity were high enough even at high capacity and even after cyclic tests. After that, continuous operation of the 0.5 MWth chemical looping combustor was performed using natural gas as fuel. In this presentation, the results and experiences will be discussed.

571233 Pulsed Electric Field (PEF) Device Equipped with a Cathode Decorated with Copper-Titania Nanostructures for Enhanced Water Denitrification

With population growth comes the increase in demand on agricultural food produce, and consequently, nitrogen-based fertilizers. This is causing an upsurge in the nitrate content of groundwater due to runoff or seepage, ushering in some damaging environmental and health impacts such as eutrophication and methemoglobinemia. Previous electrochemical reduction reports on nitrate reduction of ground water or wastewater have posted (1) low nitrate reduction rate, (2) poor selectivity towards nitrogen (preferred product), and (3) energy intensiveness, as limiting factors of their treatment systems. This led to studies which utilize electrodes with micro/nano-sized structures to increase electroactive area and improve mass transfer.

This study investigated the inclusion of copper-titania nanostructures on a titanium plate cathode for the electrochemical reduction of nitrate to nitrogen. The cathode was incorporated in a batch-type undivided electrolytic cell, which allows simultaneous reduction and oxidation reactions—a process that may be detrimental to reduction efficiency since reduced nitrate can be re-oxidized near the anode but can be an opportunity to improve selectivity towards the favored harmless nitrogen. The system was operated under pulsed electric field which greatly decreased the needed electrical energy input to reduce a unit concentration of nitrate in the synthetic contaminated ground water. Parametric electric field simulations via COMSOL Multiphysics validated the improved electric field strengths brought by the addition of copper-titania nanostructures and directed the eventual design of a flow-through denitrification device.

571258 High-Intensity Narrow-Wavelength (Hi-NW) Light-Based Technology for Disinfection of Mdro in Environment

Hospital-acquired infections (HAI) are becoming important causes of mortality and morbidity in hospitals. Multi-drug Resistant Organisms (MDROs) cause the majority of nonsocial infections. Hi-NW light-based Technology provides an alternative technology, which is safe and effective against a wide spectrum of microorganisms without the emergence of microbial resistance. This technology based on the combined use of focused LED lighting sources to assemble a light array to target multiple sites within the microbe for rapid in decontamination of MDROs commonly found in hospital environments. Multiple cellular targets and different inactivation pathways ensure that microbial tolerance and resistance are avoided. High intensity 405nm, 470nm and 275nm LED are employed and assembled into 3*3 LED array. The bactericidal rate of E.*coli*, P.*aeruginosa*, S.*aures* and E.*faecalis*., which are as surrogate for their MDRO counterparts, are 4-log (99.99%) reduction within 10mins. The morphology of testing bacteria was analyzed by scanning electron microscope. Recent experiments revealed that the bactericidal effect on MDROs (i.e., MRSA, MRPA, VRE, and CRE) were 4-log (99.99%) reduction within 10mins as well. This technology can provide high bactericidal effect on MDROs and complement existing hospital practice to address an important public health problem.

Title: Bacteriophage recovery from environmental samples using functionalized silica

Authors: Charlimagne M. MONTEALEGRE[1,2], Wei HAN[1,3], King Lun YEUNG[1]

[1]Department of Chemical and Biological Engineering, The Hong Kong University of Science and Technology, Kowloon Hong Kong SAR

[2]Department of Chemical Engineering, University of the Philippines Diliman, Quezon City 1101 Philippines

[3]Division of Environment and Sustainability, The Hong Kong University of Science and Technology, Kowloon Hong Kong SAR

Abstract

The prevalence of antibiotic resistant bacteria is driving attention to alternative antimicrobials including bacteriophages. Bacteriophages are bacterial viruses which are specific to bacteria and abundant in nature. A typically used method for isolation is by membrane filtration which can separate viruses from a large amount of liquid sample and backwash them as a concentrate. However, this method has limited scalability. Previous studies showed that bacteriophages are capable of adsorption into solids. This study used silica and aluminum hydroxide functionalized silica as a bed of solids to adsorb bacteriophages from environmental samples. Using this unit operation allows possible scale up of the process provided that the effect of various factors on the efficiency of the process is determined. Samples with high solid content were pretreated by centrifugation at 5,000×g for 10 minutes, followed by pretreatment of solids by washing and finally, dilution. The dilute sample was then passed through a bed of solids. Solids to fluid ratio and number of passes were kept constant while varying column parameters and configuration. The solid used and pretreatment method highly influenced the performance of the column. Pretreatment using a solution of glycine, broth and tryptone was very effective in solids removal but greatly increases the pressure drop across the column. This limits the flowrate through the bed, beyond the capability of the pump used. Bacteriophage recovery was determined using plaque assay. Aluminum hydroxide functionalized silica performed better in adsorbing phages relative to silica. However, in a fixed bed of solids the combination of silica with glycine-tryptone-broth solution resulted into better overall recovery relative to aluminum hydroxide functionalized silica and water.

Introduction

Bacteriophages are viruses which only infect bacteria. They are found in a wide range of environments including solar lakes and mammalian gut. Their presence in a wide range of environment and their small size makes them difficult to isolate from large quantities of liquid or in a solid matrix. This study employs a continuous process to isolate bacteriophages from environmental samples especially sludge. A continuous adsorption process will enable processing of a large volume of water and will statistically increase the chance of isolating a large number and variety of bacteriophages. Various materials for adsorption were studied in the literature (Dowd & Pillai, 1997; Sinton et al., 2012). Most of these studies focus on attenuation of viral particles in water samples yet the same principle may be used for recovering viruses including bacteriophages from water samples. Silica was among the most prevalent material found in sand filters used in early studies but Li et al. showed that modifying silica with $Al(OH)_3$ resulted into a material that recovers viruses effectively (Li et al., 1998). A challenge to continuous processes is the solids content of the matrix. The solids may clog the equipment, increase pressure drop and eventually result to a shut-down of the continuous process. This study employs a desorption solution shown to be effective in recovering adsorbed bacteriophages as a pretreatment to remove solids prior to adsorption.

Methodology

Sludge samples were collected from a wastewater treatment facility in Stonecutters Island, Hong Kong S.A.R. These were refrigerated at 5^{0}C prior to solids removal by centrifugation at 5,000×g for 10 min. Washing was repeated three times to desorb adsorbed phages from the solid matrix. Distilled water and a mixture of glycine-tryptone-broth (Liu et al., 2017) were used as desorption solution. The resulting clarified sludge was spiked with *Escherichia coli* bacteriophage T3 (ATCC® 11303-B3™). A liter of the spiked solution was passed through a column packed with 1 g of Silica (Si) or Al^{3+}-modified silica (ASi) particles (Li et al., 1998). Adsorbed phages were then desorbed from the solids using glycine-tryptone-broth solution according to the procedure of Li et al. The pH of the desorbed phage solution was adjusted to neutral and a phage purification was performed (Clokie & Kropinski, 2010).

Bacteriophage concentration was quantified using a standard plaque assay with *E. coli* (ATCC® 11303™) as the host strain (Clokie & Kropinski, 2010).

Results

Bacteriophage T3 spiked tap water was passed through a bed of the adsorbents at an average flow rate of 0.508 ± 0.036 mL/sec. Plaque assay was performed from the initial solution and the desorbed and purified bacteriophage solution.

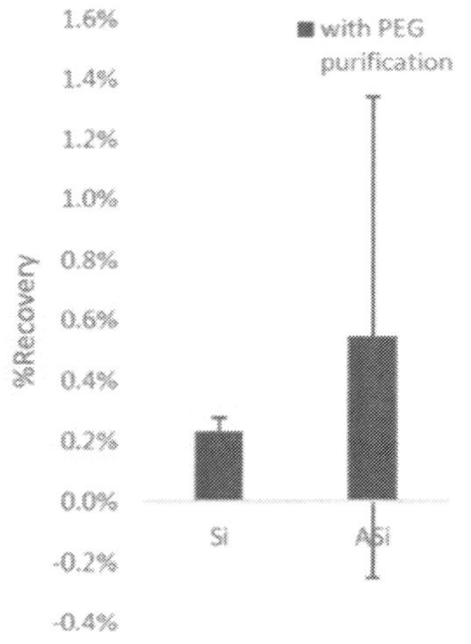

Figure 1. Bacteriophage recovery in tap water after desorption, precipitation and purification of adsorbed bacteriophages.

Figure 1 shows a bacteriophage recovery that is less than 1%. More noticeable is a very significant standard deviation especially when Al^{3+} functionalized silica was used.

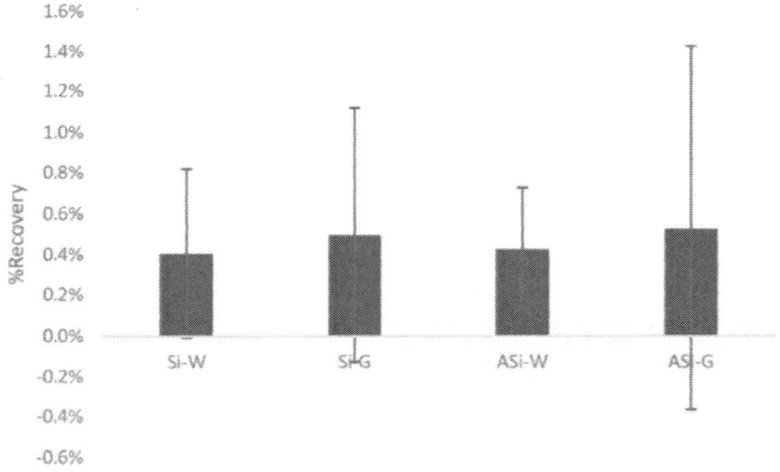

Figure 2. Bacteriophage recovery of Silica and Al3+ functionalized silica in clarified sludge.

The same observations can be drawn from figure 2. The percent recovery of the materials are statistically similar with adsorption in tap water.

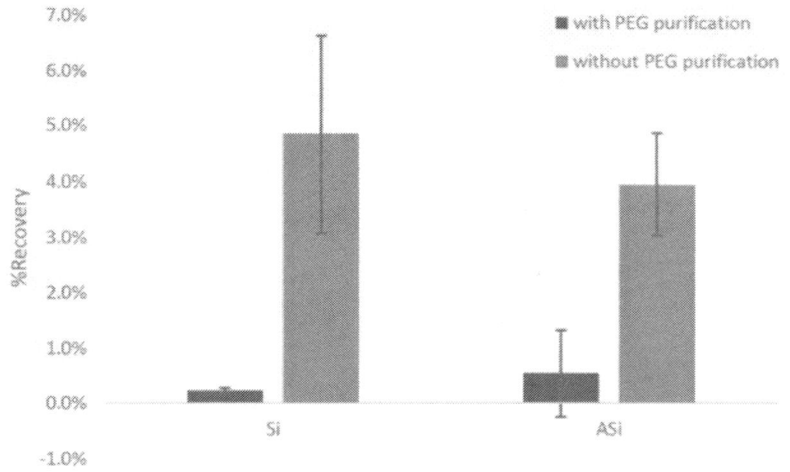

Figure 3. A comparison of bacteriophage recovery by silica and modified silica with and without purification.

Figure 3 shows a comparison of bacteriophage recovery when the purification step was not performed. A one order of magnitude increase in percent recovery was observed.

Discussion

Purification of bacteriophage containing samples was a standard practice when working with bacteriophages as this step ensures the removal of viable bacteria from the environment which can interfere with further assays. Protein purification is sensitive to pH and the optimum pH for the precipitation of target proteins, in this case the bacteriophages, is at the isoelectric point. This study used a pH 10.5 desorption solution which was effective in desorbing phages from silica. For the bacteriophage purification step, this required adjustment of the pH from 10.5 to 7 using 0.1 M H_2SO_4. This can result into localized low pH regions which can disturb the protein structure and inactivate the bacteriophage. The large standard deviation reflects how variable and unpredictable the effect of pH.

Conclusion

As this brief study has shown, there may be some benefits into working with unpurified samples. Omitting the purification step allowed an increase in the recovery of the model bacteriophage. This translates into an increased probability of obtaining a large number and variation of bacteriophages from environmental samples.

References

Clokie, M. R. J., & Kropinski, A. M. (Andrew M. B. (2010). *Bacteriophages : methods and protocols. Vol. 1, Isolation, characterization, and interactions*. Humana Press.

Dowd, S. E., & Pillai, S. D. (1997). Survival and transport of selected bacterial pathogens and indicator viruses under sandy aquifer conditions. *Journal of Environmental Science and Health . Part A: Environmental Science and Engineering and Toxicology*, *32*(8), 2245–2258. https://doi.org/10.1080/10934529709376680

Li, J. W., Wang, X. W., Yi Rui, Q., Song, N., Zhang, F. G., Ou, Y. C., & Chao, F. H. (1998). A new and simple method for concentration of enteric viruses from water. *Journal of Virological Methods*, *74*(1), 99–108. https://doi.org/10.1016/S0166-0934(98)00078-0

Liu, W., Li, C., Qiu, Z.-G., Jin, M., Wang, J.-F., Yang, D., ... Shen, Z.-Q. (2017). Development of a novel and highly efficient method of isolating bacteriophages from water. *Journal of Microbiological Methods*, *139*, 143–149. https://doi.org/10.1016/j.mimet.2017.05.019

Sinton, L. W., Mackenzie, M. L., Karki, N., Dann, R. L., Pang, L., & Close, M. E. (2012). Transport of Escherichia coli and F-RNA Bacteriophages in a 5-M Column of Saturated, Heterogeneous Gravel. *Water, Air, & Soil Pollution*, *223*(5), 2347–2360. https://doi.org/10.1007/s11270-011-1029-9

571479 New Insights of Carbon Chain Length in Cross-Linked PEI Adsorbents on CO_2 Capture

Recently, many countries and international society are seriously concerned about the global warming caused by CO_2 emissions from the industry and are carrying out realistic regulatory policies. For this reason, lots of researches on CO_2 capture, utilization and storage (CCUS) have been carried out. In particular, the CO_2 capture process is the most expensive step that should be preceded before the storage and utilization in CCUS, so technological efforts have been made to reduce its operation cost. Amine-impregnated adsorbent is one of considerable options to reduce the CO_2 capture cost, however, stubborn problems such as the formation of urea and low desorption performance in practical regeneration condition have been hurdles steadily. The formation of urea and poor regeneration performance can be directly connected to lower the working capacity of the adsorbents. Amine cross-linking technique could be adopted to solve these problems. In this study, cross-linked amine-impregnated adsorbents were manufactured by using polyethyleneimine (PEI), diepoxy cross-linkers having different carbon chain length and mesostructured silica gel for support. Scanning electron microscopy (SEM) and ^{13}C nuclear magnetic resonance (NMR) spectroscopy were carried out to characterize adsorbents and to investigate molecular structure of cross-linked PEI, respectively. Cross-linked PEI showed less 1° amine ratio compared to pristine PEI and the morphologies of adsorbents were almost similar to the morphology of silica support. The formation of urea was diminished in cross-linked PEI samples and desorption performance was improved as well, and it was verified via thermogravimetric analyzer (TGA) and Fourier transformed infrared (FT-IR) spectroscopy. Especially, with longer carbon chain length of cross-linker, samples figured out better working capacity and less urea formation. It is thought that the carbon chain of cross-linkers acts as a spacer which can weaken the interaction force between CO_2 molecules and amine active sites, consequently, a forward reaction which causes the formation of urea could be restrained and desorption of CO_2 also could proceed well.

572205 Theoretical Modelling and Experimental Validation of Neutralization Reaction of Acidic Spillover By Calcium Hydroxide [$Ca(OH)_2$] and Sodium Bicarbonate [$NaHCO_3$]

In modern society, industrial manufacture and distribution of various chemicals flourished our life and economy. However, when hazardous chemical is accidentally released from the storage, the consequence can inflict casualties and huge economic damage. Therefore, it is of importance to prepare possible accidents with appropriate safety manual. In this study, we theoretically model and experimentally validate neutralization reaction of four archetype acids (hydrogen chloride, hydrogen nitrate, sulfuric acid, and hydrogen fluoride) using two alkaline agents (calcium hydroxide [$Ca(OH)_2$] and sodium bicarbonate [$NaHCO_3$]). The batch reaction was modelled using PHREEQC interfaced within MATLAB and equilibrium solution was simulated as alkaline agent is incrementally titrated. Total heat and final temperature from the modelling were compared with the experimental results: Neutralization 3 M of acid solutions with equivalent amount of alkaline agent. The modelling result showed remarkable accuracy of R^2=0.96 with respect to experiment. $NaHCO_3$ rather lowered the solution temperature because of its characteristic neutralization heat and endothermic dissolution and CO_2 evolution which imparts $NaHCO_3$ as almighty alkaline agent for any acid. This work can be further applied to higher concentration where experiment is restricted due to violent reactivity and aid the safety remedy for chemical accident involving acid chemicals.

573182 Simultaneous Transformation of Hexavalent Chromium and Carbon Tetrachloride Using Atomized Zerovalent Iron in Co-Contaminant Aqueous Medium

Hexavalent Chromium (Cr(VI)) and Carbon tetrachloride (CCl_4) are two of the most toxic and prevalent contaminants in many of the hazardous waste sites, including DOE commissioned superfund sites such as Hanford site in Washington and Savannah River site in South Carolina, USA. Cr is a heavy metal used extensively for its anti-corrosive properties, whereas CCl_4 is a chlorinated organic solvent historically used extensively as a metal degreasing agent in industries. This paper focuses on the simultaneous transformation of Cr(VI) and CCl_4 in a co-contaminant aqueous medium by atomized Zerovalent Iron (ZVI). ZVI being a nontoxic, cheap, abundant and easy to produce reducing agent has proven to remediate hosts of several chlorinated organic compounds, heavy metals, nitrates and other contaminants. The experiments show that the both the contaminants undergo reduction at much higher rates at low pH values; hence, a pH of 2.5 was used to study the degradation. ZVI proved to be an effective reducing agent for both Cr(VI) and CCl_4

contaminants did not effects the degradation of each other in the co-contaminant media. Both the contaminants followed a pseudo-first order degradation kinetics, with the rate of Cr(VI) degradation being higher (K_{obs} =13.038 hr^{-1}) than the degradation of CCl_4 (K_{obs} =3.29 hr^{-1}) for same reaction conditions. At a pH of 2.5, nearly 100 % degradation was achieved for both the contaminants using excess ZVI within a 2-hour period, in separate as well as in a co-contaminant media.

573618 Cr(VI) Reduction By Fe(II) and Subsequent Regeneration of Fe(II) from Fe(III) By *Shewanella Oneidensis* MR-1

Hexavalent chromium, Cr(VI) is the predominant contaminant in the subsurface vadose zone and groundwater of Department of Energy's waste sites such as Hanford, WA, and has detrimental impacts on the environment, ecology, as well as human health. This contaminant can be reduced to nontoxic trivalent chromium, Cr(III) by reacting with Ferrous ion, Fe(II). During that reaction, Fe(II) is oxidized to Ferric, Fe(III), which has no ability to reduce Cr(VI) and the remediation process terminates. But, metal-reducing bacteria, *Shewanella oneidensis* MR-1, regenerates Fe(II) from Fe(III) with the help of an organic electron donor in an appropriate bacterial media. Thus, in the Cr(VI) chemical remediation process, Fe(II)-Fe(III) conversion is reversed in biological process regenerating Fe(II) and the Cr(VI) chemical remediation process continues in a cyclic fashion. This paper presents the results from regeneration of Fe(II) from pure Fe(III) as well as from the reaction products of Cr(VI) and Fe(II). Both pure Fe(III) and reaction product Fe(III) exhibited the same reduction behavior. Fe(II) production increased with the increase of the initial concentration of Fe(III); almost 80% Fe(II)-regeneration was achieved. In the experimental system, Cr(VI) was reduced first followed by Fe(III) by *Shewanella oneidensis* MR-1 when both of these species are present. So, Fe(II) regeneration with metal-reducing bacteria has a good potential for application in groundwater and wastewater remediation in reducing Cr(VI) to Cr(III).

575832 Magnetically Recoverable Carbon-Coated Iron Carbide for Organic and Arsenic Adsorptive Removal

Magnetic particles, generally nanostructured and magnetite-based, have been used extensively to remove drinking water contaminants. Compositions alternative to Fe_3O_4 could address long-standing issues of magnetic recoverability and materials integrity in real waters. The stability, magnetic separability, and adsorptive properties of nanostructured carbon-coated iron carbide (Fe_3C@C) were compared to those of Fe_3O_4 and other common iron oxide-based nanomaterials. Experimental results show that (i) Fe_3C@C is chemically stable in simulated drinking water, (ii) can be separated from water magnetically under flow with greater than 99% recovery, and (iii) has a 100′ higher adsorption capacity for methylene blue (39.8 mg/g) than Fe_3O_4 and a comparable adsorption capacity for arsenic (168 µg/g). These properties suggest Fe_3C@C as a viable magnetic engineered nanomaterial (ENM) for the removal of organics and oxo-anions with further development.

576727 Biological Transformation of Hexavalent Chromium and Carbon Tetrachloride in Aqueous Medium By *Shewanella Oneidensis* Strain MR1 and *Cellulomonas* Strain ES6

Hanford Site, operated by the U.S. Department of Energy, which was a nuclear production complex operated between 1943 and 1998, is contaminated with heavy metals, chlorinated organics, inorganics and radioactive wastes. This site poses an environmental risk because of the presence of hexavalent chromium, Cr(VI), introduced as a result of the use of chromate (a corrosion inhibitor) in the reactor cooling water. Chromium migration to the water table can result in a rapid transfer to the Columbia River, which is a main source of drinking water for Native American population in the area and is one of the largest salmon spawning water body in the United States. Therefore, the migration of Cr(VI), which has the Federal drinking water standard of 100 μg/L, could cause severe health risks to both humans and aquatic lives. Currently, the U.S. Government is spending more than $1 billion per year to remediate Hanford Site and to clean up contaminated soil and groundwater, reactors, tanks, chemical processing plants, and ancillary facilities. On the production complex, about 6,300 lbs. of chromium and 38,000 lbs. of carbon tetrachloride have been removed through "Pump and Treat Remediation". Our overall project is focused on combined chemical/biological degradation of Cr(VI) and chlorinated organics. In this paper, biological degradation of Cr(VI) and CCl_4 will be presented. *Shewanella oneidensis* strain MR1 and *Cellulomonas* strain ES6 were used to biologically reduce Cr(VI) to Cr(III). The experiments were conducted using an End-On-End reactor and a mini Incubator Shaker. The batch reactions were conducted under nongrowth anaerobic conditions in the presence of hexavalent chromium and carbon tetrachloride. After 7 days *Shewanella oneidensis* strain MR1 was able to achieve 82% reduction of Cr(VI) without any competing electron acceptors and 67% reduction in the presence of carbon tetrachloride. On the other hand, *Cellulomonas* strain ES6 reduced chromium by 43% without any electron acceptor and 35% in the presence of carbon tetrachloride.

576788 An Excel Tool for Economic and Environmental Analysis of Waste-to-Energy Technologies Processing Urban and Rural Waste

The average amount of waste generated in the United States is 4.4 lbs./day/person [1], and most of it is landfilled. The enormous amount of energy contained in these wastes presents an opportunity for renewable energy production. Renewable energy production methods that are more environmentally friendly are needed to address concerns about global warming. Therefore, it is necessary to explore opportunities to implement waste to energy technologies that will reduce landfilling and minimize environmental impact. An Excel-based calculator has been developed to compare various options for waste to energy conversion. In this poster, the methodology used to create this tool and some case studies that illustrate its use will be presented.

The options for converting waste to energy include anaerobic digestion (AD), incineration and gasification. In addition, landfilling followed by landfill gas collection and conversion is also often practiced. These landfill gases can be purified and upgraded to be used in energy recovery methods like combined heat and power units (CHP), Fischer-Tropsch Synthesis (FTS) to produce liquid fuel or the production of Compressed Natural Gas (CNG).

Most of these technologies are new and minimal data has been reported for the operation of these plants. A limitation encountered when trying to decide which of these technologies should be implemented is that there is no set tool that compares and analyses the techno-economic data and fuel-cycle environmental impact of these plants. Analyzing data from plants currently operating can provide data used to estimate cost and GHG emissions for the screening of these alternative projects. The goal of this project is to create a software tool that compiles all this data to assist decision makers such as municipalities and waste management companies on the selection of waste to energy technologies.

Using this tool, the user can select the technology desired: AD, Gasification, Incineration or Landfilling, as well as the energy recovery method: CHP, liquid fuel or CNG production. The user will also provide the feedstock type (Municipal Solid Waste (MSW), Food Waste, Animal Manure, and Farm Waste), feed properties and plant capacity. The software tool is set to provide the user with the capital and operational cost estimates, biogas production rate and its composition, and where applicable, liquid and solid production rates. The calculator also provides detailed fuel-chain analyses of the environmental impact and GHG emission savings for the use of each technology compared against fossil fuel sources and landfilling of waste.

In this presentation, we will discuss the underlying principles and methods used in building this calculator. In addition, we will present results for a case study involving the waste disposal options for Hillsborough County located in Florida.

References

1. (2013, March 27). Municipal Solid Waste. Retrieved from https://archive.epa.gov/epawaste/nonhaz/municipal/web/html/

578661 Enhancing Nitroarene Reduction with Magnetically Recoverable AuFe Nanoparticles and Localized Heating Produced By an Alternating Magnetic Field (AMF)

Gold nanoparticles are active reduction catalyst and are known to hydrogenate a variety of organic molecules (e.g., alkenes, carbonyls, nitroarenes, etc.) in the presence of a reducing agent (i.e. H_2) and are catalytically active under mild conditions (i.e., ambient temperature, etc.). Additionally, the activity, selectivity, and surface deactivation/fouling resistance can be enhanced by combining gold with secondary metal (e.g., bimetallic catalysts). A typical problem associated with gold catalysts is that under ambient conditions (i.e. room temperature and atmospheric pressure) reaction kinetics can be slow. Additionally, elevated temperatures are often used to enhance the catalytic activity of precious metal catalysts as well as reverse catalysts deactivation and surface fouling (e.g., regeneration). Using nitroarene reduction – a reaction seeded with environmental impact implications – as a probe reaction and an AuFe bimetallic catalyst, we hypothesize that applying an alternating magnetic field (AMF) will increase nitroarenes reduction reaction rates, via localized particle surface heating. Generated temperatures at the catalyst surface (e.g., the extent of localized heating) can be approximated with a temperature sensitive reaction - benzyl alcohol oxidation. Additionally, exploring different AMF exposure durations and provide a better understanding of operational effects on SAR.

561419 Removal of Heavy Metals from Contaminated Waters By Insoluble CMC Bio-Polymer

In this project, insoluble carboxymethyl cellulose adsorbent was used for removal of heavy metals including cadmium and Pb from aqueous environments. The results showed that the maximum removal of Pb and cadmium metal ions by insoluble carboxymethyl cellulose, was observed at reaction time of 24 hours, pH 2, and initial concentration of contaminant 200 ppm. While the temperature increasing did not significantly effect on the removal efficiency of cadmium and Pb ions. Also, insoluble carboxymethyl cellulose whit the same conditions is more capable for removing cadmium ions than Pb ions.

Keywords: Chemical Water Pollution, Pb, cadmium, insoluble carboxymethyl cellulose, Water Purification

566249 Thermal Treatment of the Aqueous Phase Resulting from the Hydrothermal Liquefaction of the Microalga *Chlorella vulgaris*

Microalgae are one of the most promising biofuel sources, because, among other characteristics, they capture between 10 and 50 times more CO_2 than terrestrial plants. Since the water content of these microorganisms can reach 95% by weight, hydrothermal liquefaction (HTL), which uses water at supercritical conditions, has emerged as a promising technology to convert wet biomass into biofuel. However, current biofuel yields of about 50% result in the generation of an aqueous phase with high concentrations of organic compounds that could be toxic, which constitute the main residue of the process. This work intends to evaluate the hydrothermal gasification as an alternative treatment of the aqueous phase coming from the HTL of the microalga *Chlorella vulgaris*, in order to recirculate it to the culture bioreactors.

In the first place, the algal biomass to be liquefied was obtained and biochemically characterized. Then, the HTL was carried out at constant reaction conditions (350ºC, 20 minutes). Subsequently, the aqueous phase obtained was characterized and an experimental plan was designed to establish the adequate reaction conditions (time and temperature) to perform the proposed hydrothermal treatment. Finally, the treated water was characterized, and was diluted to different concentrations to carry out comparative microalgae cultures, in order to determine the reuse potential of the process water.

At the end it was found that the aqueous phase resulting from the HTL contained nutrients necessary for the growth of the microalga, in higher concentrations than those used for the culture medium. However, the presence of toxic compounds such as phenol, which limit and inhibit the growth of microalgae, limited the recirculation of water as large dilutions factors were required. The hydrothermal gasification, used to treat the water, allowed the removal of more than 50% of the phenolic compounds present in the water. Therefore, a major recovery of water and nutrients was achieved, due to a lower dilution factor of the aqueous phase.

567487 Treating Liquid Effluents from Anaerobic Digestion: Semi-Empirical Modelling of a Multi-Stage Flash Evaporator

Anaerobic digestion is a widespread method for the treatment of waste biomass and energy crops. In 2016, almost 18,000 biogas plants with a total installed capacity exceeding 9 GW were operating in Europe. The main product of anaerobic digestion is biogas, which is subsequently combusted for the generation of electricity. A liquid by-product of anaerobic digestion is the so-called **digestate**, whose annual output ranges between 10,000 and 30,000 tons per MW of installed capacity. Digestate is an organic fertilizer, but with a very low nutrient concentration (N, P, K). Up to 95% of the digestate volume is represented by water. **Storage, transportation, and application of this fertilizer are very costly**. This is particularly true in areas with intensive agricultural production (e.g. North-western Europe), where there is a long-term problem with the accumulation of fertilizers (especially nitrogen) in soil and groundwater. Consequently, there are many legislative constraints and operators are forced to export digestate to distant areas. Transport distance may even exceed 100 km in some areas.

Vacuum evaporation is one of the technologies for **digestate volume reduction**. The evaporators allow separation of water from digestate and reduce its volume several times. The advantage of vacuum evaporation is its reliability and the ability to use low-potential waste heat (<90 ° C), which is generally in excess in biogas plants. Integrating thickening technology into a biogas plant assumes a reasonable return on investment. The material and energy balance of the technology is the key information for a techno-economic evaluation. However, digestate evaporation modelling is complicated for many reasons. The main problems in this regard include the complicated properties of the digestate (e.g. non-newtonian character, variable composition, high viscosity), the complex nature of the evaporation process (phase conversion, foaming, release of dissolved gases), and lack of verified operational data in general.

This paper presents the current state of a model, which aims to reliably predict the mass and energy balance of a **multi-stage flash evaporator** when concentrating digestate from anaerobic fermentation. The model is based on semi-empirical relationships derived from a series of experiments on a pilot plant evaporator. The acquired knowledge is partly transferable to other types of evaporation devices. The most important modelled parameters include boiling point elevation and overall heat transfer coefficient in the process of digestate evaporation and heating respectively. The model also considers input parameters such as acid dosage and anti-foaming agent addition.

571472 Pig Slurry Detoxication and Usage

Pig slurry is a valuable but unwanted by-product of industrial scale agriculture. The vicinity of industrial livestock farming suffers from odour nuisance during manure discharge. In the low-vegetation period manure discharge (autumn and winter) causes undesirable soil- and groundwater pollution through conversion of ammonia into nitrate.

The increasing consumption of meat does cause a steadily increasing surplus of pig slurry on national and international level. As a rule of thumb a feeding pig has a mean life span of 4 months and it produces about 0.6 m^3 of pig slurry with an ammonia content of up to 10000 mg/L.

Detoxication and processing of pig slurry seems to be difficult due to the variety of ingredients, such as ammonia and amines, carboxylic acids, aldehydes, alcohols and hydrogen sulfide based ethers, briefly specified as odour-intensive ingredients. The challenge of finding appropriate technologies lies in the development of suitable treatment steps that can be directly applied at the source (farm) under economically feasible conditions.

The aim of this project was the development of a process for the sustainable solution of the "pig slurry problem" by applying simple technical measures. Through combination of "high impact" technologies, a flexible treatment process for ammonia based nitrogen removal, odor control and effluent management without waste products had to be developed. All products must be suitable for agricultural and horticultural use and the aqueous effluent must at least suffice the quality for irrigation even in low vegetation periods.

The solver for the problem was to neutralize odorous ingredients such as skatol, cresols, phenols, aldehydes and above all butyric acid by alkalizing pig slurry and by stripping ammonia from the liquid slurry to a maximum residual concentration of 100 mg/L and to convert it into ammonium sulfate. Caustic lime and wood ash are recommended for alkalizing large quantities of pig slurry. Ammonium sulfate is a high-quality storable mineral fertilizer for agricultural and horticulture use.

Biochar can completely decolourize the detoxified slurry and it helps improve solid-liquid separation by sedimentation and filtration. The solid residue from phase separation has great potential for use in soil meliorisation. The meanwhile completely transparent filtrate can be used together with the clear run from the sedimentation stage for irrigation/fertilization purposes. In case of necessity the filtrate can be split into a retentate with fertilizer quality and clean permeate by reverse osmosis.

The process, as specified above, was tested in bench scale and in pilot scale and it will go into operation for technical scale validation in autumn 2019. The process is also applicable in cattle manure detoxication and management.

575598 Systematic Synthesis of Wastewater Treatment Networks Using the P-Graph Approach

Systematic synthesis of wastewater treatment networks using the P-graph approach

Kirti Yenkie[1], Jean Pimentel[2 3], Akos Orosz[4], Ferenc Friedler[2]

[1]Department of Chemical Engineering, Henry M. Rowan College of Engineering, Rowan University, NJ - 08028, USA

[2]Institute for Process Systems Engineering and Sustainability, Pazmany Peter Catholic University, Budapest, Hungary

[3]Grupo de Procesos Quimicos y Bioquimicos, Department of Chemical and Environmental Engineering, Universidad Nacional de Colombia, Bogota, Colombia

[4]Department of Computer Science and Systems Technology, University of Pannonia, Veszprém, Hungary

Wastewater treatment is an area of utmost importance to address the challenge of decreasing freshwater resources. Traditionally, wastewater treatment consists of three or four sequential stages: preliminary, primary, secondary and tertiary [1]. Preliminary treatment is intended to facilitate the removal of contaminants in the following stages. Primary treatment mainly involves the withdrawal of settleable solids. The secondary treatment is capable of removing pollutants such as chemicals and metals. In some cases, a tertiary treatment might be used for further purification if the required purity limits are not achieved during the first three stages. To create this systematic approach for designing wastewater treatment networks, we rely on the principles of mathematical modeling to generate an exhaustive enumeration of all the possible technologies and their connections during the early stages of design. This enables the generation of a maximal structure which consists of all the treatment structures. Some of these structures would be non-intuitive and could include reprocessing, and multiple technologies in parallel or series to remove the same contaminant, which would not have been obtained via typical design heuristics. Thus, the sequence of treatment technologies predicted using this maximal structure would be based on holistic consideration and would not lead to suboptimal solutions.

A representative case study of a wastewater stream containing settleable solids and chemicals is considered as the motivating example where the minimum purity requirements for water reuse have been specified. Primary and secondary stages were studied considering three types of technologies for the former, and four types of technologies for the latter. In primary treatment, the technologies are based on the particle size: sedimentation, filtration, and microfiltration; whereas, in secondary treatment, the technologies are based on the ability to remove chemicals, such as adsorption, disinfection, advanced oxidation process and rotating biologic contactors. The contaminant removal efficiencies and cost associated with the operation of these technologies were provided through information collected from design literature [2], [3] and simulation software [4]. The primary and secondary treatment stages were studied by means of the P-graph approach, which has been extensively deployed to evaluate problems of combinatorial nature, such as supply chains, evacuation routes design, and process network synthesis (PNS) [5]–[7]. This advantage is that not only the best structure in terms of cost is obtained, but also a ranked list of all feasible near-optimal solutions is created. This additional information can be relevant when considering additional performance criteria, such as controllability, operability, flexibility, sustainability, etc.

For the primary stage, an initial structure was obtained by joining the input and output materials with the technologies. The water rich streams leaving the primary stage can be considered as potential input for the next stage. The secondary structure is constructed for each of the water-rich leaving streams from the primary stage by joining the plausible materials and technologies of the secondary treatment stage. The resultant overall network consisting of the initial and secondary structure is used as the input for the P-graph algorithm, which determines the most cost-effective treatment process for a certain set of parameters. Subsequently, a parametric study is carried out by changing the feed water composition, with the objective of generating cost-effective process structures which are also capable of satisfying the treatment requirements under different inlet water conditions. Therefore, this work aims at providing guidelines for the design of efficient wastewater treatment networks depending on the inlet water composition and treated water requirements.

Key words: Wastewater treatment, Process Network Synthesis, exhaustive enumeration, P-graph

References:

[1] M. & E. Inc, G. Tchobanoglous, F. L. Burton, and H. D. Stensel, *Wastewater Engineering: Treatment and Reuse*, 4th edition. Boston: McGraw Hill Higher Education, 2002.

[2] G. Towler and R. K. Sinnott, *Chemical Engineering Design, Second Edition: Principles, Practice and Economics of Plant and Process Design*, 2 edition. Boston, MA: Butterworth-Heinemann, 2012.

[3] G. D. Ulrich and P. T. Vasudevan, *Chemical Engineering Process Design and Economics: A Practical Guide*. Process Pub., 2004.

[4] D. P. Petrides, *Intelligen, Inc.: SuperPro Designer, Batch Process Simulation, Environmental Impact Assessment*. Intelligen Inc, 2013.

[5] I. Heckl, F. Friedler, and L. T. Fan, "Solution of separation-network synthesis problems by the P-graph methodology," *Computers & Chemical Engineering*, vol. 34, no. 5, pp. 700–706, May 2010.

[6] H. Cabezas, A. Argoti, F. Friedler, P. Mizsey, and J. Pimentel, "Design and engineering of sustainable process systems and supply chains by the P-graph framework," *Environ. Prog. Sustainable Energy*, vol. 37, no. 2, pp. 624–636, Mar. 2018.

[7] L. Vance, H. Cabezas, I. Heckl, B. Bertok, and F. Friedler, "Synthesis of Sustainable Energy Supply Chain by the P-graph Framework," *Ind. Eng. Chem. Res.*, vol. 52, no. 1, pp. 266–274, Jan. 2013.

576694 Innovative Water Recycle Strategies for Reducing Water and Energy Use in Pulp and Paper Industry

Innovative water recycle strategies for reducing water and energy use in pulp and paper industry

The pulp and paper industry is one of the largest consumers of water in the US industrial sector. Application of water in the industry ranges from a chemical reactant, solvent and use in physical application as a conveyor of final or intermediary material. In recent years, there has been an increase in the industry for adoption of waste minimization strategies covering various aspects and avenues in mill operations. Process modelling and simulations is an effective approach for identifying innovative strategies for improving the industry's water efficiency and productivity while strengthening its economic sustainability. In this study, WinGEMS (Windows General Energy and Mass balance System) is used as the software tool for modelling bleaching and the paper machine sections of an integrated bleached Kraft mill in the United States. The project entails modelling the replacement of fresh water use for final bleaching stage shower washer with white water from paper machine and assess the change within the bleaching and paper machine operations. Current study simulates the bleaching section of an Elemental Chlorine Free (ECF) sequence D_0EopD1EpD2 for a softwood Kraft pulp kappa number 30. The chlorine dioxide delignification reaction is a combination of fast reaction followed by a slow reaction and the asymptotic value of kappa number linked to the floor lignin (asymptote for both kappa number and brightness). The simulation included the effect of carry-over from brownstock washers (BSW) and extraction stage washer. Table below shows the data for the base simulation of ECF bleaching modelled in WinGEMS.

	Pulp Kappa Number	Bleaching Sequence	Extraction Stage			D2 Stage Final Brightness (% ISO)
			Kappa No.	Brightness (%ISO)	COD (Kg/Tn)	
Industry Data	30	$D_0EopD_1EpD_2$	5.1	50	53	88
Simulation Data	30	$D_0EopD_1EpD_2$	5.4	50.9	53.5	89.6

The values of brightness, Kappa number and COD are in close proximity with the actual industrial values. Future work includes modelling the base case for paper machine and integrating it with bleaching process capturing non-process elements (NPEs) flows in adequate detail and examine its effects on the final paper brightness and paper machine operations.

Key Words: WinGEMS, modelling, simulation, white water, bleaching, paper machine, brightness

577454 The Synergistic Emulsification of Cactus Mucilage with d-Limonene and Silica Nanoparticles for Oil in Water Emulsion

It is common knowledge that crude oil does not dissolve easily in water. Therefore, crude oil floats on the ocean surface and hardly dissolve after an oil spill. Spilled oil can produce extensive damage to marine and wildlife habitats and tourism industries. As a rapid response method, dispersants can remove spilled oil from the sea surface and disperse it as tiny oil/dispersant/water droplets. However, the application of extremely high amounts of chemical dispersants could be detrimental to marine life and the costal environment. In this research, we are proposing to use a natural surfactant from cactus plant to stimulate the dispersion process of oil in water (O/W). This natural dispersant is extracted from the *Opuntia ficus indica* cactus plant. Compared with the traditional chemical dispersants, cactus mucilage extract has the advantages of being biodegradable, non-toxic, and can be obtained through sustainable agriculture. The cactus plant-based mucilage can diffuse to the interface when applied to O/W emulsion. Then, tiny oil droplets are surrounded by surfactants and dissolved into water. By the mixing action of the waves, spilled oil and water can be combined in two ways: natural dispersion and water-in-oil emulsification. Cactus mucilage extract can push the balance between natural dispersion and water-in-oil emulsification strongly towards natural dispersion. This dispersion process can improve the opportunity for biodegradation of the spilled oil by increasing the contact area between water and oil. The properties of cactus mucilage extract as a dispersant were studied on the surface tension, droplet size and stability of O/W emulsions, toxicity, and dispersion effectiveness. Synthetic seawater and the surrogate oil from British Petroleum (properties similar to the Macondo rig) were used. The synergistic emulsification mechanism among the surfactant, organic solvent, and silica nanoparticles have been studied. A food grade solvent, d-limonene, and silica nanoparticles have been applied with the cactus mucilage extract to improve its dispersion ability. It is found that the O/W emulsion had a lower surface tension due to the application of the solvent, which can push the balance strongly from O/W emulsion towards dispersion. Furthermore, adding silica nanoparticles increased the stability of the O/W emulsion. Baffled flask test was used to quantify to dispersion effectiveness. The factors and the levels of each factor were as follows: DOR (mg/mg) (control, 1:200, 1:100, 1:75, 1:50; 1:35, 1:20, and 1:10). The baffled flask test showed that with the synergy of both d-limonene and silica nanoparticles, the dispersion effectiveness of the cactus mucilage has been significantly improved. The ratio of cactus mucilage extract/d-limonene/SiO_2 has also been optimized through baffled flask test. The average droplets size of O/W emulsion was smaller in the systems with cactus mucilage extract/d-limonene/SiO_2 when compared with the systems using the conventional dispersant COREXIT EC9500A by measuring with optical microscope. The smaller droplets size implies that cactus mucilage extract/d-limonene/SiO_2 would provide a better stability to O/W emulsion.

561760 Selective Oxygen Reduction Using Methane for CO_2 Purification from a Simulated Flue Gas

Catalytic Oxygen Reduction using Methane for CO_2 Purification from a Simulated Flue Gas

Andrew Kuhn[1], Zhitao Chen[1], Yongqi Lu[2] and Hong Yang[1]*

[1]University of Illinois Urbana Champaign, Urbana, IL 61801 (USA)

[2]Prairie Research Institute, Champaign, IL 61801 (USA)

*hy66@illinois.edu

Introduction. Efficient sequestration of carbon dioxide (CO_2) from flue gas generated from fossil fuel sources remains a critical endeavor due to the link between rising global temperatures and increasing atmospheric CO_2 concentrations. State of the art oxy-combustion practices both increase the energy efficiency and maximize the effluent purity for coal and natural gas power plants (>85% CO_2, dry basis).[1] Streams with 95% CO_2 purity can be utilized in Enhanced Oil Recovery or chemically-upgraded to value-added chemicals such as methanol. One of the main impurities, oxygen gas (O_2), prevents the direct utilization of these streams due to undesirable oxidation reactions. Because of selectivity challenges and the energy requirement, traditional separation technologies (i.e. membranes and distillation) are unpromising. A new process must be developed that is both energy efficient and technologically practical. Here, we propose to catalytically reduce O_2 from a simulated flue gas stream to produce high purity CO_2.[2] Methane was used as the primary reducing gas due to its availability and high energy density ($CH_4 + 2O_2 \rightarrow CO_2 + 2H_2O$). In order to purify flue gas, both the oxygen conversion and selectivity to CO_2 must approach 100% because of the tight specifications for O_2 and carbon monoxide (CO). Here, we study palladium-based (Pd) catalyst systems as benchmark catalysts for the purification of CO_2 from flue gas. Three variables to increase the activity of the model Pd system were conducted: the metal loading, the oxygen-methane feed ratio, and the usage of a copper-based (Cu) oxygen scavenger sequentially after the Pd bed. This novel approach to purify and capture CO_2 is achieved by careful material design.

Results and Discussion. Pd catalysts were made by impregnating the precursor onto gamma alumina (γ-Al_2O_3). The performance of the catalysts was tested using an in-house ambient fix-bed reactor system. The CO_2 feed composition was set at 85% (balance: CH_4 and O_2 at various ratios). The product compositions were measured (on a dry basis) by gas chromatography. The temperature was varied from 300 to 600 °C. Using this fixed bed reactor system, the Pd catalytic materials were tested for oxygen conversion, selectivity to CO_2, and product CO_2 purity. A stoichiometric feed to be optimal for all catalysts. A summary of the activity of three materials is shown in Figure 1. Notably, the Pd then copper-adsorber (Pd+Cu) yielded the highest in all three categories. Additionally, we discuss the outlook for using non-precious metal catalysts to achieve the performance metrics.

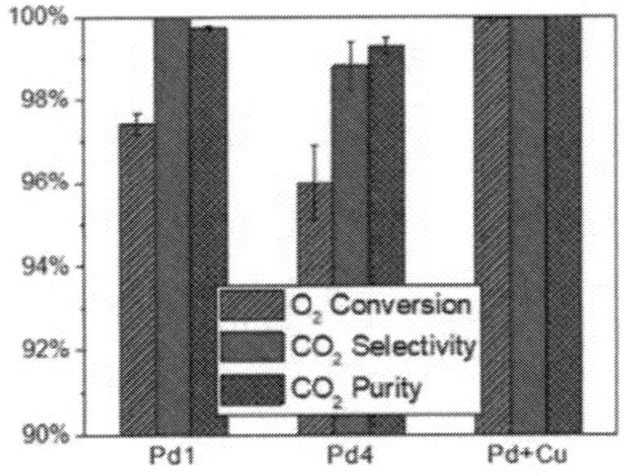

Figure 2. Oxygen conversion, selectivity to CO_2, and product CO_2 purity for 1% Pd/Al_2O_3 (Pd1), 1% Pd/Al_2O_3 (Pd4), and the sequential beds containing 1% Pd/Al_2O_3 and 13% Cu/Al_2O_3 (Pd+Cu). Ambient pressure. 450 °C. Feed: 50 mL/min (85% CO_2, 10% O_2, and 5% CH_4). WHSV: 60,000 hr^{-1}.

References

1. Buhre, B.J.P., et al., *Oxy-fuel combustion technology for coal-fired power generation*. Progress in Energy and Combustion Science, 2005. **31**(4): p. 283-307.

2. Kuhn, A.N., et al., *Sequential Oxygen Reduction and Adsorption for Carbon Dioxide Purification for Flue Gas Applications*. Energy Technology. **0**(0).

562460 Modeling Chemical Absorption of CO_2 in Highly Alkaline Solutions for Direct Air Carbon Capture Applications

Carbon dioxide is known to be one of the gases causing current global climate change. In 2015, with the signature of the Paris Agreement, nations agreed to reduce global emissions annually by 3% in order to avoid a catastrophic scenario. However this hasn't been achieved and the current path, which includes mainly treatment of point sources such as flue gas, is not enough to avoid a global temperature change greater than 2°C, rising urgency for new and innovative ways to mitigate new emissions and to reduce existing CO_2 levels in the air [1].

Although long term geological storage of CO_2 have been demonstrated as feasible, it remains unattractive from an economic perspective, as –besides applications involving enhanced oil recovery– the cost of capture and storage is significantly above the market price assigned to carbon pollution. Consequently, recent research efforts are aimed at complementing carbon capture and storage with utilization strategies that generate valuable products from CO_2 [2], [3].

The production of bioenergy and bioproducts using photosynthetic microbes is quite attractive as, theoretically, it only requires minimal energy inputs to achieve the conversion, given that most of the energy can be provided by sunlight. Furthermore, processes that rely on photosynthetic microbes can be deployed in marginal land and can capture CO_2 directly from the air.

Unfortunately, a major obstacle currently hindering the large-scale deployment of photobiological processes is their low volumetric productivity. This low productivity is partially caused by limited solubility of CO_2 in the cultivation media and very low mass transfer rates. Current approaches to overcome this obstacle rely on a continuous bubbling of CO_2 streams into photobioreactors or cultivation ponds, which is accompanied by excessive CO_2 losses and high-energy requirements.

Although several advances have been reported regarding methods suitable for capturing CO_2 directly from the air, such as metal-organic frameworks and cryogenic distillation, which obtain very high purities they are still very energy intensive. Chemical absorption, on the other hand, is a well-established technology that could be easily deployed in a sustainable way, especially if environmentally friendly and easy to regenerate solvents are used.

Here we report on the modelling of a chemical absorption process using a highly concentrated aqueous solution of sodium carbonate. The high concentration of the carbonate solution favors the mass transfer of carbon dioxide from atmospheric air into the solution, and the solution can be regenerated directly by photosynthetic microbes.

Several factors affect the suitability of direct air capture applications. First, mass transfer is limited due to the low concentration of CO_2 in the air (~400 ppm) as at these low concentrations the likelihood of interaction and therefore absorption, is reduced. Second, the stability of the nonreacting CO_2 in the liquid phase and the relatively high fugacity of the CO_2 in aqueous solutions introduce significant thermodynamic limitations [4]. Third, there are also kinetic limitations introduced by the complex reaction pathways of CO_2 and carbonate species in aqueous solutions. The kinetics of highly alkaline solutions have been studied [5] and correspond to a series of electrochemical reactions in which the absorption of CO_2 is enhanced.

To understand the absorption mechanism and the limitations of the proposed system, a mathematical model was developed in COMSOL Multiphysics, incorporating the limiting factors previously described. Two different models were developed in order to determine the controlling mechanism in the process: the falling film model, which describes a thin film of liquid falling and a bulk of gas around it, and the two film theory, which simulates the bulk of both the liquid and gas as well as the interface between them.

The two-film model was found to provide a better understanding of the reaction and its likelihood in the interface and the mass transfer of the system. From this model, it is clear that, due to the high alkalinity of the media, the hydroxide ions available for the reaction are in such abundance that their concentration at the interface enables an increased probability of interaction with the CO_2, which is present in lower concentration on air allowing chemical absorption to occur by the previously described mechanism.

References:

[1] NOAA, "Global Greenhouse Gas Reference Network," 2018. [Online]. Available: https://www.esrl.noaa.gov/gmd/ccgg/trends/weekly.html.

[2] D. W. Keith, K. Heidel, and R. Cherry, "Capturing CO2 from the atmosphere: Rationale and Process Design Considerations," *Geo-Engineering Climate Change: Environmental necessity or Pandora's box?* pp. 107–126, 2010.

[3] H. Herzog, "Air Assessing the Feasibility of Capturing CO2 from the Air," *MIT Lab. Energy Environ.*, no. October, 2003.

[4] K. S. Lackner, "The thermodynamics of direct air capture of carbon dioxide," *Energy*, vol. 50, no. 1, pp. 38–46, 2013.

[5] H. Knuutila, O. Juliussen, and H. F. Svendsen, "Kinetics of the reaction of carbon dioxide with aqueous sodium and potassium carbonate solutions," *Chem. Eng. Sci.*, vol. 65, no. 23, pp. 6077–6088, 2010.

568036 Pore Scale Investigation of CO_2 Storage and Bioconversion in Deep Saline Aquifers Using High Pressure Micromodels

CO_2 geological storage in deep saline aquifers represents a mediation solution for reducing the anthropogenic CO_2 emissions. This kind of storage required adequate scientific knowledge and tools at the pore scale to evaluate injection scenarios or to estimate reservoir capacity. In this context, high pressure / high temperature microfluidic reactors (micromodel or geological labs on chip – GloCs) turn out to be excellent tools to investigate the different mechanisms associated with CO_2 geological storage in deep saline aquifers.

This talk will first highlight the latest results obtained at ICMCB concerning the application of the GLoCs to study the invasion processes of CO_2 in water and brine saturated GLoCs. In particular, direct optical visualization and image treatments allow following the evolution of the CO_2/brine phase distribution within the pores, including displacement mechanisms and pore saturation levels. We will then present some ongoing works aiming at integrating *in situ* spectroscopy techniques in HP microreactors to get information about the dissolution and mineralization trapping. We have developed an experimental set-up to recreate 3D reactive porous media within a microfluidic channel (fixed packed bed of calcium carbonate). Thanks to X-ray laminography carried out at the ESRF, we have observed on reconstructed 2D images, the dissolution phenomena occurring during the successive injection of constant volumes of non-equilibrium solution. This proof of concept has opened new possibilities for using this methodology to acquire kinetic data on 3D reactive front phenomena in porous media.

Eventually, we will introduce the use of GLoCs as a significant tool to mimic the *in situ* biogeological reservoirs conditions to investigate CO_2 bioconversion (in the frame of the ERC project "Big Mac"). These tools could also find wider applications in geological-related studies such as Enhanced Oil Recovery, shale gas recovery or geothermal energy.

569323 Energy-Efficient Thermal Regeneration of Biphasic Solvents in a Laboratory CO_2 Desorption System

A novel biphasic solvent-enabled CO_2 absorption process (BiCAP) is under development for post-combustion carbon capture (PCC) as a promising alternative to the conventional monoethanolamine (MEA) process. The BiCAP features the regulated formation of dual liquid phases during CO_2 absorption and the regeneration of the CO_2 rich phase only with reduced mass and elevated CO_2 loading for CO_2 desorption. The process has great potential in significantly reducing the energy use compared with the MEA process.

The biphasic solvents employed in the BiCAP are a new class of amine-based water-lean blends. Two biphasic solvents, BiS1 and BiS2 were selected for this study. The CO_2 absorption capacity of either BiS1 or BiS2 is more than 30% higher than that of the benchmark 30 wt% MEA. Because of the absorbed CO_2 concentrated in the rich phase, their CO_2 desorption capacity is more than twice that of MEA.

A 10 kWe laboratory experimental system (Figure 1) consisting mainly of a flash vessel and a stripping column was fabricated to investigate the CO_2 desorption performance of biphasic solvents. The system is rated at 300 psig at 200 °C. The flash vessel is a 5-inch inner diameter (ID) and 2-ft height cylinder with a heating coil inside that uses steam as a heat source. The stripper is a 2-inch ID and 9-ft height column with stainless steel Pro-Pak random packing, attached to a sump section with a heating coil inside that uses steam to reboil the solvent. The CO_2 desorption system can be operated under two modes: the single stripper mode, under which the flash is bypassed; and the sequential flash and stripper mode.

In the CO_2 desorption experiments, for either BiS1 or BiS2, the CO_2 rich phase was separated from the biphasic solvent mixture loaded with the required amount of CO_2, and then used as the feed solution. The 30 wt% aqueous MEA solution was tested as a reference. The MEA solution is monophasic and was used directly for desorption experiments after being loaded with the required amount of CO_2.

In the single stripper experiments, the stripping of either BiS1 or BiS2 solvent was conducted at 120, 135, and 150 °C. At these temperatures, the stripping pressures were controlled at 160-280 kPa, 260-470 kPa, and 450-860 kPa, respectively, resulting in the rate of CO_2 recovery varying from approximately 90% to 30%. The stripping temperature as high as 150 °C was adopted for the biphasic solvents because they have demonstrated extremely high thermal stability in our previous studies. For the comparison, the MEA solution was tested at either 110 or 120 °C to reflect its typical stripping temperature range with the pressure controlled at 120-160 kPa or 140-260 kPa.

The heat duty in the single stripper generally decreased with increasing stripper pressure at the constant reboiler temperature, mainly because the stripping heat associated with water vapor loss in the product CO_2 stream was reduced at a higher pressure. The estimated heat duty ranged between 2,500 and 3,900 kJ/kg CO_2 for BiS1 and between 2,000 and 2,800 kJ/kg CO_2 for BiS2. In comparison, the MEA required the heat duty of 4,200 to 4,800 kJ/kg CO_2 under comparable conditions. The results of heat duty obtained at 150 °C are displayed in Figure 2(a). The low heat duty for the biphasic solvents is a result of low sensible heat use due to reduced solvent mass and low stripping heat due to reduced water vapor loss in the product CO_2 stream.

Under the sequential flash and stripper test mode, the stripper reboiler temperature remained at 120, 135, or 150 °C for BiS1 and BiS2, and 110 or 120 °C for MEA. The flash temperature was controlled at approximately 10 °C lower than that of the reboiler. Because of the relatively high level of CO_2 loading in the flash vessel, the flash pressures for BiS1 or BiS2 could be elevated to at least 0.6 bar higher than their respective stripping pressures, which were controlled to be comparable to those in the single stripper mode. Throughout the flash vessel and the stripper, the total CO_2 recovery rate achieved ranged between 40% and 80% in the different experiments. Among the total CO_2 recovery, 21% to 57% could be attributed to the flash vessel depending on the test conditions.

The heat duty in either the flash or stripper decreased with increasing operating pressure when the operating temperature remained unchanged. The heat duty in the flash was lower than that in the stripper, primarily because it was operated at higher pressure and lower temperature and thus lowered the stripping heat associated with water vapor loss. The overall heat duty weighted over both devices was estimated at 2,500-3,100 kJ/kg CO_2 for BiS1 and 2,200-2,700 kJ/kg CO_2 for BiS2, compared with the heat duty of 3,500-5,400 kJ/kg CO_2 for MEA. Figure 2(b) shows the results of heat duty obtained under 140/150 °C (for BiS1 and BiS2) or 110/120 °C (for MEA) flash/stripping. As mentioned above, the low heat duty for the biphasic solvents was attributable to their low sensible and stripping heat uses.

Compared with the single stripper, the overall heat duty incurred in the sequential flash and stripper system was comparable. However, because of the elevated flash pressure, the CO_2 stream from the flash required less compression work, which thus reduced the overall work requirement.

The experimental results revealed that the selected biphasic solvents could lower the heat duty of CO_2 desorption by 30%-45% compared with the benchmark MEA solution. In particular, the biphasic solvent BiS2 could reduce the regeneration heat duty to as low as 2,000-2,200 kJ/kg CO_2 desorbed. Additional work is ongoing to investigate different desorption process configurations to further reduce the energy use of CO_2 capture for the biphasic solvents.

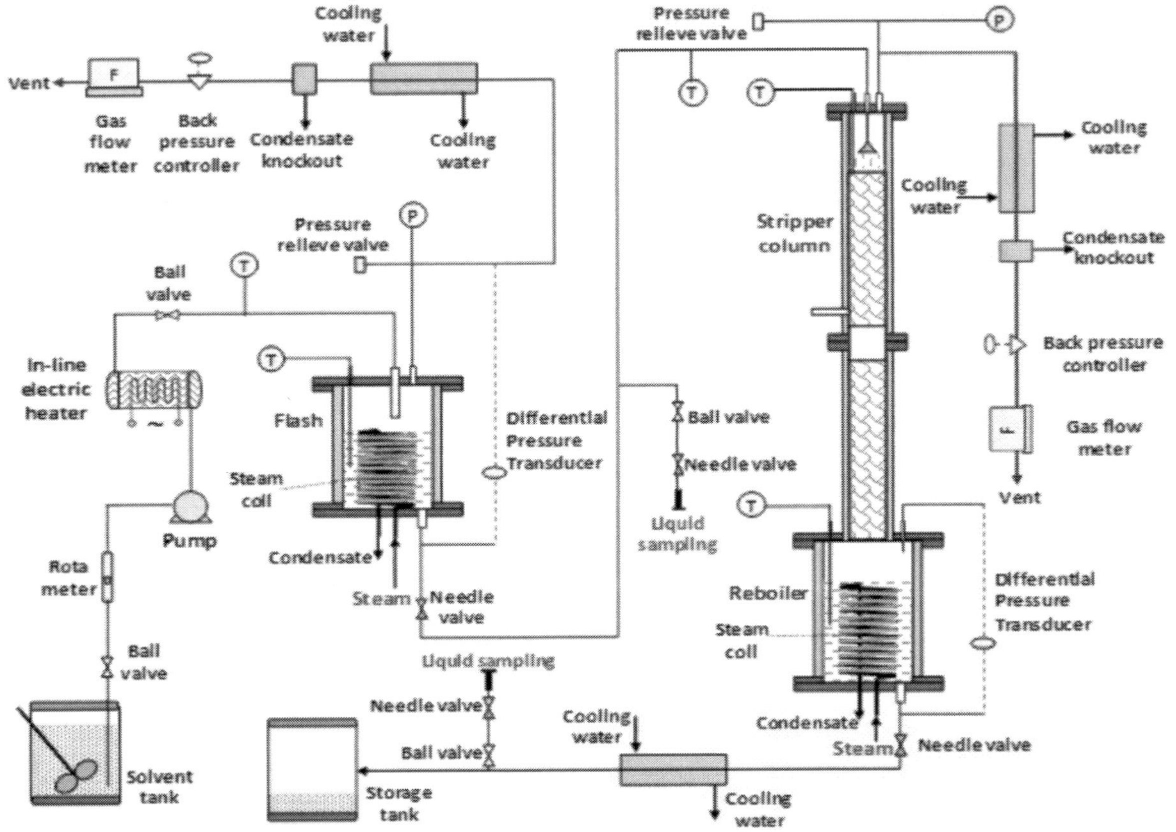

Figure 1. Schematic diagram of the laboratory flash and stripping system for CO_2 desorption

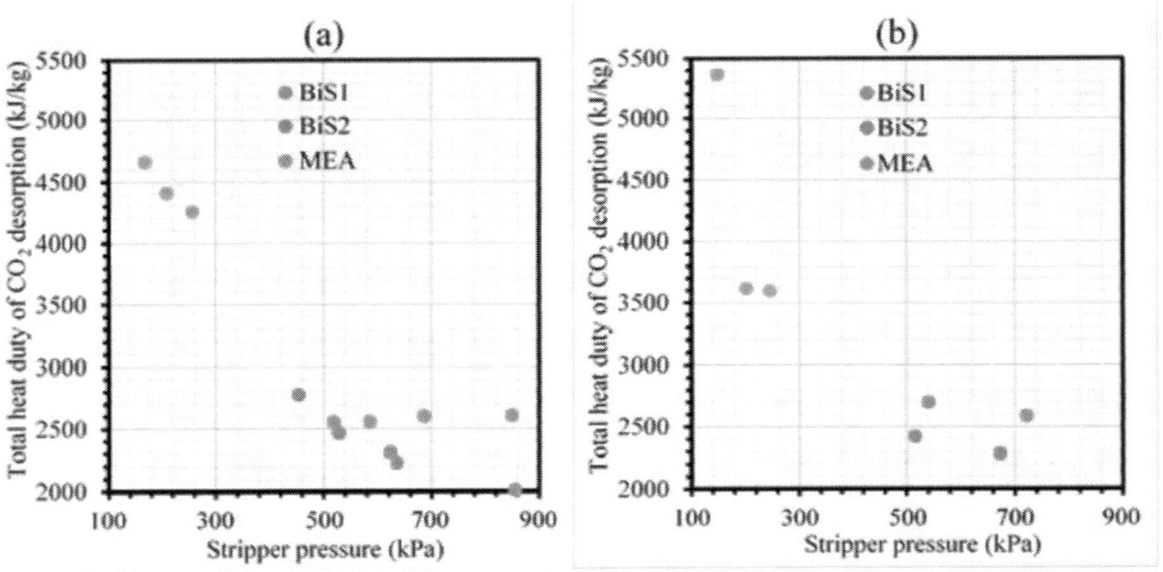

Figure 2. Heat duty of CO_2 desorption operating under (a) single stripper mode at the reboiler temperature of 150 °C for the biphasic solvents and 120 °C for MEA and (b) sequential flash and stripper mode at the flash/reboiler temperatures of 140/150 °C for the biphasic solvents and 110/120 °C for MEA.

576306 Feasibility of Polyethylenimine-Based Silica Sorbent for Post-Combustion CO_2 Separation

Feasibility of Polyethylenimine-Based Silica Sorbent for Post-Combustion CO_2 Separation

Yao Ma, Vishnu Sriram, and Joo-Youp Lee

Amino polymer-based adsorbents have been extensively investigated for post-combustion CO_2 separation for its low energy penalty as compared to amine solutions. However, most studies have focused on improving the CO_2 capacity, and have not reported the feasibility of separating CO_2 from the spent sorbents during the desorption phase.

In this study, the adsorption/desorption cyclic performances were investigated using polyethylenimine (PEI)-impregnated mesocellular silica foam (MCF) in a thermogravimetric analyzer (TGA) and a fixed-bed reactor. The sorbent could reach a working capacity of ~3.8 mmol/g (75 °C, 100% CO_2 for adsorption and 110 °C, 100% N_2 for desorption) using a multiple cyclic performance test. However, for the practical desorption phase, either temperature-vacuum swing or temperature swing need to be used. The adsorption capacity at ~75-80 °C is almost twice as much as that at 50 °C, which means the temperature control of the adsorber is critical for the maximum utilization of the sorbent. The temperature control of the reactor during the adsorption and desorption phases will be critical to achieving high CO_2 working capacity and minimum energy requirements. In addition, based on the performance results obtained from the fixed-bed reactor, the two desorption options will be compared for a techno-economic evaluation.

578375 Physico-Chemical Activation of Biochar Using Acoustic Treatment and Amine Functionalization for Efficient CO_2 Adsorption

For the last few decades, the emission of green-house gases especially CO_2, raises serious concerns since it is the major contributor to global warming and climate change. Biochars, produced from pyrolysis of woody biomasses, have highly porous structure with high surface area that makes them potential CO_2 adsorbents. The previous studies of our group suggested that CO_2 could be chemically fixed on biochar through reductive photocarboxylation under ambient conditions that can be considered an effective way to capture CO_2. These studies have also suggested that ultrasound irradiation can exfoliate the biochar structure, create new pores, and open the blocked pores, thus enhancing the biochar's chemical functionalization efficiency and subsequent CO_2 capture. Besides, the biochar surface consists of reactive oxygen functional groups (such as -COOH, C=O, -OH) that are susceptible to amine functionalization. Amination is the technique to incorporate nitrogen functionality (such as amine) to biochar structure that has important applications in CO_2 capture. The amine has a strong affinity for acidic CO_2 due to its basic nature and can react through acid-base interaction and boost adsorption efficiency significantly. Thus, the present study aims to maximize the CO_2 capture capacity of biochar by improving its carbonaceous structure through physical activation with low-frequency ultrasound and chemical functionalization using amines. The initial studies were conducted with pine-wood derived commercial biochar activated physically for 30 s under ultrasonic irradiation and then functionalized chemically with individual and blended 1°, 2° and 3° amines (diethanolamine (DEA), monoethanolamine (MEA), polyethylenimine (PEI), piperazine (PZ) and tetraethylenepentamime (TEPA)) subjecting to two different groups of activating agents (N-(3-dimethylaminopropyl-N'-ethylcarbodiimide hydrochloride)-HOBt (hydroxybenzotriazole and KOH). The work was further extended to investigate the role of different biomass origin on CO_2 capture employing Miscanthus, Switchgrass, Corn Stover, Sugarcane Bagasse, Sorghum, Wheat Straw, Rice Straw and Rice Husk as different biomass sources. The ultrasono-blended amination study revealed that biochar samples activated with EDC-HOBt-TEPA, EDC-HOBt-TEPA-MEA, KOH-MEA possess maximum adsorption capacities 2.04, 1.91 and 1.62 mmol/g respectively at 0.10 atm. CO_2 partial pressure and 70 °C compared with raw biochar (0.3 mmol/g). Comparing these results with subsequent biomass origin study, it was obtained that biochars prepared in-house from Miscanthus has highest possible adsorption capacity of 2.53 mmol/g (under similar experimental conditions as before) due to its increased specific surface area, high carbon and low ash content.

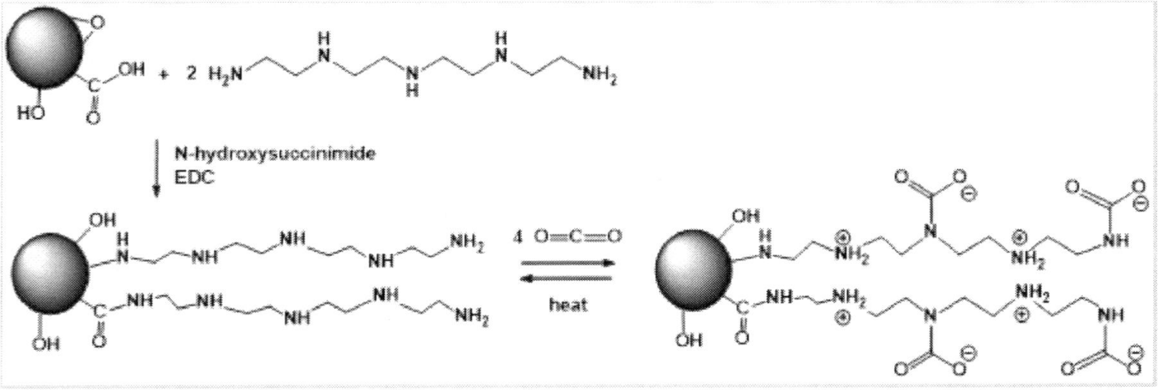

553095 Wastewater Treatment By Catalytic Wet Air Oxidation Process Based on Response Surface Methodology

Catalytic wet air oxidation (CWAO) is one of the most economical and environmental-friendly for oxidation decomposition of organic refractory materials, such as; phenol's compounds with aid of heterogeneous catalysts. In the present work an experimental work and simulation have been carried out to study the oxidation process for phenol's degradation in the effluent wastewater from the small paint manufacturing plant is located in the Iraq. Experimental set-up was designed and constructed to get reliable data. Lab-scale packed bed reactor was made of stainless steel with diameter of 25.0 mm and height of 150.0 mm.The catalyst is particles of activated carbon with diameter 0.5 mm. Pressure and temperature were monitored by high accuracy Transducers, while flow rate of air and water were measured by Rotameters. Gas Chromatography-Mass Spectrometer (GC-MS) was used to determine concentration of oxygen and phenol .Stochastic and deterministic methods were used to study reaction mechanisms and developing reaction kinetics models.Catalyst deactivation was taken into account. The selected process variables which affecting reactor performance are: initial concentration of phenol, reactor temperature, reactor pressure, air flow rate and water flow rate. Response Surface Methodology (RSM) has proven reliable search for selecting the optimal operating conditions.The oxidation process behaves as pseudo first order reaction with respect to phenol concentration. It has found that phenol's degradation could enhance by increasing of: temperature, pressure, and gas flow rate, while initial concentration of phenol and liquid flow rate give different images. High removal of phenol obtained is 95% at optimum operating conditions of: LHSV=7.0 h^{-1}, temperature=155°C, oxygen partial pressure= 1000.0 KPa and phenol concentration=10.0 mg/L. Down-flow regime is more effective than up-flow mode for improving degradation of phenol. Reasonable agreements have been found when compared the simulated results with the experimental data.

556358 Application of Ann and RSM Techniques in Optimal Parameter Evaluation for the Coag-Flocculation of Abattoir Effluent Using Crabshell Chitin

Crab-shell chitin has been applied in the coag-flocculation of abattoir waste water. The influence of process variables (pH, initial concentration, dosage, Temperature, and settling time) on the effluent final turbidity was investigated. Also, the accuracies of two modelling techniques (Response surface methodology, RSM and Artificial neutral network, ANN) in predicting the non-linear nature of the system were compared. The optimization result indicates a final turbidity of 4.96 mg/L (corresponding to 97.23 % removal efficiency) at pH = 6.7, dosage = 1.003 g/L, initial conc. = 182.2 mg/L, coagulation temp. = 345K and settling time of 36 min. Meanwhile, effluent pH was spotted as the most significant variable, with p-value <0.01%. Furthermore, the error analysis result portrayed the supremacy of ANN over RSM in data prediction accuracy as it signified lower error values (Mean square error, MSE = 13.11 and Absolute average relative deviation, AARD = 1.43%) when compared to those of RSM (MSE = 37.78, AARD = 5.93%). Thus, it was demonstrated that ANN is a better tool for optimization study of the present system.

*** Formatting Issue - Best Available Paper/Graphic ***

559760 Magnetic Graphene Oxide Modified with Chitosan and Cysteine Catalyst Was Synthesized for Removing (Pb^{2+}) Ions from Aqueous Solution

Synthesis and identification of modified magnetic Carbon Nanoparticle and study of its application in removing Lead ions (pb^{2+}) from aqueous solution

Ghazaleh. Ramezani [a], Bizhan. Honarvar [a,b*], Masoomeh. Emadi [a], Sayyed Esmaeil. Moradi [a]

[a] Department of Chemical Engineering, Marvdasht Branch, Islamic Azad University, Marvdasht, Iran

[b] Department of Civil Engineering , University of Texas at Arlington, Arlington, Texas

Abstract

magnetic graphene oxide modified with Chitosan and Cysteine (GO/Fe3O4/chi/cys) was synthesized for removing (pb^{2+}) ions from aqueous solution. The structural properties were analyzed by different techniques such as FE-SEM, FTIR and VSM. Physicochemical analysis such as effect of pH, contact time, adsorbent dosage and initial concentration of pb^{2+} was also studied. The results showed that the maximum capacity of GO/Fe3O4/chi/cys in Lead ions adsorption (Equilibrium concentration of 120 ppm) occurred at =5.75, =30min and Adsorbent dosage=0.1 gr. Maximum empirical adsorption capacity) was calculated 85.4 . The equilibrium data were analyzed using Langmuir and Freundlich models. The experimental data could be well described by the Langmuir isotherm . Also, the adsorption kinetics followed the mechanism of the pseudo-second-order equation ; So Pb^{2+}ions adsorption onto GO/Fe3O4/chi/cys may be monolayer and chemisorption.

Keywords:

Water treatment; Removal of pb^{2+}; Nano composite; magnetic graphene oxide; Adsorption Isotherms; Kinetics

1.INTRODUCTION

Lead is one of the most toxic metal ions and it's found in +2 oxidation state (pb^{2+}) in the inorganic form. Its presence in drinking water resources, can damage to the brain, blood composition, liver, kidney, nervous and reproductive system [1].So, removal of that, from aquatic systems has attracted the

attention of researchers. The limit of lead concentration in drinking water specified 0.05 from the world Health Organization [2]. Many processes were developed to treat water pollution with metal ions, such as membrane separation process, chemical precipitation, ion exchange, reverse osmosis and adsorption [3]. Among these commonly methods, adsorption processes are more popular than the others due to their high removal efficiency and the simplicity of the purification process [4].efficiency of adsorption processes have directly depended on the kind of adsorbent. So In recent decades, Nano materials based on carbon allotropes specially graphene based materials, such as graphene oxide (GO) and reduced graphene oxide(R-

GO) have been used as efficient and affordable adsorbents in adsorption processes [5]. Unique properties of graphene like; mechanical strength, great flexibility, water solubility and particularly large specific surface area are reasons that cause graphene ,as the famous material of the 21st century, to be used in synthesis various Nano composites[6,7].

In this paper, at first we synthesized magnetic graphene oxide modified with Chitosan and Cysteine (GO/Fe3O4/chi/Cys) as a novel adsorbent .The characterization of Nano composite were analyzed carefully by Field Emission *Scanning Electron Microscope* (FE-SEM), Fourier Transform Infrared Spectroscopy(FT-IR), and Vibrating Sample Magnetometer (VSM).then, we used it to remove lead ions from synthetic wastewater in a batch system.in the next step.

2. Experimental

2.1. Materials

Natural graphite powder, Pure ethanol , , Glacial acetic acid, , , ,sodium hydroxide and were purchased from merck,Germany.Acetic acid, glutaraldehyde, , , , , dapcco and Ethanolic were obtained from sigma-Aldrich, Germany. Ammonia (25% by weight) with a molecular weight of 17.03 was bought from Arvin Chem Co. Chitosan powder (CS, Mw 1.3 × 105 Da, 90% deacetylation) was supplied by Sinopharm Chemical Reagent Ltd (Shanghai, china).metal solutions were prepared by, Double distilled water from Aban Distillery Pasargad Co and Lead nitrate salt (Pb (NO3) 2) from Company Products Merck.

2.2. Preparation of GO/ Fe3O4/Chi/CS Nano composite

In this study, Go was prepared via hummers method [24].at first; 20 gr graphite powder and 40 gr $NaNo_3$ were added to 500 ml sulfuric acid solution at . While the mixture was stirred well, 120 gr KMno4 was added to the mixture gradually. Afterwards by using co-precipitation method GO was changed to GO/ Fe_3O_4 [25].For this purpose, 150 gr and 60 gr were dispersed in 100 ml Hcl solution (0.4 M). the solution that contained iron ions (with molar ratios between Fe2+ and Fe3+ 2:1) was added to 1100 ml the graphene oxide solution (0.01818 g/ml) Which was synthesized in the previous step. The mixture was stirred for 2 hours and then added immediately to 1.5 liters of ammonia solution (0.7 M). The ammonia solution was degasified farther with Argon gas For 15 minutes. The reaction system was mixed at 60 ^{oc} for 2 Hours and then stirred at room temperature for 12 hours. The black precipitate was collected and washed with Double distilled water until its PH reached about 7.the Magnetic graphene oxide(GO/ Fe3O4) was kept wet for the next step. For the last step of surface modification of Nano composite with cysteine; at first, glutaraldehyde (25% wt.) and Glacial acetic acid were suspended in pure ethanol and heated on the magnetic stirrer to ^{0c}. After that, added to pure ethanol and heated to 120 ° C. This solution was added to the first solution. They were mixed well. The resultant mixture was kept for 24 hours under reflux condition in an oil bath at ^{0c}.the green precipitate was filtered and washed with 50 ml hot ethanol and then dried. The prepared Nano composite (GO/ Fe3O4/Chi/CS) was used as adsorbent for removing of lead in a batch system.

3. Results and discussion

3.1. Characterization of GO/ Fe3O4/Chi/CS

3.1.1. Field Emission Scanning Electron Microscope (FE-SEM) Analysis

In order to study the morphology and structure of synthesized Nano composite at different stages of surface modification, Field Emission Scanning Electron Microscope (FE-SEM) Analysis Was investigated. For this purpose, FE-SEM instrument (The SIGMA VP model) was used in 25 kv and the Results was shown in **Fig.1.** As seen in **Fig.1.a**; Graphene oxides sheets had a completely porous and irregular structure with Particle size smaller than 200 nm. according to **Fig.1.b**, Fe_3O_4 nanoparticles with the size of about 60 nm were take placed successfully on the graphene oxides sheets.**Fig.1.c** represents a thin coating of

*** Formatting Issue - Best Available Paper/Graphic ***

chitosan at the adsorbent surface which causes a fundamental change in the adsorbent morphological structure. At this stage; the adsorbents surface seems relatively rough. According to **Fig.1.d** the final structure of the Nano composite is Non-uniform and heterogeneous.

Fig.1. FE-SEM images of GO/ Fe3O4/Chi/CS at different stages of surface modification

4. Conclusion

In this paper, a novel adsorbent of GO/ Fe_3O_4/Chi/CS composite was successfully prepared and investigated for the removal lead ions from aqueous solutions. GO/ Fe_3O_4/Chi/CS is a Super-Paramagnetic adsorbent and it can be separate by an external magnetic field so easily. The optimal conditions for adsorption of lead (II) ions were achieved at =5.75, =30min and Adsorbent dosage=0.1 gr. Maximum empirical adsorption capacity) was 85.4 . The adsorption isotherms could be explained by the Langmuir model and the adsorption kinetics data were well fitted by the pseudo-second order model (R^2 = 0.9996) Therefore the adsorption of lead (II) ions onto GO/Fe_3O_4/Chi/CS Nano composite is a monolayer, favorable (0 1). and chemical adsorption.

Acknowledgments

This research work was supported by the Islamic Azad University, Marvdasht, Iran.

References

[1] Hessam R., Mehdi., Afife K., Graphene-Based Polymer Nanocomposites, Polymerization Quarterly , Pages 86-107

[2] Mishra, P. C., and R. K. Patel. "Removal of lead and zinc ions from water by low costadsorbents." Journal of Hazardous Materials 168, no. 1 (2009): 319-325.

[3] Judd S., Jefferson B., Handbook of Membranes for Industrial Wastewater Recovery and Re-us, Elsevier Science & Technology Books, Pages 7-9.

[4] Mehta, D., Mazumdar, S., & Singh, S. K. (2015). Magnetic adsorbents for the treatment of water/wastewater—a review. *Journal of Water Process Engineering*, *7*, 244-265.

[5] Ye, N., Xie, Y., Shi, P., Gao, T. and Ma, J., 2014. Synthesis of magnetite/graphene oxide/chitosan composite and its application for protein adsorption. Materials Science and Engineering: C, 45, pp.8-14.

[6] Novoselov, K.S., Fal, V.I., Colombo, L., Gellert, P.R., Schwab, M.G. and Kim, K., 2012. A roadmap for graphene. Nature, 490(7419), pp.192-200.

[7] Novoselov, K.S., Geim, A.K., Morozov, S.V., Jiang, D., Zhang, Y., Dubonos, S.V., Grigorieva, I.V. and Firsov, A.A., 2004. Electric field effect in atomically thin carbon films. science, 306(5696), pp.666-669.

[8] Dreyer, D.R., Park, S., Bielawski, C.W. and Ruoff, R.S., 2010. The chemistry of graphene oxide. Chemical Society Reviews, 39(1), pp.228-240.

[9] Lerf, A., He, H., Forster, M. and Klinowski, J., 1998. Structure of graphite oxide revisited‖. The Journal of Physical Chemistry B, 102(23), pp.4477-4482.

[10] Pei, S. and Cheng, H.M., 2012. The reduction of graphene oxide. Carbon, 50(9), pp.3210-3228.

567356 Degradation of N-Containing Pollutants By a Fenton-like Reaction with Cu-Based Catalysts

Fenton reaction, which can generate hydroxyl radical (•OH) with strong oxidative capacity to decompose bio-refractory contaminants, is attracting more and more attention. Comparing with conventional homogeneous Fenton reaction (Fe^{2+}/H_2O_2), Cu based heterogeneous catalyst is a more promising candidate for practical application for no iron sludge and wider operation pH range.

In this work, $Cu@SiO_2$ with a plum pudding-like core shell structure was designed and reduced at different temperature to identify the catalytic roles of Cu^0, Cu^+ and Cu^{2+} species. $Cu@SiO_2$ reduced at 200 ¡ãC ($Cu@SiO_2$-R200) exhibited the highest performance, and showed a 100% rhodamine B (10 ppm) removal within 30 min. Meanwhile, the characterization results of CO-DRIFTS, UV-vis DRS, XPS and XANES suggested that Cu^+ concentration on the catalysts surface reached its maximum after reducing at 200 ¡ãC, which indicated that the Cu^+ was the primary active site and produced ·OH.

Comparing with Cu based catalysts supported on SiO_2, Cu/Al_2O_3 could exhibit higher degradation efficiency at the same reaction condition. However, Cu-doped mesoporous alumina showed almost no degradation effect on nitrobenzene, which is an important material in chemical industries related to pesticides, explosives, dyes and paper. We found that bimetallic Fe-Cu catalysts could exhibit high performance in degradation of nitrobenzene. 5Fe2.5Cu-Al_2O_3 showed a 100% nitrobenzene (100 ppm) removal within 1 h, which was superior to the nitrobenzene degradation efficiency over monometallic Fe and Cu catalysts. With combination of the catalytic performance and the characterization results of XPS (Fig. 1) and H_2-TPR, we proposed that the synergistic effect between Fe and Cu species played a vital role in promoting the reduction of Fe^{3+} to Fe^{2+}, thus enhancing the generation of hydroxyl radicals (•OH) and the degradation efficiency (Fig. 2).

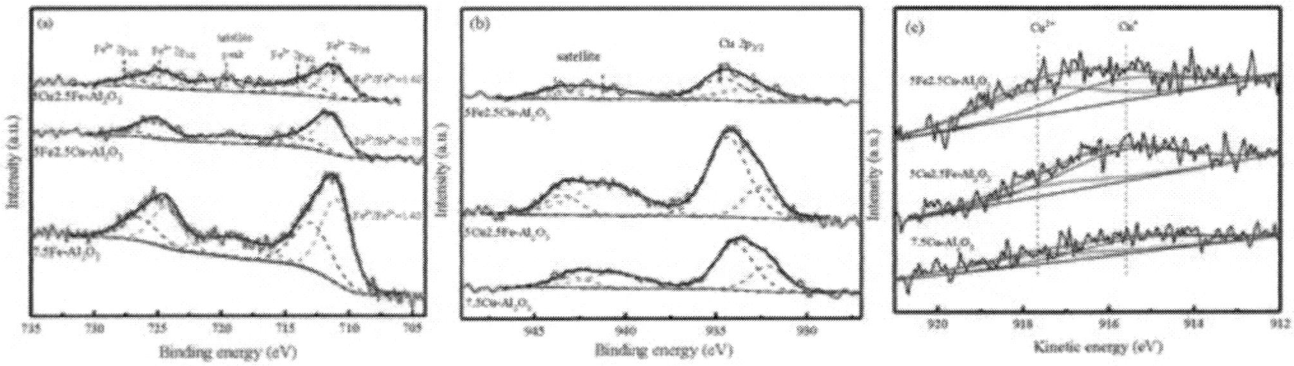

Fig. 1. XPS spectra of synthesized samples. (a) Fe 2p, (b) Cu 2p, (c) Cu LMM.

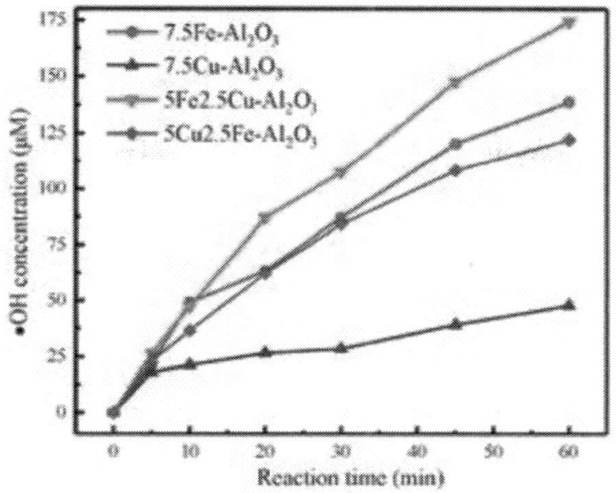

Fig. 2. The concentration of generated •OH using different catalysts (Reaction conditions: 323 K, 1000 ppm H_2O_2, 1 $g{\cdot}L^{-1}$ catalyst, 1500 ppm benzoic acid).

Since the Fenton-like reaction is usually followed by biological treatments, the biocompatibility and biodegradability of the intermediate is of vital importance in the whole remediation process. To clarify the reaction pathway, we carried out an on-line analysis of the intermediate during NB degradation in 30 min using a GC-MS system. Combining our results with previous works, a plausible reaction pathway for NB degradation for the Fe-Cu bimetallic Fenton-like system was illustrated by Fig. 3. The electrophilic attack on the benzene ring of NB was the predominant step. Besides, UV-Vis spectra (Fig. 4)and GC-MS results confirmed that no stable highly toxic 1,3-dinitrobenzene existed throughout the reaction, indicating that this catalytic system can also be readily coupled with a biodegradation process in the practical application.

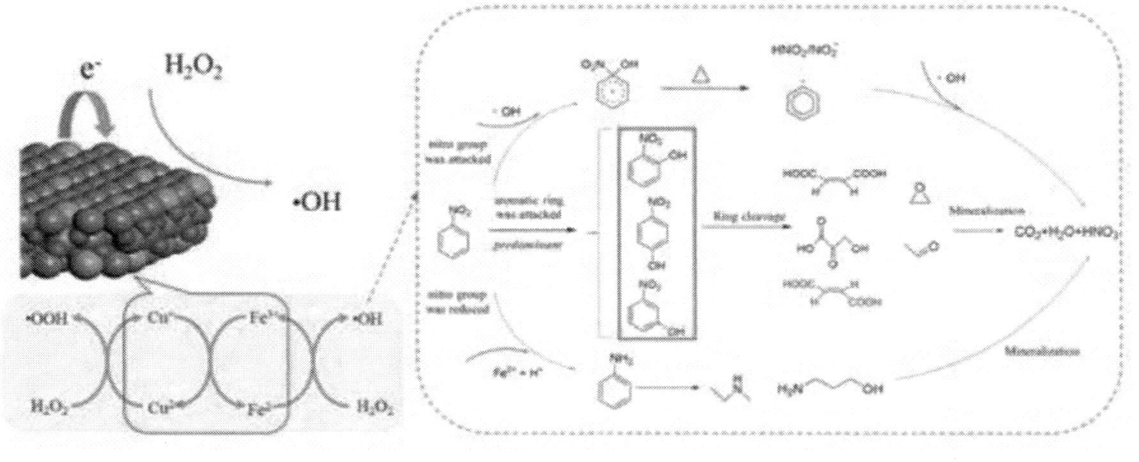

Fig. 3. Proposed possible mechanism of NB degradation over $5Fe2.5Cu\text{-}Al_2O_3$ in the presence of H_2O_2.

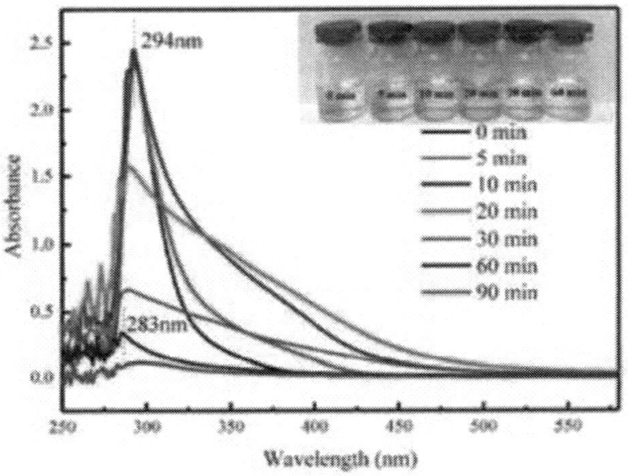

Fig. 4. UV-Vis absorption spectra over $5Fe2Cu\text{-}Al_2O_3$ during the reaction time

This work is supported by National Natural Science Foundation of China under grant No. 21676157.

567484 Modeling and Kinetics Study of Methyl Orange Degradation over a Cu/γ-Al_2O_3 Fenton-like Catalyst in Continuous Fixed Bed Reactor

Catalytic wet peroxide oxidation (CWPO) is recognized as one of the most promising and environmental-friendly techniques for reclamation of wastewater containing toxic and bio-refractory compound. It takes advantage of the great oxidative capacity of the hydroxyl radical (HO·) driving from hydrogen peroxide (H_2O_2) at moderate conditions (< 373 K, 0.1 - 0.5 MPa). Compared with the batch reactor, continuous fixed bed reactor (FBR) was a superior configuration for industrial application of CWPO process [1, 2]. However, a lack of study on the diffusion and kinetics for this liquid-solid system in the continuous FBR results in poor understanding of this process, which further hinders its scale-up and commercial use.

In this work, copper supported on alumina pellets (Cu/¦Ã-Al_2O_3) were synthesized by impregnation method, and the average diameter of pellets size was 2.3 mm. The as-prepared catalyst was denoted as 2.3Cu/¦Ã-Al_2O_3. The morphology of pellets exhibits good sphericity with a rough surface (Fig. 1(a)), however, the hardening phenomenon was more serious for the surface comparing with the inside of the pellet (Fig. 1(b - d)). In addition, Al, O and Cu elements are well-dispersed for the whole pellet (Fig. 1(e)).

A series of performance tests were conducted in an up-flow FBR for methyl orange (MO) degradation. The 2.3Cu/¦Ã-Al_2O_3 exhibited high activity and excellent leaching resistance during 200 h testing (leaching amount of Cu < 0.6 mg/L). The conversion of MO and COD equal to 97% and 92% was achieved over 2.3Cu/¦Ã-Al_2O_3 catalyst (reaction conditions: $pH_0 = 8.2$, flow rate = 2 mL/min, [catalyst] = 9.9 g, $[MO]_0 = 30$ ¦ÌM, $[H_2O_2]_0 = 29$ mM, and 50¡æ). The apparent reaction kinetics was investigated on the basis of the pseudo-first order simplification for the MO degradation and H_2O_2 decomposition, and the apparent E_a were calculated to be 21.0 and 13.8 kJ/mol, respectively. Compared with the intrinsic Ea studied for MO degradation (57.8 kJ/mol) and H_2O_2 decomposition (47.7 kJ/mol) in batch reactor, the mass transfer was significant for MO/H_2O_2/2.3Cu/¦Ã-Al_2O_3 system in the FBR.

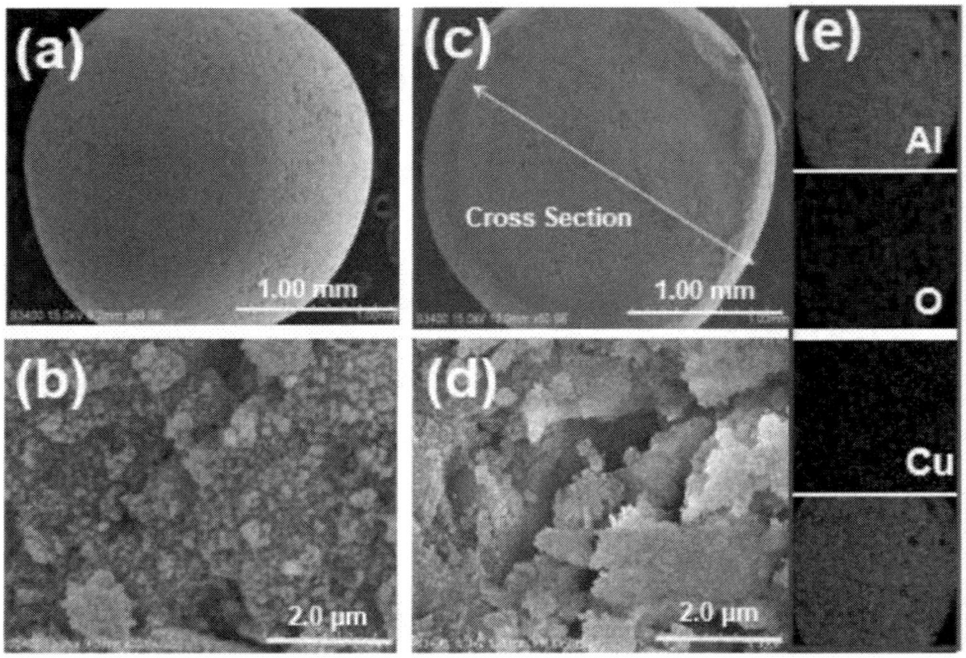

Fig.1 SEM and EDX images of 2.3Cu/¦Ã-Al_2O_3 catalysts: (a - b) SEM images of 2.3 Cu/¦Ã-Al_2O_3 pellet and corresponding enlarged image, (c - b) SEM images of cross-section for 2.3Cu/¦Ã-Al_2O_3 pellet and corresponding enlarged image, (e) EDX mappings of Al, Cu and O elements for cross-section of 2.3Cu/¦Ã-Al_2O_3 pellet.

In addition, the apparent kinetic study for MO oxidation and H_2O_2 decomposition were modeled with account for the external and internal diffusion limitation, which were quantified by measuring the apparent kinetic rate constant by changing the pellets sizes (mean d_p = 2.3, 3.5, 4.2 mm) and flow rate (1 - 20 mL/min). As shown in Fig. 2, modified experimental correlation ($Sh = 1.5Re^{1.10}Sc^{0.42}$) was proposed for the calculation of external mass transfer coefficient. Additionally, intrinsic kinetics was studied in the batch reactor to determine the effectiveness factor (¦Ç) in FBR to provide accuracy description of the experimental data.

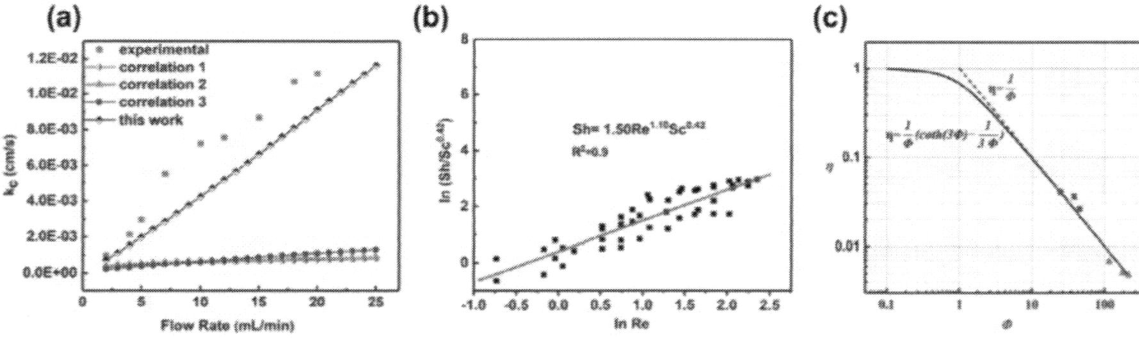

Fig. 2 (a) Values of k_c for MO oxidation, ($[H_2O_2]_0$ = 29 mM, $[MO]_0$= 30 ¦Ìm, flow rate = 2 - 20 mL/min, mean d_p = 2.3 mm, and [Catalyst] = 9.9 g). (b) Fitting results of the relationship between Sh, Re and Sc. (c) Calculated ¦Ç and ¦µ for H_2O_2 decomposition and MO oxidation (blue square: H_2O_2, red triangle: MO)

In summary, Cu/¦Ã-Al_2O_3 pellet catalysts prepared by impregnation method showed outstanding activity and stability against leaching for H_2O_2 decomposition and MO degradation in the FBR. Quantified analysis showed that H_2O_2 decomposition and MO degradation in the FBR was both controlled by the external and internal mass transfer resistance, especially the latter. The geometric size of the pellet catalysts exert a pronounced effect on the apparent kinetic of the process.

For the predication of apparent kinetic rate in the FBR, a combined diffusion-kinetic model was proposed. A modified empirical correlation for the external mass transfer at low Re was determined by fitting the experimental data in FBR and the effective factor (¦Ç) for internal mass transfer resistance was calculated according to the intrinsic kinetics measured in the batch reactor. This diffusion-kinetic model provided a reasonable description of the experimental data and can be expanded to other pollutants degradation, which provided a good solution for the CWPO process analysis and reactor design.

References

[1] Ioannou LA, Puma GL.Fatta-Kassinos D. Treatment of winery wastewater by physicochemical, biological and advanced processes: A review. *J. Hazard. Mater.* 2015; 286,343-368.

[2] Esteves BM, Rodrigues CSD.Madeira LM, in Applications of Advanced Oxidation Processes (AOPs) in Drinking Water Treatment, eds. Gil, Galeano,Vicente, Springer International Publishing, Cham, 2019,211-255.

(562bb) Semi-Pilot Plant for Tertiary Treatment of Domestic Wastewater using Algal Photo-Bioreactor, with Artificial Intelligence

Ahmed Karam[1], Mohamed K. Mostafa[2], Abdelsalam Elawwad[3], Khaled Zaher[3], Ahmed S. Mahmoud[4] and Robert W. Peters[5]

[1]Civil, Infrastructure Engineering and Management Program, Nile University, Egypt

[2]Faculty of Engineering and Technology, Badr University in Cairo (BUC), Cairo, Egypt

[3]Sanitary and Environmental Engineering Dept., Cairo University, 12613 Giza, Egypt

[4]Sanitary and Environmental Institute (SEI), Housing and Building National Research Center (HBRC), Egypt

[5]Department of Civil, Construction and Environmental Engineering, University of Alabama at Birmingham, Birmingham, AL, USA

Abstract

This study attempted to investigate the removal of biological oxygen demand (BOD), chemical oxygen demand (COD), total suspended solids (TSS), ammonia–nitrogen (NH_4–N), and total phosphorus (TP) from secondary treated domestic wastewater using algal photo-bioreactor. A semi-pilot plant was constructed and operated for 112 days under continuous flow conditions at Zenin wastewater treatment plant, Giza, Egypt (WWTP) which consists of an algal photo-bioreactor with an effective volume of 188 litters and a lamella settler. The removal of the studied parameters was studied at different hydraulic retention times (HRTs) and mixed liquor suspended solids (MLSS) concentrations. A continuous illumination of the photo-bioreactor was maintained using sunlight in the morning and incandescent lamps at night. The best overall organic and nutrients removal efficiency was recorded at HRT of 16.1 hours. Artificial neural network (ANN) with a structure of 3–10–1 was used to predict the BOD, COD, TSS, NH_4–N, and TP removal efficiencies. It was revealed that the ANN model adequately predicted the studied parameters removal efficiencies with r^2 greater than 90%.

Key words: Algal photo-bioreactor, Artificial Neural Networks (ANN), Domestic wastewater treatment, Tertiary treatment.

1. Introduction

Domestic wastewater can be defined as used water with dissolved and suspended solids discharged from residential units. It always contains a wide range of potential contaminants in different concentrations with high organic matter and may additionally contain harmful mixes and pathogenic microorganisms (Mara 2013). The conventional secondary treatment methods able to improve municipal wastewater attributes by reducing total solids (TS), chemical oxygen demand (COD), and biochemical oxygen demand (BOD). However, a shortcoming is potentially signified by less efficient performances in nutrient (Nitrogen and Phosphorus) removal. Nitrogen in wastewater exists in the form of ammonia, which is toxic to most aquatic life at high concentrations and can cause many environmental problems such as aquatic hypoxia and eutrophication (Camargo and Alonso 2006). As well as, excessive phosphorus can cause health problems and also speed up eutrophication(Liu et al. 2008, Correll 1998). The conventional treatment techniques based on physical and chemical methods may present either uneconomical or incompetent alternatives for nitrogen (N) or phosphorus (P) removal from wastewater (Elawwad et al. 2017). Therefore, searching for low cost eco-friendly method for nutrient removal is essential for further cleaning for the secondary effluents. Numerous studies have well documented the removal of nutrient from wastewater by microalgae (Oswald and Gotass 1957, Doran and Boyle 1979, Hashimoto and Furukawa 1989, Boelee et al. 2011, Abinandan and Shanthakumar 2015, Christenson and Sims 2011). Microalgae have the ability to accumulate toxic compounds and heavy metals into its biomass and found very effective for removing N and P from wastewater by consuming them for their growth (Zhu et al. 2013). Throughout the photosynthesis process, microalgae consume great amounts of carbon dioxide (CO_2) to produce glucose ($C_6H_{12}O_6$) and oxygen (O_2) which is required for aerobic bacteria to biodegrade organic contaminants and ammonia removal in an algal-bacterial integrated ecosystem. The use of photo-bioreactor algal-bacterial ecosystem as a post treatment for domestic wastewater can achieve great potential. It will decrease land requirement, aeration cost and produce high biomass for more biofuel production (Pittman et al. 2011). Nevertheless, the extreme growth of microalgae could constrain the growing of nitrifying bacteria (Choi et al. 2010). The microalgal growth is considerably affected by various factors such as pH, light, temperature, mixing, and concentration of dissolved oxygen while the hydraulic retention time (HRT) has a vital impact on biomass concentrations in the algal photo-bioreactor (Xu et al. 2015). Alternatively, biomass recycling can diminish the washout rates and empowers the system to be worked at shorter HRTs (Elawwad et al. 2017, Valigore et al. 2012). Artificial neural network (ANN) was used to drive relationships between input data (HRT, flow rate, and mixed liquor suspended solids (MLSS)) and output data (removal efficiencies of the tested parameters: BOD, COD, TSS, NH_4–N, and TP). The main aim of this study is to investigate the feasibility of using a semi-pilot plant for tertiary treatment of real secondary treated wastewater using algal photo-bioreactor in continuous feed conditions without mechanical aeration. Furthermore, the effect of HRT and loading rates of nutrient. The predominant microalgae species toward the end of the pilot investigation were identified. At last, ANN was utilized to signify the relationship between the input and output data.

2. Materials and Methods

A semi-pilot plant was made of 6 mm transparent acrylic material and installed in "ZENIN WWTP", Giza, Egypt. It consists of feeding tank (500 liters), algal photo-bioreactor (188 liters), and lamella settler (100 liters) with two biomass recycling pumps. Figure 1 show photographs of the semi-pilot plant. The algal photo-bioreactor was operated by the aid of two overhead stirrers worked at 120 rpm interchangeably. The light required for algae growth in the photo-bioreactor was obtained from sunlight during the day and a fluorescent light operated only during night. The lamps were associated with a photocell for "turning on" at evening time. From literature, it has reported that the average light intensity for algal growth should be not less than 200 μmol m^{-2} s^{-1} (Cheirsilp and Torpee 2012). The secondary treated wastewater produced from Zenin WWTP is feeding firstly by a submersible pump to the 500 L feeding tank, then into the photo-bioreactor by gravity after that discharged to the lamella settler for algal biomass settling. The average characteristics of the secondary treated wastewater were COD (60-70) mg/L, ammonia-nitrogen NH_4-N, (10-12) mg/L, and total phosphorus TP of (3-5) mg/L. After settling, the settled biomass is recycled back to enhance the biomass growth in the photo-bioreactor as shown in Figure 1c.

2.1. Experimental Work

2.1.1 Start-up operation

The algal photo-bioreactor was filled up with secondary treated wastewaters discharged from Zenin WWTP and cultivated with algal biomass from another high-rate algal pond experiment. The added algal biomass was mixed with the secondary treated wastewater inside the photo-bioreactor for 10 days before beginning the continuous feed.

Figure 1. Semi-pilot plant treatment tanks and settled algal-bacterial biomass: a) algal photo-bioreactor, b) lamella settler, and c) settled algal-bacterial biomass in the lamella settler

2.1.2. Continuous operation

The semi-pilot plant was operated in different HRTs 23.8, 16.1, 12.1, 8.5, and 3.9 hrs for 16, 18, 15, 20, and 22 days, respectively with fixed solids retention time (SRT) of 15 days as reported by (Karya et al. 2013, van der Steen et al. 2015). The total working days of continuous operation were 112 days at ambient temperature.

2.1.3. Sampling program

Samples were taken three times a week from the inlet and outlet of the semi-pilot plant to analyze different parameters. The collected samples were stored refrigerated (4°C) until tested on the same day in the National

Research Centre (NRC), Giza, Egypt. The analyzed parameters were COD, BOD, total suspended solids (TSS), total Kjeldahl nitrogen (TKN), ammonia-nitrogen (NH_4-N), and total phosphorus (TP). All analysis methods were performed based on the standard method for the examination of water and wastewater (Apha 1995). The algal species identification was carried out using a ZEISS AXIO Imager A2 microscope.

2.2. Artificial intelligence

An ANN model was developed with three main layers: an input, hidden, and output layers, to predict the removal efficiencies of the tested parameters (BOD, COD, TSS, NH_4-N, and TP). The input layer receives records from three experimental factors (HRT, MLSS, and flow rate), and the hidden layer consisted of ten neurons. The ANN model was structured as 3–10–1 (Figure 2). Equation 1 represents "tansig" transfer function which has been selected for the hidden layer, while Equation 2 represents "purlin" activation function which has been selected for the output layer. A cascade-forward backprop algorithm was used for data classification (Mahmoud et al. 2018). A comparison was conducted between the target and the output through calculating the mean squared error (MSE). Equation 3 was employed to adjust the weights and biases, where the MSE was passes back from the output node to the input layer. The network was trained using Levenberg–Marquartd method (trainlm), which is appropriate for small to medium-size networks(Mahmoud et al. 2019).

$$f(x) = x, \quad -\infty < f(x) < +\infty \qquad (1)$$

$$f(x) = \frac{e^x - e^{-x}}{e^x + e^{-x}}, \quad -1 \le f(x) \le 1 \qquad (2)$$

$$MSE = \frac{\sum_{i=1}^{N}(t_i - a_i)^2}{N} \qquad (3)$$

Where: N is the number of measured data, and a_i and t_i are the outputs and predicted target, respectively.

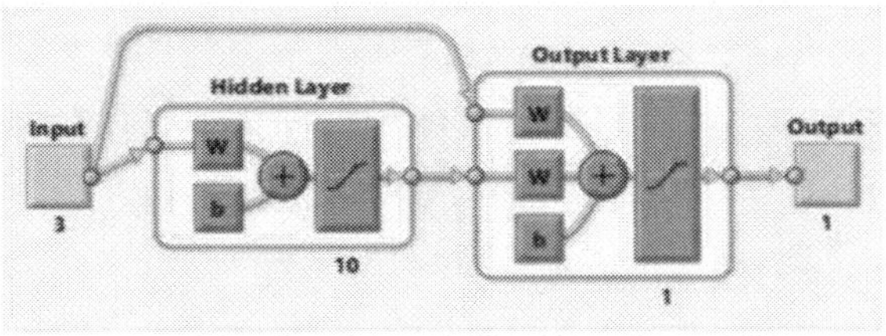

Figure 2. ANN of 3 – 10 – 1 structure used for the prediction of studied parameters removal efficiency

3. Results and Discussion

The results were taken and showed in different tested phases numbered from 1 to 5, matching to HRTs of 23.8, 16.1, 12.1, 8.5, and 3.9 h, respectively. After 21 days of continuous operation, the system of treatment has reached to the steady state. Table 1 presents the results of influent, effluent and average removal efficiencies of BOD_5, COD and TSS with MLSS for the five different phases. From phase 1 to 4, the removal efficiency of BOD_5 and COD was ranged between 55.8 to 67.5%, 51.4 to 67.3, respectively. However, the removal efficiency of TSS was ranged between 64.6 to 66.9%. During phase 5, the removal efficiency of BOD_5 and COD was significantly decreased to about 45.5 and 44.6%, respectively, as a result of the algal biomass wash-out and provided by the low concentration of MLSS as presenting in Table 1. The same can be deduced for the TSS results. During phase 5, the TSS removal efficiency decreased to 61.2% which reflect the poor ability of the lamella settler to settle the algal biomass in low HRT (3.9 h).

Table 1: MLSS concentrations, influent and effluent BOD_5, COD, and their average removal at the different phases

Phase	HRT	MLSS	$BOD_{5(inf)}$	$BOD_{5(eff)}$	Average BOD_5 removal efficiency	COD_{inf}	COD_{eff}	Average COD removal efficiency	TSS_{inf}	TSS_{eff}	Average TSS removal efficiency
Unit	Hours	g/l	mg/L	mg/L	%	mg/L	mg/L	%	mg/L	mg/L	%
1	23.8	1.32	46.7-62.1	21-27	55.8	60-74.4	25.7-32.9	57.9	27-30.3	9-11.1	65.3
2	16.1	1.14	46-48.3	13.6 -17.1	67.5	60-61.5	18.5-21.6	67.3	28-30.1	9.2-11	65.9
3	12.1	118	47-73	17.8-21.9	66.0	60-90	21.2-25.2	54.9	28.2-32.6	10-11	66.9
4	8.5	1.1	46.4-51.9	16.9-21.8	60.3	58.2-66	19.9-25.7	51.4	25.6-30.3	9.2-11	64.6
5	3.9	0.56	46.3-52.3	20.0-34.4	45.5	60-66	26.0-42.4	44.6	21-30.1	8.8-12	61.2

Figure 3 shows the average removal efficiencies of BOD_5, COD and TSS for the five different tested phases. The MLSS concentrations were regularly increased till reaching around 1200 mg/L by the end of the start-up period. Operating the system at SRT of 15 days was effective to sustain the MLSS concentrations while reducing HRTs down to 8.5 h, as shown in Table 1. At phase 5, the influent flow was increased to 1.16 m^3/d on average, the MLSS declined by almost half to 558 mg/L. This reduction may be due to the algal biomass wash-out. Data from this study can be comparable while Van der Steen et al. (2015) has operated SBR photo-bioreactor at 15 days SRT and 24 h HRT with biomass VSS concentration of 736 mg/L. The ammonia and phosphorus influent and effluent concentrations at the diverse tested phases were mentioned and displayed in Table 2 and Figure 4. For phases 2 and 4, the ammonia and phosphorus removals were above 85% and 60%, respectively. However, during phase 5, a significant decrease in the ammonia removal efficiency was observed to be 63.1% due to the algal biomass wash-out. The algal photo-bioreactor showed best performance at phase 2 (HRT 16.1 h) with an average removal efficiency of 90.2% and 85.6% for ammonia and phosphorus, respectively. The system accomplished great outcomes for phosphorus removal as the phosphorus effluent concentrations did not surpass 2 mg/L.

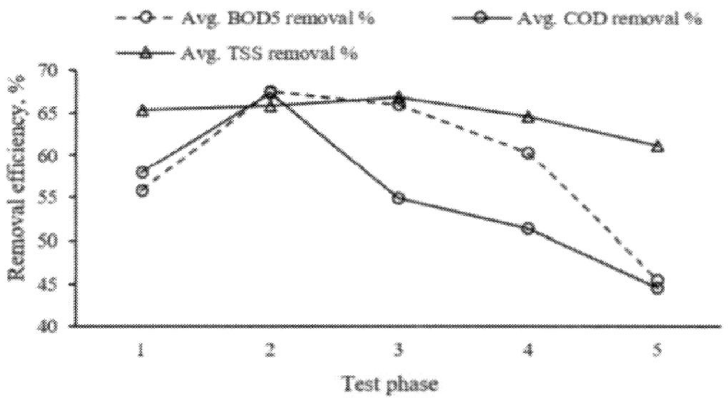

Figure 3. Average BOD_5, COD and TSS removal efficiencies

Table 2: Influent and effluent ammonia and phosphorus and their average removal at different phases.

Phase	HRT	Influent NH_4-N	Effluent NH_4-N	Average NH_4-N removal efficiency	Influent P	Effluent TP	Average TP removal efficiency
Units	Hours	mg/L	mg/L	%	mg/L	mg/L	%
1	23.8	10.8-12.4	1.6-3.7	78.9	3.7-5.8	0.9-1.8	71.6
2	16.1	6.5-10.4	0.6-1.1	90.2	3.0-3.7	0.2-0.7	85.6
3	12.1	8.2-12.5	1.1-1.7	86.3	3.2-5.6	1.4-1.8	61.2
4	8.5	7.4-11.7	1.2-1.6	85.7	2.8-5.6	0.5-1.9	72.1
5	3.9	7.9-10.8	2.3-4.8	63.1	2.6-5.2	0.5-2.5	59.7

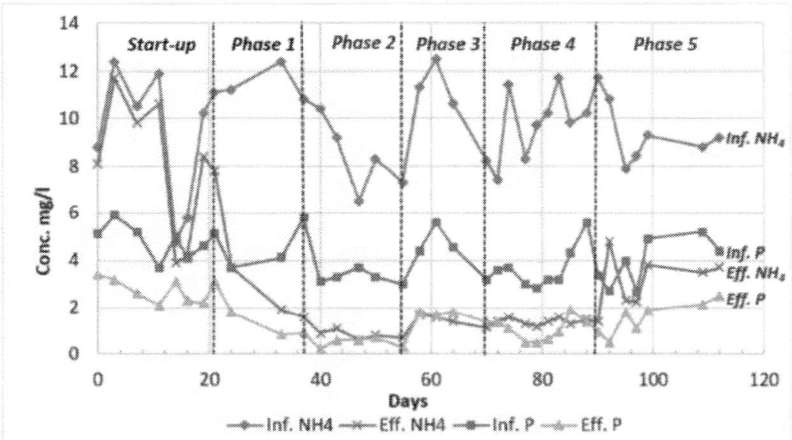

Figure 4. Influent and effluent ammonia and phosphorus concentrations in the influent and effluent of photo-bioreactor (Units are in mg N/l and mg P/l for ammonia and phosphorus, respectively)

At the end of the study, samples were taken from the semi-pilot photo-bioreactor for the identification of the predominant algal species. Scenedesmus, Spirulina, Spirulina and Anomoeonels were observed as shown in figure 5. It was mentioned that Scenedesmus sp. has quick development rates and dominates in continuous microalgal-based treatment systems (Munoz and Guieysse 2006).

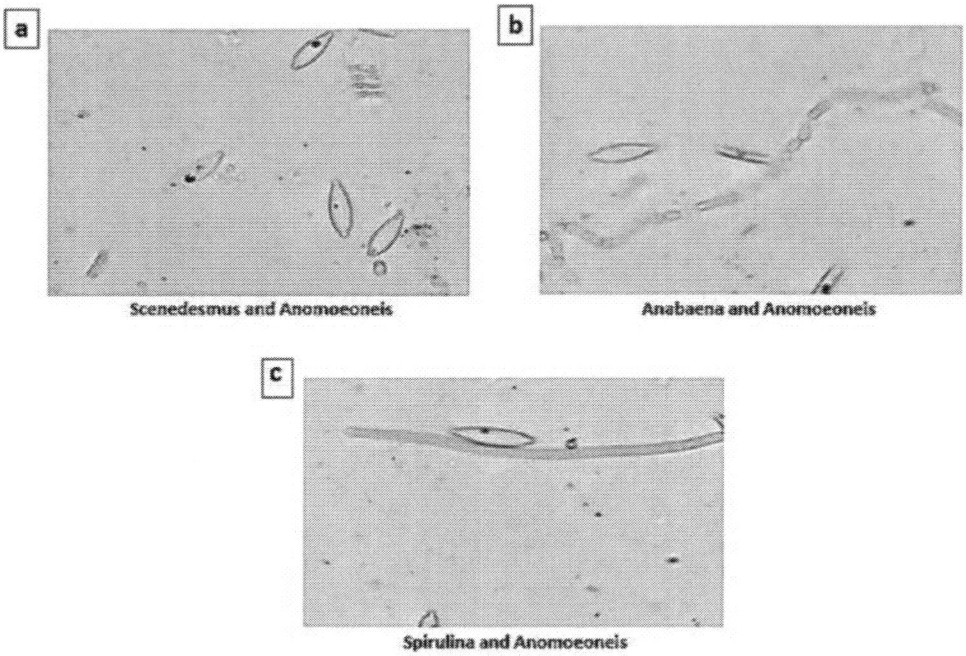

Figure 5. Dominant algal species at the end of the study: a) scenedesmus and anomoeoneis, b) anabaena and anomoeoneis, and c) spirulina and anomoeoneis (40x magnification)

3.2. Artificial intelligence

3.1.1. Training and validation performance

The magnitudes of the gradient for all tested parameters (BOD, COD, TSS, NH4-N, and TP) ranged from 0.39279 to 23.6078 (Figure 6). Since, these values exceeded the least error level of (1 x 10 – 5), the training process was terminated after 6 epochs. For all tested parameters, the MSE of the training phase decreased with the increase in epochs (Figure 7). The MSE of the validation phase increased after epoch 0, indicating the data overfitting. The test and validation curves increased in an analogous way with the number of iterations. The best validation performance was 10.6526, 5.9537, 0.98633, 4.6835, and 32.7335 at epoch 0 for BOD, COD, TSS, NH4-N, and TP, respectively.

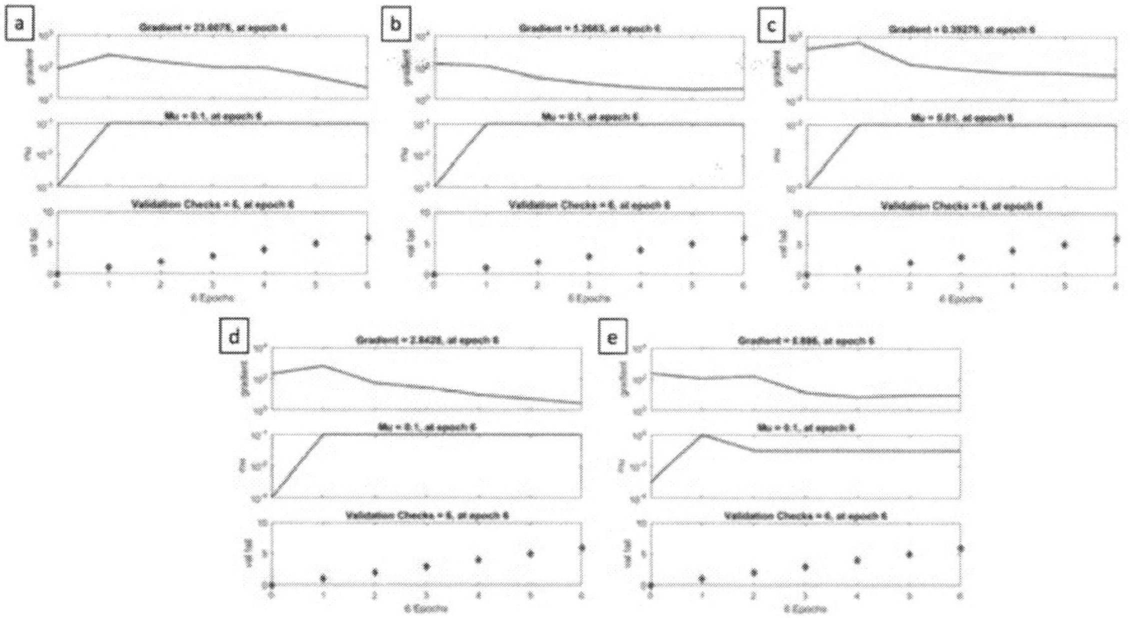

Figure 6. Training performance: (a) BOD, (b) COD, (c) TSS, (d) NH_4-N (e) TP

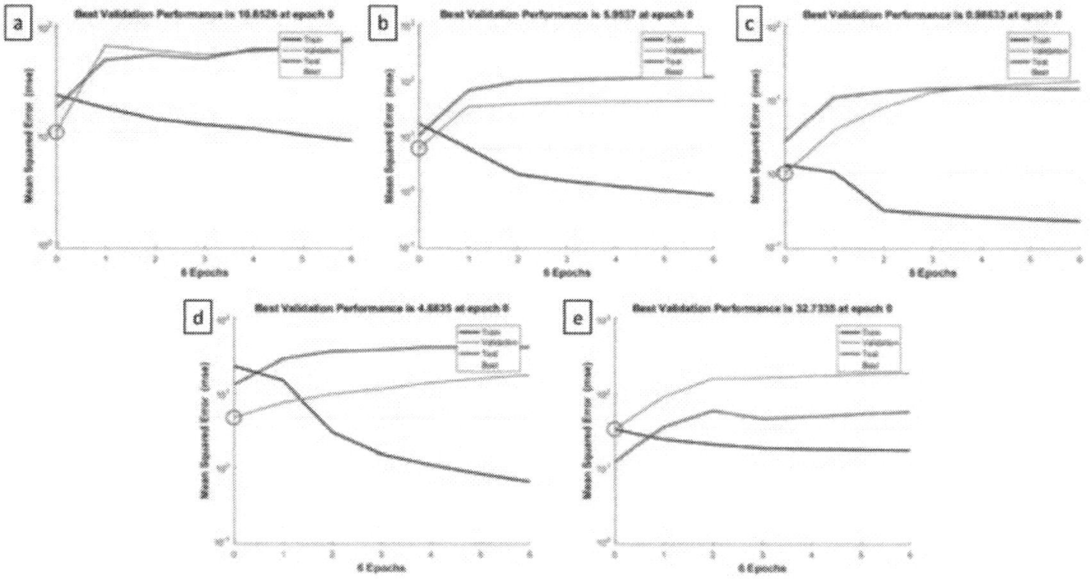

Figure 7. Best validation performance: (a) BOD, (b) COD, (c) TSS, (d) NH_4-N (e) TP

3.1.2. Regression plot

Figure 8 shows a relation between the coefficient of determination (r^2) and outputs and targets of the generated network. For all tested parameters, a good fit was detected between the output and the target data for the three phases: training, validation and test, with r^2-values ranged from 0.90 to 0.97, 0.93 to 0.99, and 0.90 to 0.98, respectively. The overall r^2-value for all tested parameters ranged from 0.91 to 0.98, indicating the model reliability.

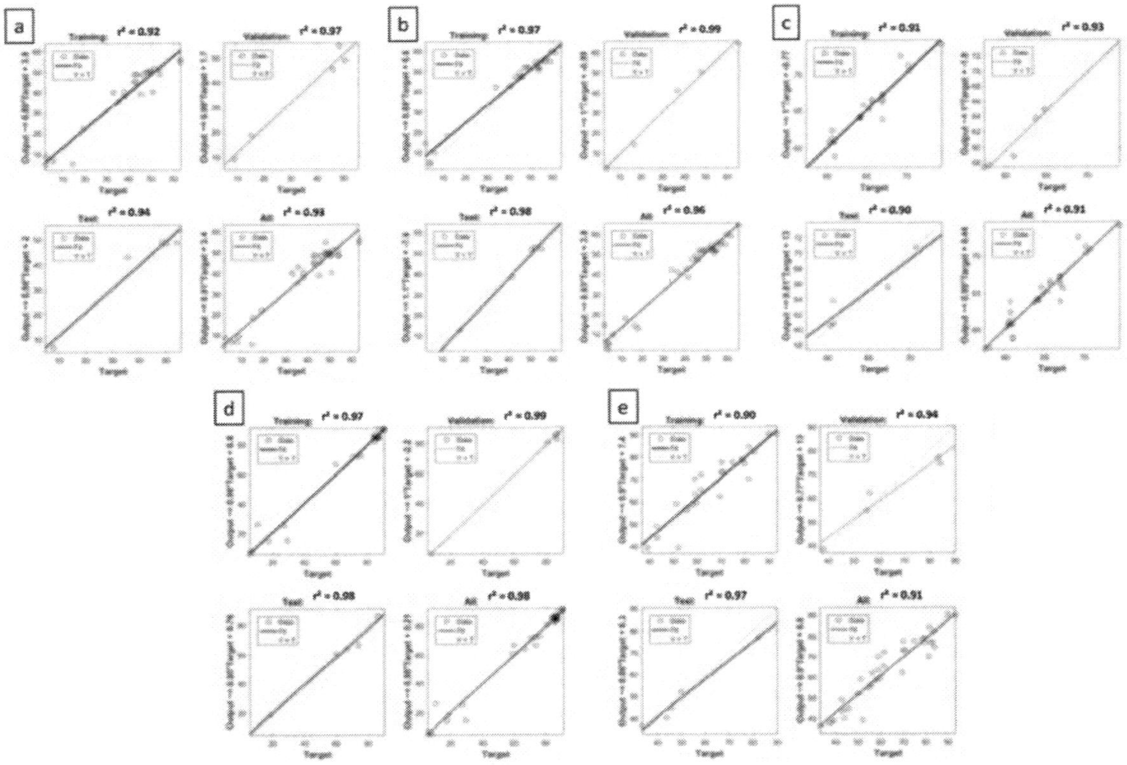

Figure 8. Regression plot between target and output: (a) BOD, (b) COD, (c) TSS, (d) NH_4-N (e) TP

4. Conclusions

This paper investigated the applicability of using a semi-pilot algal photo-bioreactor for tertiary treatment of secondary treated wastewater in a real WWTP. The system was tested at a fixed SRT of 15 days and at different HRTs. The best performance was observed at HRT of 16.1 h for both nutrient and organic removal, where the BOD, COD, TSS, NH4-N, and TP removal efficiencies reached 67.5, 67.3, 65.9, 90.2, and 85.6%, respectively. However, for a more economical design, HRTs can be shortened to 8.5 h while maintaining removal efficiencies above 60, 51, 64, 85, and 72%, respectively. It was observed that Scenedesmus was the most predominant species. Under the studied conditions, ANN was able to accurately predict the removal efficiency of tested parameters with r^2 value ranged from 0.91 to 0.98. In conclusion, algal-bacterial reactors proved to be effective in domestic wastewater treatment, so future research should concentrate on system optimization and scaling-up.

References

Mara, D. (2013) Domestic wastewater treatment in developing countries, Routledge.
Camargo, J.A. and Alonso, Á. (2006) Ecological and toxicological effects of inorganic nitrogen pollution in aquatic ecosystems: a global assessment. Environ Int 32(6), 831-849 - 0160-4120.
Liu, Y., Villalba, G., Ayres, R.U. and Schroder, H. (2008) Global Phosphorus Flows and Environmental Impacts from a Consumption Perspective. Journal of Industrial Ecology 12(2), 229-247.
Correll, D.L. (1998) The Role of Phosphorus in the Eutrophication of Receiving Waters: A Review. Journal of Environmental Quality 27(2), 261-266.
Elawwad, A., Karam, A. and Zaher, K. (2017) Using an Algal Photo-Bioreactor as a Polishing Step for Secondary Treated Wastewater. Polish Journal of Environmental Studies 26(4), 1493-1500.
Oswald, W. and Gotass, H. (1957) Photosynthesis in sewage treatment. Trans. Amer. Soc. Civil Engrs.;(United States) 122.
Doran, M.D. and Boyle, W.C. (1979) Phosphorus removal by activated algae. Water Research 13(8), 805-812 - 0043-1354.
Hashimoto, S. and Furukawa, K. (1989) Nutrient removal from secondary effluent by filamentous algae. Journal of Fermentation and Bioengineering 67(1), 62-69 - 0922-0338X.
Boelee, N.C., Temmink, H., Janssen, M., Buisman, C.J.N. and Wijffels, R.H. (2011) Nitrogen and phosphorus removal from municipal wastewater effluent using microalgal biofilms. Water Research 45(18), 5925-5933 - 0043-1354.
Abinandan, S. and Shanthakumar, S. (2015) Challenges and opportunities in application of microalgae (Chlorophyta) for wastewater treatment: a review. Renewable and Sustainable Energy Reviews 52, 123-132 - 1364-0321.
Christenson, L. and Sims, R. (2011) Production and harvesting of microalgae for wastewater treatment, biofuels, and bioproducts. Biotechnology advances 29(6), 686-702 - 0734-9750.
Zhu, L., Wang, Z., Shu, Q., Takala, J., Hiltunen, E., Feng, P. and Yuan, Z. (2013) Nutrient removal and biodiesel production by integration of freshwater algae cultivation with piggery wastewater treatment. Water Research 47(13), 4294-4302 - 0043-1354.
Pittman, J.K., Dean, A.P. and Osundeko, O. (2011) The potential of sustainable algal biofuel production using wastewater resources. Bioresource technology 102(1), 17-25 - 0960-8524.
Choi, O., Das, A., Yu, C.P. and Hu, Z. (2010) Nitrifying bacterial growth inhibition in the presence of algae and cyanobacteria. Biotechnology and bioengineering 107(6), 1004-1101 - 0006-3592.
Xu, M., Li, P., Tang, T. and Hu, Z. (2015) Roles of SRT and HRT of an algal membrane bioreactor system with a tanks-in-series configuration for secondary wastewater effluent polishing. Ecological Engineering 85, 257-264 - 0925-8574.
Valigore, J.M., Gostomski, P.A., Wareham, D.G. and O'Sullivan, A.D. (2012) Effects of hydraulic and solids retention times on productivity and settleability of microbial (microalgal-bacterial) biomass grown on primary treated wastewater as a biofuel feedstock. Water Research 46(9), 2957-2964 - 0043-1354.
Cheirsilp, B. and Torpee, S. (2012) Enhanced growth and lipid production of microalgae under mixotrophic culture condition: effect of light intensity, glucose concentration and fed-batch cultivation. Bioresource technology 110, 510-516 - 0960-8524.
Karya, N., Van der Steen, N.P. and Lens, P.N.L. (2013) Photo-oxygenation to support nitrification in an algal–bacterial consortium treating artificial wastewater. Bioresource technology 134, 244-250 - 0960-8524.
van der Steen, P., Rahsilawati, K., Rada-Ariza, A.M., Lopez-Vazquez, C.M. and Lens, P.N.L. (2015) A new photo-activated sludge system for nitrification by an algal-bacterial consortium in a photo-bioreactor with biomass recycle. Water Science and Technology 72(3), 443-450 - 0273-1223.
Apha, A. (1995) WEF, 1998. Standard methods for the examination of water and wastewater 20.
Mahmoud, A.S., Mostafa, M.K. and Nasr, M. (2018) Regression model, artificial intelligence, and cost estimation for phosphate adsorption using encapsulated nanoscale zero-valent iron. Separation Science and Technology 54(1), 13-26.
Mahmoud, A.S., Ismail, A., Mostafa, M.K., Mahmoud, M., Ali, W. and Shawky, A.M. (2019) Isotherm and kinetic studies for heptachlor removal from aqueous solution using Fe/Cu nanoparticles, artificial intelligence, and regression analysis. Separation Science and Technology, 1-13.
Munoz, R. and Guieysse, B. (2006) Algal–bacterial processes for the treatment of hazardous contaminants: a review. Water Research 40(15), 2799-2815 - 0043-1354.

569663 *In Situ* Treatment of Cyanotoxins in Water Sources Using Fiber-Immobilized TiO_2/Fe Nanoparticle Catalysts

Due to climate change, nutrient runoff from agriculture, and worsening eutrophication of water sources, harmful algal blooms (HABs) are globally increasing in frequency and distribution.[1,2] HABs are cyanobacteria that accumulate biomass and produce cyanotoxins such as microcystin-LR; 25 to 75% of HABs are estimated to be toxic.[3,4] Cyanotoxins, like microcystin-LR, are highly toxic and have a median lethal dose (LD_{50}) of about 50 μg/kg of body weight, which is equivalent to that of sarin gas.[4] Because of nutrient pollution, cyanobacteria and cyanotoxins are becoming more prevalent and are severely damage ecosystems; cyanobacteria have negative impacts on ecosystem functions, such as organism relationship disturbances, biodiversity changes, light conditions, and oxygen concentrations.[4] Several animal studies have shown evidence of microcystins exhibiting tumor promoting properties.[4] Cyanobacteria and cyanotoxins adversely impact human health and promote negative health effects like liver damage, immunotoxicity, and neurotoxicity.[5] Annually, the U.S. alone spends $2.2-4.6B on methods, including filtration, flocculation, coagulation, or sedimentation, to battle the effects of HABs.[1,3,6] However, these methods are only temporary solutions, and key disadvantages associated with these techniques include detrimental environmental impacts and inefficiency.[1]

Currently, there is no known method that adequately mitigates both HABs and cyanotoxins present in water sources. Studies have shown that both titanium dioxide (TiO_2) and iron nanoparticles can effectively inactivate cyanobacteria while also reducing available nutrients for algal bloom formation.[1,7] For cyanotoxin degradation, iron and titanium dioxide nanoparticles have primarily been studied. Although studies have not focused on cyanotoxin degradation by the combination of iron and TiO_2, research has shown synergistic effects between iron and TiO_2 for photocatalytic degradation of organic pollutants such as methylene blue and hydroquinone.[8–10] One concern is freely-dispersed nanoparticles in water sources and the effects on aquatic organisms. While freely-dispersed TiO_2 can cause harmful effects on some desirable aquatic species, iron nanoparticles have minimal toxicity effects towards beneficial organisms.[1,7] Therefore, a novel in situ treatment approach could be the immobilization of TiO_2/Fe nanoparticle catalyst on a retrievable fiber, which would allow light activation of nanoparticle catalyst for the deactivation of cyanobacteria and degradation of cyanotoxin.

This study will focus on the degradation of cyanotoxins. Continued work for the following will be discussed specifically: testing commercially available and experimentally synthesized TiO_2 and Fe nanoparticles for cyanotoxin degradation, immobilizing nanoparticles on fiber, and the evaluation of in situ fiber treatment of cyanotoxins.

References

1. Marsalek, B. *et al.* Multimodal action and selective toxicity of zerovalent iron nanoparticles against cyanobacteria. *Environ. Sci. Technol.* **46,** 2316–2323 (2012).

2. He, X. *et al.* Toxic cyanobacteria and drinking water: Impacts, detection, and treatment. *Harmful Algae* **54,** 174–193 (2016).

3. Meng, X., Savage, P. E. & Deng, D. Trash to Treasure: From Harmful Algal Blooms to High-

Performance Electrodes for Sodium-Ion Batteries. *Environ. Sci. Technol.* **49,** 12543–12550 (2015).

4. Bláha, L., Babica, P., Maršálek, B. & Ludek Bláha, A. Toxins produced in cyanobacterial water blooms-toxicity and risks. *Interdisc Toxicol* **2,** 36–41 (2009).

5. Tucek, J. & Zboril, R. Multimodal Action and Selective Toxicity of Zerovalent Iron Nanoparticles against Cyanobacteria. (2012).

6. Meglic, A., Pecman, A., Rozina, T., Leštan, D. & Sedmak, B. Electrochemical inactivation of cyanobacteria and microcystin degradation using a boron-doped diamond anode — A potential tool for cyanobacterial bloom control. *J. Environ. Sci. (China)* **53,** 248–261 (2017).

7. Bessa Da Silva, M., Abrantes, N., Nogueira, V., Gonç Alves, F. & Pereira, R. TiO 2 nanoparticles for the remediation of eutrophic shallow freshwater systems: Efficiency and impacts on aquatic biota under a microcosm experiment. *Aquat. Toxicol.* **178,** 58–71 (2016).

8. Mazille, F., Schoettl, T. & Pulgarin, C. Synergistic effect of TiO2 and iron oxide supported on fluorocarbon films. Part 1: Effect of preparation parameters on photocatalytic degradation of organic pollutant at neutral pH. *Appl. Catal. B Environ.* **89,** 635–644 (2009).

9. Wang, Q., Xu, S. & Shen, F. Preparation and characterization of TiO2 photocatalysts co-doped with iron (III) and lanthanum for the degradation of organic pollutants. *Appl. Surf. Sci.* **257,** 7671–7677 (2011).

10. Wang, X., Tang, Y., Leiw, M. Y. & Lim, T. T. Solvothermal synthesis of Fe-C codoped TiO2 nanoparticles for visible-light photocatalytic removal of emerging organic contaminants in water. *Appl. Catal. A Gen.* **409–410,** 257–266 (2011).

572305 Degradation of Pharmaceutical Compound Diclofenac in Aqueous Solution By Horizontal Continuous-Flow Dielectric Barrier Discharge

Degradation of pharmaceutical compound diclofenac in aqueous solution by horizontal continuous-flow dielectric barrier discharge

Yinhai Yang, Lecheng Lei, Bin Yang *

Key Laboratory of Biomass Chemical Engineering of Ministry of Education, College of Chemical and Biological Engineering, Zhejiang University, Hangzhou, P. R. of China, 310027

keyangb@zju.edu.cn

Diclofenac (DCF)£¬a recalcitrant pharmaceutical residue is difficult to be removed by traditional treatments in water. This paper investigates degradation of diclofenac in a horizontal continuous-flow dielectric barrier discharge (DBD) system. Some factors, such as peak voltage, discharge distance and initial concentration were studied to explore influence mechanism for degradation efficiency and yield of diclofenac. As the peak voltage increased from 12 to 15kV, the degradation rate increased, while yield increased first then decreased. While the degradation rate increased with the reduction in discharge distance. At lower initial concentration, the degradation rate is higher with same time treatment. After 30 min discharge treatment, removal of DCF can reach 97.6% and the degradation yield was 0.245g/kWh for the initial concentration of 1mg/l. The decomposition of DCF followed first-order reaction and the rate constant respectively were $0.0893 min^{-1}$, $0.0604 min^{-1}$, $0.0476 min^{-1}$, $0.0428 min^{-1}$ at the initial concentration of 1mg/l, 10mg/l, 20mg/l, 40mg/l. Degradation intermediates during discharge treatment was identified by LC-MS and the possible degradation pathway was proposed. The total organic carbon (TOC) remained high and only a fraction of DCF was mineralized when DCF was almost removed completely. The results confirmed DCF can be removed efficiently and mineralized inefficiently, which means that DBD is expected to be a promising water treatment technology.

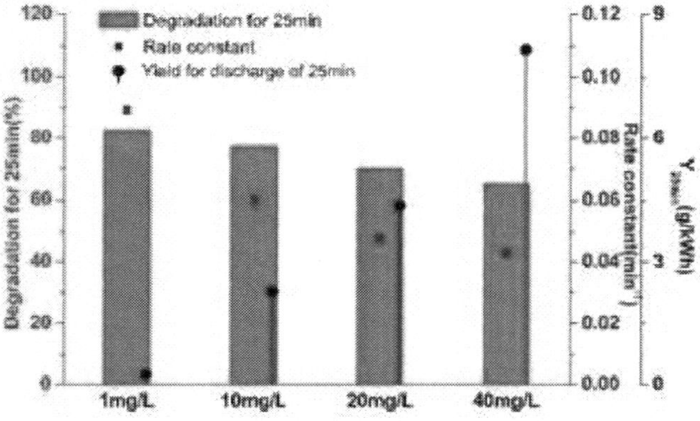

Acknowledgement:

This work was supported primarily by the National Nature Science Foundation of China (NSFC) (No. 21878271, and 21878270).

References:

1. Magureanu, M. . (2015). Degradation of pharmaceutical compounds in water by non-thermal plasma treatment. *Water Research, 81*, 124-136.

2. Rodriguez-Narvaez, O. M. Peralta-Hernandez, J. M., (2017). Treatment Technologies for Emerging Contaminants in water: A review. *Chemical Engineering Journal*, **323**, 361-380.

579897 Experimental Study on the Removal of Polycyclic Aromatic Hydrocarbons Contaminants from Surface Water

Polycyclic aromatic hydrocarbons (PAHs) are emerging water contaminates. Activated carbon has been proven to be advantageous as a suitable absorbent for polycyclic aromatic hydrocarbons (PAHs) removal from surface water. However, published PAHs-AC adsorption isotherm data are spare and not included all 7 priority PAHs in EPA standard. In this research, adsorption isotherm (24 hours) for 16 priority PAHs, including all 7 priority EPA PAHs, from 2 mg dissolved organic carbon (DOC)/L solution onto granular activated carbon (GAC) at 25°C were performed by batch shake experiments. Adsorption isotherm at 24 hours data were fitted with Brunauer–Emmett–Teller (BET) isothermal models in the range of PAHs concentrations studied (0.1-0.8 ppb). In addition, the oxidation effects of different doses of potassium permanganate (PM)/ sodium bisulfite (BS) on PAHs removal were investigated. The research results provide valuable guidance for the removal of PAHs from water.

582971 Kinetics and Equilibrium Study of New Activated Carbon Materials for Acetaldehyde Removal

Kinetics and equilibrium study of new activated carbon materials for acetaldehyde removal

Simão P. Cardoso[1], Inês S.S. Ferreira[1], Inês Matos[2], Maria Bernardo[2], Márcia Ventura[2], Isabel M. Fonseca[2], Anabela Valente[1], Carlos M. Silva[1]

[1] *CICECO, Department of Chemistry, University of Aveiro, 3810-193 Portugal*

[2] *Departamento de Química, CQFB, Universidade Nova de Lisboa, Faculdade de Ciências e Tecnologia, 2825-114 Caparica, Portugal*

**simaocardoso@ua.pt*

Introduction

Volatile organic compounds (VOCs) are well-known hazardous pollutants, present in air and water, produced from various industrial sources, including the petroleum industry, internal combustion engines, and paints. Hazardous VOCs are subjected to stringent legislation since concentrations as low as 100 ppmv or less may cause respiratory diseases or critical health problems such as cancer, liver and kidneys failure [1]. Concerning to indoor air quality, part of air pollution is related to VOCs, namely from paints, wood preservatives, aerosol sprays, building materials cleansers and disinfectants. For this reason, it is imperative to remove or manage VOC gases in indoor environments where people spend long periods of time [2].

Among the numerous methods for the elimination of VOC, adsorption has been a best option due to its efficiency and relatively low cost, its adsorbate recovery capacity and the applicability to low concentrations. Various adsorbents have been continuously investigated for VOC abatement, including activated carbon, zeolites, polymers or silica [3].

The excellent properties, low cost and large specific surface areas of porous carbonaceous materials, such as activated carbons, make these types of adsorbents, usually in the form of pellet or granules, most attractive for practical applications [3]. The progressive adsorption of pollutants on the surface of the activated carbon leads to a gradual reduction in its adsorption capacity until adsorption is no longer possible. The main goal of regeneration is to remove these pollutants and recover the original adsorption capacity. At indoor level, few researchers have explored ways to make adsorption-based treatment setup reversible [2]. Common techniques employed for the regeneration include conductive heating, hot gas purging, solvent extraction, and acid-based treatment. Thermal regeneration is widely used at industrial scale and it is the most advantageous for indoor air purifiers [4].

The present study deals with the development and testing of new carbonaceous materials with high adsorption capacity, in a reversible adsorption-desorption process for indoor air purification. Experimental data obtained from an adsorption unit was modelled to gain insights into kinetic and equilibrium phenomena.

Adsorption measurements

A schematic representation of the adsorption unit is given in Figure 1. The flow rate of the feed with the desired acetaldehyde concentration is obtained by mixing the C_2H_4O/He stream with pure He in the mixing chamber. The adsorption measurements were performed in a continuous flow fix-bed quartz column placed in a temperature-controlled furnace, with the main objective of determining the corresponding breakthrough curves. Before the assays, the adsorbent was thermally treated at 180 °C for 4 h under He flow, and then the column was cooled to 25 °C. Subsequently, the gaseous mixture VOC/He was fed at 50 mL min^{-1} and 2 bar, and the outlet gas concentration was analyzed in a gas chromatograph equipped with Barrier Ionization Discharge detector (GC-BID). For regeneration, the adsorbent was heated at 180 °C for 6 h under He flow.

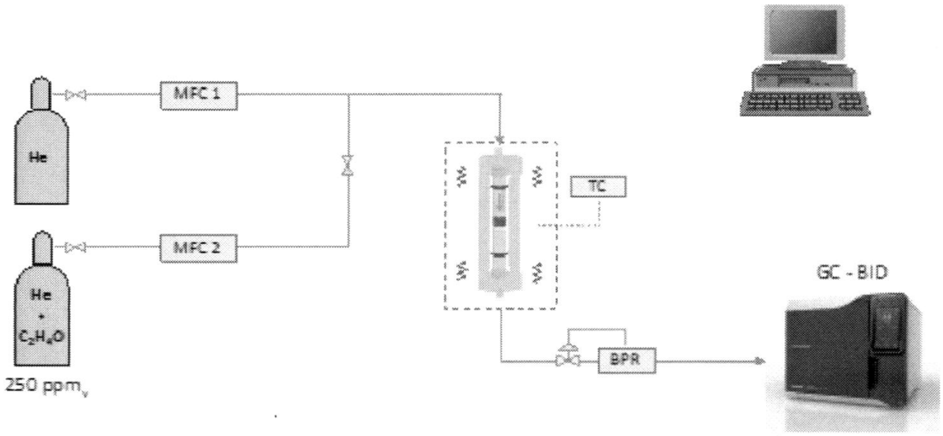

Figure 1. Layout of the experimental apparatus for VOC adsorption-adsorbent regeneration cycles.

New activated carbon materials

Biomass derived highly microporous carbons were prepared by chemical activation. The adsorbent RHC (S_{BET}=2610 m^2/g; pH pzc=6.92) was prepared from rice husk, using KOH as chemical activating agent. Adsorbent RCB3 (S_{BET}=2100 m^2/g; pH pzc=3.0) derived from a lignin-rich plant, namely conteira (*Hedychium gardnerianum*) - an invasive plant in azores island - and subjected to chemical activation using phosphoric acid. RHC presented a somewhat neutral surface, whereas RCB3 presented an acid pH pzc.

k-carrageenan is a sulfated anionic polysaccharide extracted from edible red algaes such as the specie *Chondrus crispus*, also known as Irish moss. *k*-Carrageenan aerogel was prepared by dissolution of the polymer into the ionic liquid [bmim][Cl]. The ionic liquid was totally removed from the gel by successive washing steps with ethanol being further dried with supercritical CO_2. The BET specific surface area (S_{BET}) was 216 m^2/g and the pH pzc was *ca.* 4.

Results and discussion

Preliminary results were obtained using a commercial activated carbon and a feed stream composed of 50 ppm acetaldehyde in He. From the measured breakthrough curves it was possible to determine, for each sorption cycle, the following information: the breakthrough time (based on 10 % of the inlet concentration), stoichiometric time, adsorption capacity and regeneration efficiency. The results listed in Table 1 were compared with data from the literature for acetaldehyde adsorption over carbon adsorbents, being possible to conclude that this material presents a reasonably good adsorption capacity for the first cycle [5]. Nonetheless, it was observed that along the various regeneration cycles the adsorbent decreased significantly its capacity, corresponding to 16-22 % regeneration efficiency, making this material not promising for several reuse cycles. Currently, new materials are being prepared to achieve better performance in terms of capacity, stability and reusability. Such results will be presented during the conference.

Table 1. Breakthrough time, adsorption capacity and regeneration efficiency of each cycle.

Cycle	Breakthrough time (h)	Adsorption Capacity (mmol g^{-1})	Regeneration efficiency (%)
1st	11.2	1.48	-
2nd	3.6	0.32	21.88
3rd	3.2	0.23	15.97

Acknowledgements

This work was developed within the scope of the project CICECO-Aveiro Institute of Materials, FCT Ref. UID/CTM/50011/2019, financed by national funds through the FCT/MCTES, and the Smart Green Homes Project POCI-01-0247-FEDER-007678, a co-promotion between Bosch Termotecnologia S.A. and the University of Aveiro. These projects are financed by Portugal 2020 under the Competitiveness and Internationalization Operational Program and by the European Regional Development Fund (FEDER). Simão P. Cardoso also thanks the post-doc Grant (BPD/CICECO/5247/2017) financed through the Smart Green Homes Project.

Keywords: activated carbon; Volatile organic compounds; adsorption

References

[1] J.M. Kim, J.H. Kim, C.Y. Lee, D.W. Jerng, H.S. Ahn, Toluene and acetaldehyde removal from air on to graphene-based adsorbents with microsized pores, J. Hazard. Mater 344 (2018) 458-465.

[2] R. Xiao, J. Mo, Y. Zhang, D. Gao, An in-situ thermally regenerated air purifier for indoor formaldehyde removal, Indoor Air 28 (2018) 266-275

[3] W.K. Pui, R. Yusoff, M.K. Aroua, A review on activated carbon adsorption for volatile organic compounds (VOCs), Int. Rev. Chem. Eng. 35 (2018) 1-19.

[4] F. Salvador, N. M.-Sanchez, R. S.-Hernandez, M.J. S.-Montero, C. Izquierdo, Regeneration of carbonaceous adsorbents. Part I: Thermal Regeneration, Micropor. Mesopor. Mater. 202 (2015) 259-276.

[5] H. Wang, B. Wang, J. Li, T. Zhu, Adsorption equilibrium and thermodynamics of acetaldehyde/acetone on activated carbon, Sep. Purif. Technol. 209 (2019) 535–541.

582974 The use of a new Fe Substituted Niobium Silicate (NS91) for Hg(II) ion exchange from aqueous solutions

The use of a new Fe Substituted Niobium Silicate (NS91) for Hg(II) ion exchange from aqueous solutions

Z. Lin[1], E. Fabre[1,2], S.P. Cardoso[1], E. Pereira[2,3], C. M. Silva[1]

[1] *CICECO, Department of Chemistry, University of Aveiro, 3810-193 Portugal*

[2] *CESAM, Department of Chemistry, University of Aveiro, 3810-193 Portugal*

[3] *LAQV-REQUIMTE, Department of Chemistry, University of Aveiro, 3810-193 Portugal*

**simaocardoso@ua.pt*

1. Introduction

Many industrial activities are responsible for the release of mercury into the aquatic bodies, such as battery and lamps production, mining and metallurgical processes, and chlor-alkali, petrochemical and paint industries [1]. The persistent character of mercury allows its accumulation in the living organisms and magnification along the food chain causing several impacts on environment and human health [2]. The European Union lists mercury as a priority hazardous substance and encourages its cessation or phasing out of discharges and emissions by 2021 [3]. Moreover, the Agenda for Sustainable Development of United Nations of 2030 in one of its goals, promotes the development of new efficient and low-cost technologies to improve the quality and reuse of water. In line with that, ion exchange has been widely applied for water treatment due to its ease of operation and efficacy to treat large volumes of dilute solutions [4–6]. However, the cost and recover of ion exchanger is a limiting factor, which drives the search for new materials with higher capacities and economic viability.

Microporous materials have high surface areas and may be very selectivity for target metals. In this work, a Fe Substituted Niobium silicate (NS91) has been synthesized, characterized and applied for Hg(II) removal by ion exchange. The effect of the pH of solution and increase in the mass of solid have been investigated. The framework of NS91 is negatively charged electrostatically balanced with exchangeable cations what makes this material of great interest for cationic ion exchange.

2. Experimental

The NS91 was synthesized as follow. Sodium silicate solution, sodium hydroxide, sodium chloride, potassium chloride, potassium fluoride and water were mixed. Then a mixture of niobium pentachloride and iron trichloride was added to above solution with thoroughly stirring. The resulting precursor was agitated or under ultrasonic bath alternately in a covered container for at least 2.5 hours and then treated in

Teflon in-lined autoclave at 230 °C for 7 days without agitation. Finally, the solid product was collected by centrifugation, and washed thoroughly with water, and dried at 80 °C. The obtained samples were characterized by powder XRD, SEM, EDS and TG.

The ion exchange studies were performed under batch conditions, at 22 ± 1 °C in 1 dm^3 volumetric flasks magnetically stirred at 650 rpm. The spiked solutions were prepared in high purity water (18.2 MO cm^{-1}) by dilution of the stock solution of Hg(II) in order to obtain the initial concentration of 50 μg dm^{-3}. The effect of pH was evaluated in the range of 3 - 9, adjusted with HNO_3 and NaOH and fixed masses of 7 mg dm^{-3} of NS91. The mass effect was evaluated with (7 and 10 mg dm^{-3}), at pH 6.0. Liquid samples were collected, filtered through a Millipore membrane, and analyzed afterwards by cold vapour atomic fluorescence spectroscopy (CV-AFS) to quantify Hg(II) concentration in solution. The experiments were conducted until reaching equilibrium (*ca.* 120 hours) and a control solution (solution without solid) was always run in parallel to check experimental losses.

3. Results and Discussion

From a series of syntheses with different conditions, we have obtained very similar powder XRD patterns. The SEM micrograph in Figure 1 shows that the run product consists of aggregate of small bar crystallites with ca. 50 nm. TGA in Figure 1 gives ca. 14 % mass loss between 20 and 800 °C. With the atomic ration obtained from EDS, the possible composition of the run product may be $(Na,K)_2(Nb,Fe)_2Si_2O_{10}.4H_2O$.

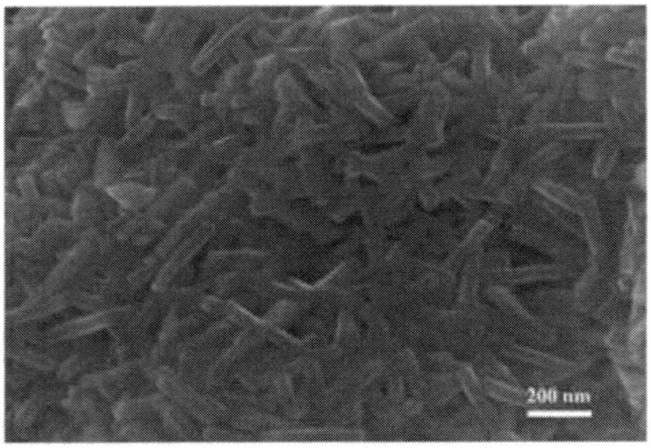

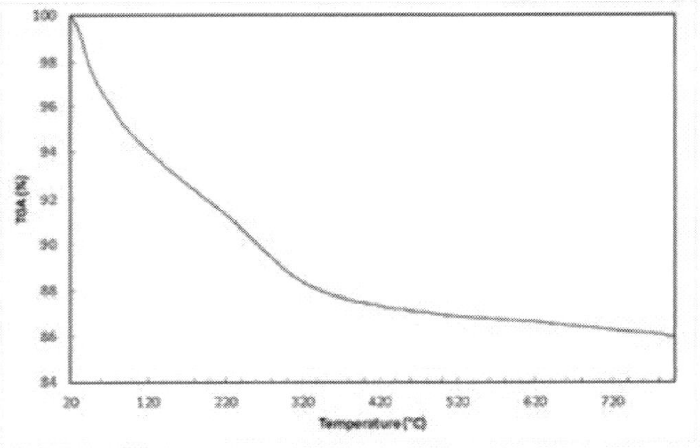

Figure 1. SEM image of obtained sample.

Figure 2. TGA of obtained sample.

After searching in XRD database, we found that the PXRD of our iron niobium silicate is very similar to that of a rubidium niobium silicate. Using the structure data of the rubidium niobium silicate, preliminary simulation gives good agreement to experimental data (see Figure 3). The proposed structure is shown in Figure 4.

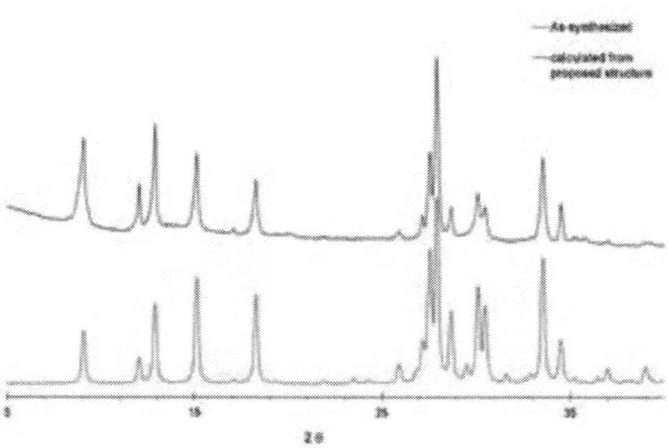

Figure 3. Experimental and calculated PXRD.

Figure 4. The proposed structure view along *a* axis.

Regarding the ion exchange, the effect of pH is one of the most important in metal removal processes. The pH impacts the degree of ionization and speciation of mercury, and the competition with coexisting ions in solution. Table 1 shows the influence of the pH of the solution on the Hg(II) elimination. It is clearly shown that this variable strongly affects the process, the mercury removal increases at pH 5 and then decreases at pH 9. As the pH increases the concentration of H^+ in solution is lower and the Hg(II) removal enhances until the optimum pH 5. The reduction on the % of removal observed in the range of 5 – 9 may be related with the adjustment of the pH with NaOH which provides more Na^+ ions in solution. The exchangeable cations of NS91 are Na^+ ions and their higher concentration in solution reduces the gradient concentration leading to weaker driving forces for mass transport and hence decreases the ion exchange.

Table 1. The influence of pH of the solution on the Hg(II) removal .

pH	3	4	5	6	7	8	9
Removal (%)	5.3	83.7	89.9	74.6	42.0	39.4	26.8

The NS91 showed high affinity to remove Hg(II) from aqueous solutions, with only 7 mg dm^{-3} of solid, 75 % of mercury was removed. These results are ascribed to the increase in the number of available sorption sites with the addition of higher mass.

Acknowledgements

This work was developed within the scope of the project CICECO-Aveiro Institute of Materials, FCT Ref. UID/CTM/50011/2019, financed by national funds through the FCT/MCTES, and the Smart Green Homes

Project POCI-01-0247-FEDER-007678, a co-promotion between Bosch Termotecnologia S.A. and the University of Aveiro. These projects are financed by Portugal 2020 under the Competitiveness and Internationalization Operational Program and by the European Regional Development Fund (FEDER). This work has also been performed with the support of CNPq *(Conselho Nacional de Desenvolvimento Científico e Tecnológico,* Brazil*)*.

Simão P. Cardoso also thanks the post-doc Grant (BPD/CICECO/5247/2017) financed through the Smart Green Homes Project.

References

[1] C. Liu, J. Peng, L. Zhang, S. Wang, S. Ju, C. Liu, Mercury adsorption from aqueous solution by regenerated activated carbon produced from depleted mercury-containing catalyst by microwave-assisted decontamination, J. Clean. Prod. 196 (2018) 109–121. doi:10.1016/J.JCLEPRO.2018.06.027.

[2] M.E. Mahmoud, A.A. Yakout, M.M. Osman, Dowex anion exchanger-loaded-baker's yeast as bi-functionalized biosorbents for selective extraction of anionic and cationic mercury(II) species, J. Hazard. Mater. 164 (2009) 1036–1044. doi:10.1016/j.jhazmat.2008.09.017.

[3] Directive 2013/39/eu of the European Parliament and of the Council of 12 August 2013 amending Directives 2000/60/EC and 2008/105/EC as regards priority substances in the field of water policy, Off. J. Eur. Union. (2013). https://eur-lex.europa.eu/legal-content/EN/TXT/PDF/?uri=CELEX:32013L0039&from=EN (accessed June 18, 2019).

and industrial wastewaters by ion-exchange method, Chemosphere. 56 (2004) 91–106. doi:10.1016/j.chemosphere.2004.03.006.

[5] C.B. Lopes, M. Otero, Z. Lin, C.M. Silva, J. Rocha, E. Pereira, A.C. Duarte, Removal of Hg^{2+} ions from aqueous solution by ETS-4 microporous titanosilicate-Kinetic and equilibrium studies, Chem. Eng. J. 151 (2009) 247–254. doi:10.1016/j.cej.2009.02.035.

[6] S.P. Cardoso, I.S. Azenha, Z. Lin, I. Portugal, A.E. Rodrigues, C.M. Silva, Experimental measurement and modeling of ion exchange equilibrium and kinetics of cadmium(II) solutions over microporous stannosilicate AV-6, Chem. Eng. J. 295 (2016) 139–151. doi:10.1016/j.cej.2016.03.007.

583788 Sustainable Strategies to Recycle Waste Electrical and Electronic Plastics by Thermochemical Processes

Electronic waste disposal and recycle are becoming a global concern as the generation of e-waste increases exponentially with time. An amount of 44.7 million metric tonnes (Mt) of e-waste was generated globally in the year 2016 and only 20% (8.9 million metric tonnes) was recycled. The presence of recoverable components, such as valuable metals (gold, silver and copper) and non-metallic fractions (plastics), makes the e-waste recycling an economically viable recycling industry particularly in the developing countries. However, e-waste contains serious toxic substances such as brominated fire retardants, heavy metals etc. [1]. These hazardous substances are responsible for the serious health issues to the people associated with the e-waste recycling and also causes pollution to the environment. Therefore, it is significant to develop sustainable technologies those can separate the toxic content and recover the valuable materials from e-waste. A significant portion of electronic waste consists of the non-metallic fractions, as most of the outer body of the small and large electronic and electrical equipment's are made from plastics[1]. A substantial amount of plastics also present in the waste printed circuit boards. The current study focusses on the development of a thermochemical process to recover toxic free plastics, polymer or the plastics derived fuel/chemicals in the most efficient manner. Solvent extraction process [2] has been used to remove trace metals and BFRs and subsequently thermochemical processes[3] such as pyrolysis will be used to recover toxic free chemicals and fuels from the e-waste plastics.

References

[1] E. Dimitrakakis, A. Janz, B. Bilitewski, and E. Gidarakos, "Small WEEE: Determining recyclables and hazardous substances in plastics," *Journal of Hazardous Materials,* vol. 161, no. 2, pp. 913-919, 2009/01/30/ 2009.

[2] X. Zhao, L. Zhan, B. Xie, and B. Gao, "Products derived from waste plastics (PC, HIPS, ABS, PP and PA6) via hydrothermal treatment: Characterization and potential applications," *Chemosphere,* vol. 207, pp. 742-752, Sep 2018.

[3] Y. Shen, X. Chen, X. Ge, and M. Chen, "Thermochemical treatment of non-metallic residues from waste printed circuit board: Pyrolysis vs. combustion," *Journal of Cleaner Production,* vol. 176, pp. 1045-1053, 2018/03/01/ 2018.

562653 A Comparison of Two Bioconversion Processes for the Production of Commercially Important C_4 Streams

Four carbon compounds like 1,4-butanediol (1,4-BDO), succinic acid (SA) have been identified as platform chemicals for producing polymers from renewable resources. The successful commercialization of 1,3-propanediol by Dupont Tate & Lyle Bioproducts and 1,4-BDO by Genomatica Inc. has established the commercial viability of green chemicals. In this talk we present the techno-economic analyses to produce four-carbon compounds (C_4s) like 1,4-BDO, gamma butyrolactone (GBL), and tetrahydrofuran (THF) from dextrose through two separate routes - (i) single step bioconversion and (ii) hydrogenation of bio-based succinic acid (SA). In the single step bioconversion process for production of 1,4-BDO, capital cost of the bioconversion section was found to contribute to around 58 % of the installed equipment cost. Hydraulic load from bioconversion section and membrane filtration was found to contribute the most to operating cost. The process was found to be economically feasible with the minimum sales price (MSP) of 1,4-BDO of $1.82/kg as compared to a market price of $2.5/kg. A mechanical vapor recompression scheme used in the evaporator section was found to benefit the process, increasing energy efficiency and cost effectiveness. In comparison, the SA hydrogenation process, called the C_4s refinery, was not economically competitive given the current state of technological progress. The C_4s refinery produces 1,4-BDO, GBL, THF and ammonium sulfate (AS) fertilizer. The capital costs of the C_4s refinery were found to be 1.8 times higher while the operating costs were found to be 2.5 times higher. An important parameter is the SA fermentation titer. Increase in SA titer values to the maximum values reported in literature were not able to reduce the price of C_4s to below the market price, demonstrating the need for progress in this area. It was concluded that for an extended process involving hydrogenation of SA, additional increases in SA titer values and productivity by genetically modifying the relevant organisms will be required to achieve viability especially in comparison to the single step fermentation process for 1,4-BDO. A secondary reason for unfavorable economics was found to be the dependence of the C_4s refinery on thermal separations which contribute to significant operating and capital costs, technological progress in separation processes such as membranes for organic water separations is therefore necessary. Overall, given the published state of technology, the direct fermentation route produces C_4s (1,4-BDO) at $ 1.82/Kg and is economically viable while the C_4s refinery produces C_4s at $ 3.67/Kg and is not competitive with current production routes.

572716 An Integrated Tool for Rapid Assessment of Chemical Manufacturing Process Emissions, Treatment and Sustainability Performance

The design and assessment of chemical manufacturing processes have been aided by a variety of tools that enable the comprehensive analysis and performance testing of both the steady and transient responses of processes with different levels of complexity. However, comprehensive tools that encompass the evaluation of process emissions, treatment simulation, and sustainability assessment of chemical plant performance are yet to be presented.

In this contribution, the tools developed by the co-authors for simulating Pollution Control Units (PCUs) [1] and performing sustainability assessment with GREENSCOPE [2] are combined for the first time. The PCUs allow for the simulation of several unit operations associated with emission treatment of gas, liquid and solid streams [1], while GREENSCOPE enables sustainability analysis considering four categories: Environmental, Mass, Energy, and Economic Efficiency. Each category has several unique metrics called indicators which provide insight into the sustainability performance of any chemical process. The new integrated tool thus allows for the analysis of scenarios including the impact of current and future models of unit operations on process economics, environmental sustainability, and emissions.

This work therefore presents a unified software framework that provides the opportunity to simulate and perform emission control analysis, pollution control unit sizing and sustainability assessment that could be used to support risk evaluation for different chemical manufacturing scenarios. The developed tool is illustrated via appropriate sections of a biomass/coal gasification process for syngas production with the end goal of acetic acid manufacturing.

References

1. Li S, Feliachi Y, Agbleze S, Ruiz-Mercado GJ ,Smith RL, Meyer DE, Gonzalez MA, Lima FV. A process systems framework for rapid generation of life cycle inventories for pollution control and sustainability evaluation. Clean Technol. Envir., 20(7):1543-1561, 2018.
2. Ruiz-Mercado GJ, Gonzalez MA, Smith RL. Sustainability indicators for chemical processes: III. biodiesel case study. Ind. Eng. Chem. Res. 2013;52(20):6747–60.

577369 Exploring the Potential of Chemical Looping Systems to Provide Alternative H_2 Production Systems Solutions Using Process Simulations and Techno-Economic Analysis

The iron-based moving bed based chemical looping technology developed at the Ohio State University (OSU) is a promising candidate to yield cost benefits in natural gas utilization for a carbon-constrained scenario. OSU has developed four commercially relevant, distinct approaches for high efficiency H_2 production from natural gas[1,2,3]. The first approach focuses on a three-reactor scheme that produces CO_2 and H_2 in separate reactors, providing process intensification benefits[1]. The second approach utilizes a membrane based separation scheme to surpass the thermodynamic limits associated with the H_2 production system in the first approach[2]. The third approach retrofits the methane to syngas based co-current reaction system in a steam methane reformer and considers co-feed of steam and CO_2 for high efficiency H_2 production[3]. A fourth approach focuses on modifying the first three reactor approach and producing syngas from the reducer in addition to the Hydrogen from the oxidizer. This presentation will initially examine, how the application of these four chemical looping systems can lead to process intensification and integration opportunities. The thermodynamic performance will be further quantified by a cold-gas efficiency and an energy efficiency comparison between the three approaches and a baseline Steam Methane Reforming case with 90% Carbon capture. Relevant experiments that include high-pressure kinetic data and sub-pilot scale experimental demonstration will be used as an input for a base techno-economic analysis model for the four approaches. The integrated process and techno-economic analysis conducted with guidance from industrial partners useful and surprising insights into how chemical looping Hydrogen production approaches provide solutions under different adoption and carbon-capture scenarios.

References

1. Kathe, M. V., Empfield, A., Na, J., Blair, E., & Fan, L. S. (2016). Hydrogen production from natural gas using an iron-based chemical looping technology: thermodynamic simulations and process system analysis. *Applied Energy, 165*, 183-201.
2. Nadgouda, S. G., Kathe, M. V., & Fan, L. S. (2017). Cold gas efficiency enhancement in a chemical looping combustion system using staged H 2 separation approach. *International Journal of Hydrogen Energy*,Â *42*(8), 4751-4763.
3. Kathe, M., Empfield, A., Sandvik, P., Fryer, C., Zhang, Y., Blair, E., & Fan, L. S. (2017). Utilization of CO_2 as a partial substitute for methane feedstock in chemical looping methaneâ €“steam redox processes for syngas production. *Energy & Environmental Science*.

570829 Economic Analysis of Battery (NMC) Cathode Material Production in Flame Spray Process with Sustainable Solvents

The cost of cathode materials contributes approximately 32.7% to the cell construction cost of lithium-ion batteries, which has a significant impact on the price of battery packs. The synthetic routes of cathode material ($LiNi_{1/3}Mn_{1/3}Co_{1/3}O_2$ (NMC333)) based on the solid-state method or co-precipitation method involve complex mixing, drying, and sintering processes, resulting in a material cost too high to reach the U.S. Department of Energy battery price target of $125/kWh. To reduce the cathode material manufacturing cost, our lab has developed a flame spray process that utilize sustainable solvent glycerol as solvent to manufacture the cathode material. The economic analysis of the new green chemical process was studied based on discounted cash flow method. The major economic indicator used in this study is the minimum cathode material selling price (MCSP), whereas the co-precipitation method for NMC333 production was selected as the reference for economic analysis. Results show that compared with the co-precipitation method, the new flame spray process has the potential to reduce the operation cost by 31% with comparable capital investment and material cost. Moreover, MCSP of the new process is $19.1/kg, which is 83% of that of the co-precipitation method. When the flame spray process is integrated with the sintering process inline, the MCSP is as low as 15.6 $/kg, which demonstrates that the new process is an attractive manufacturing process for cathode material production. Furthermore, sensitivity analysis and uncertainty analysis were used to examine the effects of plant scale, raw material price, and analysis parameters on the MCSP and evaluate the certainty interval of the MCSP of the flame spray process.

574815 Soybean Oil As a Bio-Solvent to Sweeten Sour Natural Gas

Soybean oils are readily available bioresources composed of unsaturated fatty acids capable of binding sulfur compounds. Predictive models offer an opportunity to develop cost-effective methods using bio-based solvents to sweeten natural gas. While increases in fracking have increased availability of natural gas – a fuel that when combusted produces less greenhouse gases than coal or gasoline – natural gas is rich in hydrogen sulfide, rendering it sour gas. Hydrogen sulfide is corrosive to processing equipment and poses risks to human health and the environment. Predictive models such as COSMO-RS, the conductor-like screening model for real solvents, can be used to examine interactions at a molecular level, screen solvents, and choose process parameters. COSMO-RS was used to simulate the partitioning of hydrogen sulfide between soybean oil, high oleic soybean oil, free fatty acids, and gas phases. Predicted partition coefficient (*K*) values ranged from 0.16 to 0.70 for soybean and high oleic soybean oils at temperatures from 10 - 100°C; *K* values between 0 – 1 are indicative of hydrogen sulfide's preference for the liquid phase over the gas phase. Experimental results demonstrate the feasibility of removing H_2S from gas using soybean oils and are compared to the predicted partitioning. This study provides proof-of-concept for using high oleic soybean oil as a bio-based solvent as a new method for cleaning sour gas.

574828 Reducing Energy, Solvent, and Resin Demand of Sorption-Based Processes for Carboxylic Acid Recovery

Current methods such as ion exchange and distillation to separate bio-based carboxylic acids are material and energy intensive, typically generating waste streams and requiring energy intensive distillation steps to recover a pure product. Sorption-based approaches increase the sustainability of carboxylic acid separations by reducing downstream waste handling, however, current sorption-based technologies typically require large volumes of elution solvent and still result in a dilute carboxylic acid stream. Thus, to obtain a concentrated product, excess water and solvent must be removed from the elution stream via evaporation or distillation which adds a substantial operational cost to the process. This presentation will discuss recent work towards reducing solvent, adsorbent requirements and reducing downstream distillation energy demands in these processes. Specifically, four commercially available weak-base resins were screened by determining their adsorption capacity for butyric acid, a fermentable acid targeted for biofuels. Then, to recover the adsorbed acid, non-aqueous solvents were tested and their regeneration constant were determined. Findings indicate that specific properties of the resins and eluents correlate with the energy, resin, and solvent demands of the system. These results provide a general framework for optimizing sorption-based processes for recovering dilute bioproducts.

572591 Storage of Excess Wind Energy and Utilization of Waste Flue Gas CO_2 through Sustainable Methanol Production

As a larger production of wind energy and other types of renewable energies are predicted in the future, utilizing the excess energy from these sustainable energy sources can increase the energy efficiency of the wind farm by making good use of all the electricity produced. Moreover, CO2 emissions from industry, that contribute to climate change, are also an increasing problem that requires a solution. In this study, we present a novel concept, which aims at solving these two problems by integrating excess energy from wind farms and CO2 from flue gas into an innovative production process using methanol as an energy storage/carrier. The excess wind energy is used to produce hydrogen through water electrolysis and CO2 is captured from flue gas, e.g. from the cement industry. Compared with the traditional methanol production which heavily relies on natural gas as its source of hydrogen, carbon, and energy, this process has a smaller environmental footprint. The novelty of our concept relies on shifting the focus from operating a methanol production plant towards valorizing the excess renewable energy, which is wasted otherwise and considering methanol as a value-added product (as opposed to Szima et al. 2018). Therefore, the plant capacity of 30 000 metric tons per year is scaled based on the quantity of the excess wind energy of the largest wind farm in Denmark. This capacity is significantly lower than those of traditional methanol production plants, which range up to 1.9 million metric tons per year (Ferrostaal, 2015). This adds a constraint to improve the process integration efficiency of the proposed concept. Hence, a techno-economic analysis is performed to investigate the overall feasibility of the proposed concept by identifying key design decisions and essential cost drivers. As results obtained show high sensitivity to both electricity and methanol prices, an uncertainty analysis was conducted in order to determine their effects on the profitability metrics of the proposed novel process. A sustainability analysis is complementarily carried out to emphasize the added environmental benefits of the proposed concept in comparison to the traditional methanol production process.

Reference:

Szima, Szabolcs, and Calin-Cristian Cormos. "Improving methanol synthesis from carbon-free H2 and captured CO2: A techno-economic and environmental evaluation." *Journal of CO2 Utilization* 24 (2018): 555-563.

Ferrostaal. "M5000 Methanol Plant" from *www.ferrostaal.com/en/top-references-ferrostaal/m5000-methanol-plant/* (2015), visited 12.04.2019.

575320 Towards a Novel Energy Financial Security: The Texas A & M Energy Spot Price Index

The available financial instruments providing exposure for the investors to the energy markets are limited to equity based stocks (i.e. S&P 500 Energy) and futures on individual energy commodities (i.e. WTI Crude Oil, Henry Hub Natural Gas). Furthermore, these options are predominantly associated with the oil and gas sector, with almost zero emphasis placed on the renewable energy. Given the lack of a financial security to effectively capture the entire U.S. energy landscape and represent the average cost of energy, we introduce the Texas A&M Energy Spot Price Index (ESPI) as a novel framework to fulfill these needs.

The energy landscape of the U.S. is carefully depicted to determine the products that are directed to the end-use sectors of the U.S. economy [1-5]. The total energy demand of these products, together with their spot prices in the market, serve as the backbone of the Energy Spot Price Index. Since the weights of the Energy Spot Price Index constituents have a lag between two and three months, we introduce a rolling horizon based parameter estimation model to determine the current weights. The spot prices of the products are taken from Bloomberg when available. For products whose spot prices are not available, we introduce and examine three different approaches [6-8]. The methodology is illustrated on 120 months to show the capability of accurately determining the value of the Energy Spot Price Index. The proposed Energy Spot Price Index is the first of its kind that precisely quantifies the average market price of energy. As such, it can be utilized as an efficient tool for all investors who look for exposure over the average market price of energy, trading over the whole energy market, hedging their investments, as well as a benchmark for measuring the overall cost of energy. Therefore, the Energy Spot Price Index can be used as a tradable financial security (e.g. ETF) that is updated daily providing an alternative for the investors who look to diversify their portfolios.

References:

- Elia, J. A.; Baliban, R. C.; Xiao, X.; Floudas, C. A. Optimal energy supply network determination and life cycle analysis for hybrid coal, biomass and natural gas to liquid (CBGTL) plants using carbon-based hydrogen production. Computers and Chemical Engineering 2011, 35, 1399-1430.
- Elia, J. A.; Baliban, R. C.; Floudas, C. A. Nationwide energy supply chain analysis for hybrid feedstock processes with significant CO2 emissions reduction. AIChE Journal 2012, 58 (7), 2142-2154.
- Elia, J. A.; Floudas, C. A. Energy Supply Chain Optimization of Hybrid Feedstock Processes: A Review. Annual Review of Chemical and Biomolecular Engineering 2014, 5 (1), 147-179.
- Onel, O.; Niziolek, A. M.; Floudas, C. A. Integrated biomass and fossil fuel systems towards the production of fuels and chemicals: state of the art approaches and future challenges. Current Opinion in Chemical Engineering 2015, 9, 66-74.
- Floudas, C.A.; Niziolek, A. M.; Onel, O.; Matthews, L.R. Multi-Scale Systems Engineering for Energy and the Environment: Challenges and Opportunities. AIChE Journal 2016, 62 (3), 602-623.
- The Texas A&M Energy Price Index: A Predictive Framework, 2019
- Energy Information Administration (EIA), Electric Power Monthly, 2019
- U.S. Department of Energy (DOE), Green Power Markets, 2018

566273 Techno-Economic Analysis of Septic Tank Waste Treatment in a Centralized Hydrothermal Treatment Facility

Hydrothermal treatment, a thermochemical conversion process that is particularly useful for wet wastes, of septage produces valuable solid and liquid fertilizers while eliminating human pathogens. Additionally, it treats a problematic waste that is a burden for most traditional waste water treatment facilities. Operation of a centralized treatment facility would augment an area's capacity to utilize the nutrients and valuable products available in septage while also mitigating its potential environmental hazards. This facility would be alternative to landfilling, which can be detrimental to public waterways, or treatment at a traditional waste water plant. This analysis investigates the most important factors for an economically feasible facility that both treats septic tank waste and produces fertilizer products. The environmental impact of the plant is also compared to the standard methods of septage treatment.

578172 Turning a Hazard into Sustainable Revenue Streams: H_2S Valorization in Offshore Oil and Gas Productions through Process Intensification

The grand challenges of the 21st century lie squarely at the interplay of water, energy, and food— the so-called water-energy-food nexus. Add climate change into the mix, and the challenges become *more* challenging. Aggressive measures to decarbonize the energy market are necessary to tackle the threat of climate change. However, due to the ever-increasing global population, energy requirements are expected to increase 50% by 2030.[i] While ensuring that we protect the natural environment, it is also important to provide energy security. This calls for an energy supply that is consistent, reliable, sufficient, and affordable. Natural gas has arguably the least carbon footprint among fossil energy resources (carbon dioxide emission of burning natural gas is approximately half that of coal). While natural gas cannot be viewed as the permanent solution to energy security, it plays a vital role as the bridge fuel in the transition to a low-carbon economy.

Hydrogen sulfide (H_2S) is a major hazardous by-product in oil and gas production as well as many bio-based manufacturing processes, such as anaerobic digestion. Desulfurization of sour gas is essential prior to transportation or further processing due to its extremely toxic, acidic, and corrosive nature. Several offshore oil and gas production platforms are suffering from elevated levels of H_2S in recent years, and the production facilities are not designed for such high levels of H_2S. Currently, H_2S is removed using a non-regenerative scavenging process. The scavenger reacts with H_2S and produces less acidic, water-soluble compounds in an irreversible reaction.

Although the current process is rather established, it comes with significant challenges. Such challenges include high OPEX due to excessive amounts of the scavenger used in offshore productions, environmental concerns coming from the increasing levels of H_2S and increased mass of discharged spent scavenger, and fouling issues creating blockages in several downstream unit operations, such as wastewater treatment facilities.

For these reasons, design and development of more sustainable approaches that can fulfil the H_2S content threshold limit requirements and that can eliminate the disadvantages of the current non-regenerative H_2S-scavenging process are deemed necessary, specifically from an economic point of view. This work focuses on the design and techno-economic analysis of an intensified process that can be installed on an offshore platform to simultaneously remove and convert H_2S from produced natural gas to value-added product(s) to generate new revenue streams. The added benefits of this new process are that it allows for the efficient recovery of resources by turning the hazard, H_2S, into an opportunity. The proposed process is therefore well suited for the application of intensification concepts, integrating reaction and separation into a single unit.

[i] IEA, "World Energy Outlook 2012," International Energy Agency, 2012.

554760 Critical CO_2 Carnot Engine for Industrial Waste Heat Recovery and Utilization

Industrial processes regularly discard large volumes of low-grade waste heat to the environment. Harnessing this heat and transforming it into usable power would be environmentally and economically advantageous for industries. An overlooked solution to utilizing this waste heat is the use of a Carnot cycle to generate work with supercritical CO_2 as the working fluid. Supercritical $CO_2$is a novel working fluid for a Carnot engine as its critical temperature (31.1C) is much lower than the boiling point of water. This allows for the lower isotherm of the Carnot cycle to be constructed at the critical temperature, resulting in a sizable Carnot cycle with a near flat lower isotherm and upper isotherm just below the temperature of the low-grade heat. Thus, this type of Carnot cycle encompasses a larger area on a P-V diagram than any other thermodynamic conditions, thereby optimizing the amount of work that can be converted to power using a relatively small engine.We describe this approach of waste heat harnessing and illustrate ideas for utilizing the waste heat from an industrial toluene process. The impact of compression ratio, upper isotherm temperature, and pressure in the engine are investigated. Resulting savings amount to hundreds of thousands of kWh/year, thereby reducing cooler loads and electricity bills for the process, as well as reducing environmental emissions.

References:

[1] J.T. Banchero, B.D. Smith (Ed.), R.J. Hengstebeck. *Disproportionation of toluene*, 1969.

[2] Eldred H. Chimowitz Madeleine R. Laitz, F. Douglas Kelley. *Critical CO_2Carnot Cycle for Waste Heat Utilization*, 2017.

INDUSTRIAL SYMBIOSIS IN TRINIDAD AND TOBAGO: A FOCUS ON CARBON DIOXIDE

Thérèse G. Lee Chan and David A. Janes, The University of the West Indies, St. Augustine, Trinidad and Tobago

Abstract

Enterprise input-output (EIO) models are used to analyze the flows of resources, such as energy and materials, around industrial networks. The majority of the greenhouse gas, carbon dioxide (CO_2), emitted in Trinidad and Tobago (T&T) is from industrial sources. In this study, a simple representative model was developed of a network of CO_2 exchanges within an industrial park in T&T. The EIO model was developed from both process engineering and economic data. It incorporated performance indicators, which had been proposed in the literature, to calculate the level of industrial symbiosis (IS) and benefits - based on "the three pillars of sustainability" - realized in the representative industrial network. The eco-connectance, which is one measure of the level of IS, was determined as 1.3 for the representative network, thereby indicating the presence and level of IS in the industrial park. Furthermore, optimized flow networks were created, with and without a CO_2-reusing propylene carbonate plant. Multi-objective optimization was used to determine how to add the CO_2-reusing plant to the industrial network to minimize both CO_2 emissions and implementation costs. The addition of a CO_2-reuse plant reduced the CO_2 emissions by 3%, demonstrating there is scope for improving the existing IS network.

Introduction

Industrial symbiosis (IS) can be regarded as using, recovering and redirecting resource-laden wastes for reuse, in a network of separate commercial operations, so resources remain in productive use for longer. IS is considered an effective approach to waste management (Yazan, Romano et al. 2016) delivering both economic and environmental benefits. The twin-island state of Trinidad and Tobago (T&T) is rich in hydrocarbon resources, and was amongst those countries generating the highest carbon dioxide (CO_2) emissions per capita at 25.72 tonnes per head in 2016 (Knoema World Data Atlas, 2018). In 2014, 74% of the CO_2 emitted in T&T originated from industry (UNDP, 2014). The majority of the petrochemical industry in Trinidad is housed in the Point Lisas Industrial Estate (PLIE), which contains several ammonia and methanol plants.

In this paper, the exchanges of process CO_2 between various petrochemical plants in PLIE are used to assess the level of industrial symbiosis occurring on the industrial park. The aim of this work was to develop a simple representative model of this nascent symbiotic industrial system, in terms of CO_2 exchanges at PLIE. Our model was then used to perform an integrated process engineering and economic analysis to quantify the level of exchanges and benefits beneath "the three pillars of sustainability" and thence determine the scope for improvement.

In part because there are currently no legislative restrictions on CO_2 emissions in T&T, PLIE is not thought of locally as being a bastion of progressive industrial ecology. Although PLIE was not originally planned for industrial symbiosis, waste CO_2-streams have been shared amongst plants since 1983 (Boyle and Baetz, 1998) when the Trinidad and Tobago Urea Company plant was commissioned to use waste process CO_2, ammonia and some facilities from a nearby separately-run Fertrin ammonia plant to produce urea. The current "self-organized" state of sharing of waste streams developed over the years, rather like early sharing at archetypical eco-industrial park (EIP) at Kalundborg, Denmark (Ehrenfeld and Gertler, 1997), in that each additional exchange in the park was an independently negotiated business deal, which was established only if expected to be commercially beneficial.

Few studies have been conducted on the application of industrial symbiosis in T&T. In 1998, four companies in PLIE were evaluated in terms of waste management and the study (Boyle and Baetz 1998) found that there was potential for the waste reduction by reusing and recycling, although the study did not consider CO_2 as a waste resource. In 2008, Boodlal (2008) compiled an inventory of CO_2 resources in Trinidad, in which the emissions from the chemical plants at PLIE were calculated with appropriate stoichiometric ratios for the relevant reactions.

Methodology

There is no standard methodology for measuring the level of IS in an industrial network or park. Some authors use economic, environmental and social indicators to show how IS improves a park by comparing the indicators for the park with and without IS activities. However, this does not give an actual measure of the level of IS. Some of the proposed measures for industrial symbiosis are eco-connectance, by-product and recycling rate (Tiejun 2010), eco-efficiency (Park and Behera 2014), environmental impact momentum (Yilmaz et al., 2016) and technical exchange efficiency (Fraccascia, Albino et al., 2017). Yilmaz et al. (2016) published a comprehensive list of indicators that can be used when considering industrial symbiosis.

Material flow analysis (MFA), enterprise input-output (EIO) tables, life cycle analysis (LCA), eco-efficiency, eco-connectance and waste recovery are some of the tools which have been used in reducing waste in a system in a sustainable manner. MFA, LCA and eco-efficiency are not preferred since they do not consider the whole picture.

In this study, the process CO_2 emissions for selected plants were evaluated based on reported information about the various technologies employed and the plant capacity. Flue-gas CO_2 emissions were excluded from this study, because of the much lower CO_2-purity of these streams. Potential means for improving the sharing and reusing of CO_2 to create shared value, that is the IS level, were considered. Nine petrochemical plants in PLIE were selected as being representative of the industrial park. The plants considered were ammonia, methanol, urea and melamine. An EIO table for the industrial park was developed in MATLAB® (ver. R2019a), the chosen modeling environment.

The general equations and variables for an EIO system are:

$$X = Ax + f \tag{1}$$

$$r = Rx \tag{2}$$

$$w = Wx \tag{3}$$

Variables:

A – Technical coefficient matrix
x – Total output vector
f – Vector of final demands
R – Resource intensity matrix
r – Vector of primary input
W – Waste intensity matrix
w – By-products and waste vector

It is assumed that there is only one main product for each process. A, R and W are matrices of constants and are typically observed or calculated values for a specified technology for the process. Therefore, if process demand is changed, the outputs from the process would be changed by that value.

Three EIO tables were developed for a *base*, *current* and *proposed* case. The base case was one in which there was no symbiosis occurring and this was used as a baseline in the study. In the current case, the existing symbiosis was shown and in the proposed case, a CO_2-reusing plant was added to the industrial park. Material and energy flows for the base and current cases were estimated on the basis of the average production rate of the various plants over a three-year period. Various plants, which reuse CO_2, were considered, such as manufacturing processes for biofuels (Benn-Ogeer et al., 2015), propylene carbonate (Demirel, 2015) and polyethercarbonate polyols (von der Assen and Bardow, 2014). The propylene carbonate plant was selected based on the feasibility and the availability of published performance information. The data used to simulate and cost the plant in Aspen Plus® (ver. 10) was obtained from Demirel (2015). For all the cases, the direct and total coefficients were determined. Besides the evaluation of the direct and total coefficients, the level of industrial symbiosis occurring in the various scenarios was evaluated using various industrial symbiosis measures and sustainability indicators.

In this study, the focus was on the exchange of process CO_2 amongst the plants. The CO_2-rich process streams, which originated from the ammonia plants, were supplied to the sinks at an assumed CO_2-purity of 99 wt%. US dollars were used for all prices and costs. The price of CO_2 supplied to the sinks was determined from the calculation of the levelized cost of CO_2 transportation and an assumed raw material cost for CO_2, to buttress the mutualism of the IS. Consequently, pipelines from the CO_2 sources to sinks were evaluated. The equation developed by McCoy and Rubin (2008) was used to estimate the diameter of the pipelines required for the sink flows. The costs associated with purification and compression of the CO_2 were taken into consideration and together with the diameter and length of the pipeline, these were used in the determination of the levelized cost of CO_2 transportation (\$ $tonne^{-1}$ of CO_2).

Optimization of the industrial park was computed in MATLAB by taking into consideration the amount of waste CO_2 and the economics. Single objective optimization for each case was carried out to minimize waste CO_2 and to minimize the cost. Cases were considered using linear programming (LP) and mixed integer linear programming (MILP). In the MILP, binary variables were used to prevent small flow rates through the pipelines. For the LP and MILP cases, the dual-simplex algorithm and the branch and bound algorithm were respectively utilized. Finally, multi-objective optimization using the genetic algorithm in the MATLAB function *gamultiobj* was used to minimize the amount of waste CO_2 and the cost. The Pareto front for each case was generated.

Results and Discussion

Material and energy flows were calculated for the input and output of the plants in the industrial park. These flows were then used to generate the EIO tables for the base, current and proposed cases. The cases showed that from the ammonia plants, there were approximately 202,000 kg/h of nearly pure CO_2 being produced. The EIO for the current case indicated that approximately 35% of the CO_2 was reused in methanol and urea plants in the industrial park and the remainder was vented to the atmosphere. Introducing a new plant, which reused CO_2 from the industrial park, increased the CO_2 usage by approximately 3%, thereby reducing the amount of CO_2 emitted to the atmosphere. From the EIO table, the direct and total coefficients were determined. Analysis of the direct coefficients showed that the largest direct coefficient for the by-products and waste was attributed to the nearly pure CO_2 being emitted from the ammonia plants.

The core EIO model gives an idea of the material and flow exchanges within the network, but does not measure the level of industrial symbiosis. The calculated values for the performance indicators, used to measure IS in these three cases, are shown in Table 1.

Table 1. Industrial Symbiosis Performance Indicators

Industrial Symbiosis Indicator	Base	Current	Proposed
Eco-connectance	0	1.3	1.8
Environmental impact momentum	0	0.56	0.62
Fraction of waste CO_2 utilized	0	0.36	0.38

In the base case, it was assumed that there were no exchanges amongst the plants. The values obtained for the current case show that there is some level of industrial symbiosis occurring and those for the proposed case indicate that the addition of a plant which utilizes CO_2 as a raw material can lead to increased levels of industrial symbiosis.

In this study, a propylene carbonate plant was proposed for the industrial park. There are several ways in which CO_2 can be utilized, but because many of these uses are emerging technologies, it is difficult to obtain information for them. The propylene carbonate plant had a capacity of approximately 10,500 kg/h and consumed 4,500 kg/h of CO_2. This capacity was determined in accordance with the current market for propylene carbonate. The results from the model indicated that the feasibility of such a plant is highly dependent on the price of the catalyst since this was the most expensive component of the model.

One of the objectives of this study was to optimize the flow of CO_2 in the three cases. This optimization considered minimizing the waste CO_2 and the cost. The levelized cost of transporting the CO_2 was calculated and hence pipelines from the CO_2 sources to the sinks were considered. The transportation costs of CO_2 ranged from US$9 per tonne to $50 per tonne of CO_2. The higher end of the costs was attributed to smaller flows in the pipelines and hence due to a fixed cost recovery period, the price would consequently be higher. It should be noted that the costs obtained were derived from a fixed flow in the pipeline and did not take into account the change in the flow rate and the cost. This was done to ensure that the optimization model remained linear. According to literature, there is no fixed cost for CO_2 since it is dependent on the purity of CO_2, distance and the state in which the CO_2 is transported.

Various optimization cases were considered using the current and proposed data. In the first scenario, minimization of waste CO_2 was carried out. MILP and LP were used to solve the optimization models for the base and proposed cases. Results from the LP case with the current data indicated that the CO_2 from four out of the five plants should be sent to the CO_2 sinks. The MILP results indicated the same but the flows were different for the plants. However, for the proposed cases both MILP and LP showed that flows should be sent to the sinks from all the plants. It is suggested that some flows be sent from each plant since the existence of a plant venting all its CO_2 to the atmosphere would increase the environmental burden for that particular plant.

The optimal annual costs for the current and proposed cases were $12 million per year and $14 million per year respectively. It is evident that in the current case, the transportation cost for CO_2 is less since there are less sinks which require CO_2. It should also be noted that for the proposed case, there would be some additional expenditure in constructing and maintaining the propylene carbonate plant. Multi-objective optimization for minimizing the CO_2 waste and cost was also done for both cases. The annual costs for each case were $12.2 million per year and $14.6 million per year, respectively.

Sustainability indicators were employed to evaluate the optimized cases using multi-objective optimization (see Table 2). The eco-efficiency was determined as a ratio of the annual production to the amount of CO_2 used.

Table 2. Sustainability Indicators

Indicator	Sustainability Dimension	Units	Current Case	Proposed Case
Eco-efficiency	Environmental	-	9.8	9.4
CO_2 emissions	Environmental	t/h	277	273
Resource efficiency	Environmental	t/h	0	4.5
Relative change of CO_2 recycling	Environmental	%	36	38
Total Capital Investment (CO_2 transport)	Economic	$	22.5	26.5
Total Capital Investment (new plants)	Economic	Million $	-	17.2
Profit as a result of IS activities	Economic	Million $/y	0.41	(0.43+3.6)
Job Creation	Social	-	No	Yes
Creation of Industrial Symbiosis	Social	-	6	7

The new plant led to a reduction in the overall CO_2 emissions thereby increasing the amount of CO_2 being reused. Addition of the propylene carbonate plant resulted in a decrease in eco-efficiency. Nevertheless, it would lead to an increase in jobs and would create an increased level of industrial symbiosis in the industrial estate.

Further work may investigate the actual linkages amongst the plants and determine the capacities of the existing pipelines. It could also consider looking at CO_2 in flue gas and other CO_2 utilization options. Additionally, exchanges of other materials and energy from the industrial park need to be included to get a fuller picture of how industrial symbiosis is being and could be put into practice.

Conclusions

The model indicates that current partners in the on-going industrial symbiosis at the Point Lisas Industrial Estate ought to be achieving sustainable benefits. The estimated levelized cost for CO_2 transportation ranged from $9 per tonne to $50 per tonne of CO_2. The introduction of a few chemical manufacturing plants reusing CO_2 should lead to increased IS indicator values and further triple-bottom-line benefits, even though it would only make a small dent in the quantity of as yet unused process carbon dioxide. The integration of data from process engineering models into EIO models provides a means to analyze and optimize potential IS scenarios.

References

1. Benn-Ogeer S. K., D. A. Janes and R. Deo (2015). "A Life Cycle Assessment (LCA) of the Production of Biofuels in Trinidad from Waste CO_2 With Microalgae", e-Journal of the Caribbean Academy of Sciences 8(1).

2. Boodlal, D. (2008). "Trinidad and Tobago's CO_2 Inventory and Techno-economic Evaluation of Carbon Capture Options for Emission Mitigation." 3rd Tobago Gas Technology Conference, Lowlands, Tobago.

3. Boyle, C. A. and B. W. Baetz (1998). "A prototype knowledge-based decision support system for industrial waste management: Part II. Application to a Trinidadian industrial estate case study." Waste Management 17(7): 411-428.

4. Demirel, Y. (2015). "Sustainability and Economic Analysis of Propylene Carbonate and Polypropylene Carbonate Production Processes Using CO_2 and Propylene Oxide." Chemical Engineering & Process Technology 0: 6:3.

5. Ehrenfeld, John; Nicholas Gertler (1997). "Industrial Ecology in Practice: The Evolution of Interdependence at Kalundborg". Journal of Industrial Ecology 1: 67–79.

6. Fraccascia, L., V. Albino and C. A. Garavelli (2017). "Technical efficiency measures of industrial symbiosis networks using enterprise input-output analysis." International Journal of Production Economics 183: 273-286.

7. Knoema World Data Atlas. (2018). "Trinidad and Tobago – CO_2 emissions per capita". [retrieved 01/04/2019]. Available from: https://knoema.com/atlas/Trinidad-and-Tobago/CO2-emissions-per-capita.

8. McCoy, S. T. and E. S. Rubin (2008). "An engineering-economic model of pipeline transport of CO_2 with application to carbon capture and storage." International Journal of Greenhouse Gas Control 2(2): 219-229.

9. Park, H. S. and S. K. Behera (2014). "Methodological aspects of applying eco-efficiency indicators to industrial symbiosis networks." Journal of Cleaner Production 64: 478-485.

10. Tiejun, D. (2010). "Two quantitative indices for the planning and evaluation of eco-industrial parks." Resources, Conservation and Recycling 54(7): 442-448.

11. UNDP, Factor CO_2. (2014). "Projection of the Business as Usual (BaU) and measures scenarios of carbon emissions for the electrical energy generation, industry and transportation sectors". Port of Spain, Trinidad and Tobago.

12. von der Assen, N., and A., Bardow, (2014). "Life cycle assessment of polyols for polyurethane production using CO_2 as feedstock: insights from an industrial case study." Green Chemistry 16: 3272-3280.

13. Yazan, D. M., V. A. Romano and V. Albino (2016). "The design of industrial symbiosis: an input–output approach." Journal of Cleaner Production 129: 537-547.

14. Yilmaz, O., E. Yontem and E. Alkaya (2016). Industrial Symbiosis Indicators. FISSAC.

553061 Reclaiming Phosphorus from Secondary Treated Municipal Wastewater with Engineered Biochar

Phosphorus (P), a vital growth-limiting nutrient, is often lost in wastewater discharge, which may not only deteriorate water quality but also accelerate P depletion. In this study, laboratory experiments were conducted to investigate the reclamation and reuse of P from secondary treated wastewater (STWW) by engineered biochar loaded with aluminum oxyhydroxides (AlOOH). Biochar loaded with colloidal and nanosized AlOOH crystalline flakes was successfully produced through direct pyrolysis of hickory wood chips pretreated with aluminum salt. STWW samples were collected from a local wastewater treatment plant to evaluate the P sorption ability of the engineered biochar. The real wastewater was used in batch sorption experiments and the results showed that the engineered biochar effectively removed P from the STWW with relatively fast kinetics (<8 h) and good capacity (8,346 mg-P/kg), mainly through the electrostatic attraction mechanism. When the engineered biochar was packed in fixed-bed columns, it also effective removed P from the real STWW. The P removal efficiencies of the fixed-bed columns were affected by biochar dosage and flow rate. Various fixed-bed filtration models were applied to simulate the column experimental data and the model results matched the experimental breakthrough curves of P well. After the use, the engineered biochar was laden with P and thus significantly stimulated seed *(Vigna radiata R. Wilczek)* germination and early seedling growth. Findings from this study demonstrate that engineered biochar has promising applications in wastewater treatment because it can not only reduce P level in discharge to benefit the environment but also reclaim it as an agricultural fertilizer to improve P biogeochemical cycling and sustainability.

575667 Nutrient Recovery from Municipal Wastewater Using a Methanotroph-Microalgae Co-Culture

In recent times, human activity has increased the amount of nitrogen fixation in the world causing negative consequences for our environment (greenhouse effect, reducing the protective ozone layer, contaminating drinking water), the human health and the economy [1]. Thus, effective treatment of wastewater (rich in nutrient nitrogen) plays a key role in managing the nitrogen cycle and improving our ecosystems and public health.

At water resource recovery facilities (WRRFs), anaerobic digestion (AD) is the most widely employed method to reduce the amount of sludge from both primary and secondary treatment, as well as reducing pathogens and odor [2]. The nitrogen and phosphorus content in the AD effluent are usually removed by secondary treatment; however, secondary treatment consumes energy and does not always remove the nutrients to the desired degree, which often cause toxic algae bloom in water ways and costal area. In addition, the impurities (e.g., CO_2, H_2S, etc.) in biogas resulting from the AD are difficult to remove economically; thus, the value of biogas is low. As a result, 80% of WRRFs either flare the produced biogas or use it for heating, while only 20% of them use biogas for power generation or driving machinery [3].

In this work, we aim to develop a new wastewater treatment platform where we use a coculture of methanotroph-microalgae to augment the AD treatment. In the proposed framework, the methanotroph-microalgae coculture recovers nutrient from the AD effluent as well as both methane and carbon dioxide from biogas without requiring external oxygen supply. To realize the potential industrial application, the robustness of the co-culture must be determined considering the inhibitory compounds in wastewater. Such robustness will determine the level of required pretreatment of wastewater, which could significantly reduce the economic feasibility of the new platform.

In our previous work, we have assembled several synthetic methanotroph-photoautotroph cocultures that exhibit stable growth under various substrate delivery and illumination regimes [4]. In this work, using *Methylococcus capsulatus* - *Chlorella sorokiniana* as the model coculture system, we investigated the co-culture's potential in combined nutrient recovery and biogas co-utilization on pretreated AD effluent. The performance of the co-culture technology was assessed on filtered AD effluent diluted with clarifier water; where both AD effluent and clarifier water were obtained from the South Columbus Water Resources Facility. Our results show that co-culture can be maintained completely on wastewater effluents with good nutrient removal. AD effluent pretreated by filtering has shown to maintain better growth than on AD effluent pretreated by a combination of filtering and autoclaving. We have also been able to support the co-culture on AD effluent after allowing the sludge to settle. In addition, the co-culture enabled ~85% recovery of total nitrogen and up to ~99% recovery of total phosphorus in the diluted AD effluent. Our results show that the methanotroph-microalgae co-culture is a promising platform for wastewater treatment through simultaneous nutrient recovery from AD effluent and biogas conversion with the potential of producing value added products from wastes.

References

[1] Driscoll, C., Whitall, D., Aber, J., Boyer, E., Castro, M., Cronan, C., ... Ollinger, S. (2003). Nitrogen Pollution: Sources and Consequences in the U.S. Northeast. Environment: Science and Policy for Sustainable Development, 45(7), 8–22.

[2] Cushman-Roisin, B; Anaerobic Digestion of Wastewater Sludge; Introduction to Environmental Engineering; Dartmouth College; New Hampshire. 2018

[3] Water Environmental Federation. Biogas production and use at water resource recovery facilities in the United States. Alexandria, VA: Water Environment Federation; 2013.

[4] Roberts, N.; Hilliard, M.; Bahr, K.; He, Q. P.; Wang, J. Coculture of Methanotrophs and Microalgae – a Flexible Platform for Biological CH4/CO2 Co-Utilization, 2017 AIChE Annual Conference, Oct. 28 – Nov. 3, 2017, Minneapolis, MN. This work won 2017 AIChE Session's Best Paper Award.

557792 Removal of Organics, Metals, Microbes, and Algae Using Electro-Magnetic Fields

Removal of organics, metals, microbes, and algae using Electro-magnetic fields

Abstract

US Clean Water Technology, LLC (USCWT) is commercializing process intensified technology using an Electro Clean Water Unit (ECWU) to treat water from oil and gas operations, bio-diesel production, refinery operations and municipal water plant clean-up of Fats, Oils, and Greases. A 1000 barrel per day unit has demonstrated the technology at an oil field in California, cleanup of wastes at a large chemical producer in California and other locations over the last four years. USCWT is scaling this technology to 3000 bbl./day units for use in a variety of applications.

The flow through technology coagulates any insoluble organic for recovery and at the same time oxidizes soluble organics, metals, microbes and algae without any added treatment chemicals. The oxidized materials are recovered by filtration and carbon absorption (if needed). The unit is compact, automated and linked to the internet for remote monitoring. Further development of this technology is supported with a RAPID project supported by AIChE in conjunction with University of Texas, Texas A/M and University of Pittsburgh.

The proprietary technology features two stages of electrolysis. Sacrificial aluminum electrodes generate aluminum ions which accelerate coagulation of the larger insoluble molecules. Specialized electrodes then generate an electromagnetic field to both enhance coalescence and break down water to form hydrogen, oxygen, peroxides, and hydroxides. Salt in the water breaks down to form soluble chlorine compounds. Insoluble contaminants coalesce and coagulate. Hydrogen and oxygen bubbles help transport the coalesced phase to the surface where it can be skimmed and recovered for reuse or disposal.

Specialized mixed metal oxide electrodes in the second electrolysis stage drive oxidation of dissolved contaminants. The electrodes are an active type in the electrolysis, meaning that the surface of the electrode facilitates the generation of oxidizing specie, which themselves are highly reactive. A process intensification focus on the fundamental reaction mechanisms drove engineering enhancements to overcome mass transfer resistance at the electrode surface resulting in reduced residence time and improved capital efficiency. The second stage effectively oxidizes soluble hydrocarbons, bacteria, and some metals.

The overall process is adapted to meet each application. In general, the waste water passes through a pre-filter to remove sediment, vegetation, etc. The electrolysis stages follow with recovery of insoluble organics and oxidation of soluble contaminants. Post filters remove suspended solids and absorbable contaminants. Other post treatments can be added, such as salt removal, ultra-filtration, or reverse osmosis, depending on the requirements of the treated water. The technology can deliver a treated water suitable for recycle and reuse.

The unit is fully automated and can be monitored remotely so that necessary human oversight is limited. Its small footprint allows the unit to treat water at the point of origin for reuse eliminating transportation costs. Electrical operating costs are about $0.10/bbl. treated water.

Data will be presented to show what this technology can deliver, such as analytical results from oil field producer water processed through both bench and pilot scale equipment to show dramatic reductions in oils and greases, hydrocarbons, metals, BOD, suspended solids, and COD.

The picture shows frack water before and after treatment.

This innovative process intensive technology removes 90+% of organics, handles varying feeds with no intervention, features a low operating cost, requires no chemical additives to remove organics and microbes, can fit on a trailer for easy mobility, and scales readily. It provides industrial water generating industries with economical on-site waste treatment technology enabling the recycle and reuse of wastewater.

577953 Coupled Membrane Distillation-Crystallization for Water and Salt Recovery from High-Strength Brine

The need for alternative freshwater sources is increasing in the U.S. and around the world due to rising populations, expanding urbanization, and climate events. Although many coastal regions rely on seawater desalination for potable water, inland regions cannot rely on the oceans as a feed source for water treatment due to the high cost of transporting water over long distances and against an elevation gradient. Inland brackish waters are a promising alternative water source for non-coastal regions, and desalination of these waters has potential to provide high-quality product water. Unfortunately, most desalination techniques developed for coastal regions cannot be directly translated to their full extent inland due to differences in water chemistry between seawater and brackish waters. Additionally, concentrated brine disposal is a challenge in non-coastal areas as the ocean cannot serve as a repository for the brines.

This work evaluates Membrane Distillation-Crystallization (MDC) as a method for effectively treating reverse osmosis (RO) brine resulting from desalinating inland brackish waters to minimizing concentrate volume. The process combines membrane distillation (MD) with crystallization. MD is a non-pressure-driven separation process that involves the transport of water through a hydrophobic, microporous membrane. Compared to conventional distillation methods, MD requires only small temperature differences and has a simple configuration that makes it highly suitable for coupling to other processes. The driving force of MD is the vapor pressure of the brackish water, which is relatively independent of salt concentration, thus it can be used to treat brines to near saturation. When coupled with a crystallization unit, the overall system can remove solid salts crystals away from the membrane surface to maximize membrane longevity and produce high-quality distillate water.

The bench-scale system uses a novel, multi-stage, cascading crystallizer designed by the University of Nevada team in conjunction with a flat-sheet MD unit with 140 cm^2 of active membrane area. The MDC system has been validated using a 290 g/L solution of NaCl as an initial feed, with MD operating at 50 / 25 C as the feed / distillate side temperatures. After 10 hours of batch operation, the MDC system recovered 2.5 times more water than MD alone, while also capturing 150 g of solid salt in the crystallizer. When cooling was added to the crystallizer, 3.3 times more water was recovered compared to MD alone, and 195 g of solid salt were recovered. Further testing with additional salts is currently underway and will be presented.

578143 An Asymmetric Iron-Based Redox-Active System for Electrochemical Ion Separation in Aqueous Media

Chemical separations are of crucial importance and are used ubiquitously in the large-scale production of commodity chemicals and valuable species. However, as the purification steps in industrial manufacturing are often highly material and energy-intensive, there is a clear urgency for pursuing green engineering to develop and implement clean energy-efficient separation processes to reduce the economic and environmental impacts of high global energy consumption.[1,2] The flip side to the same coin is the critical relevance of separations in aqueous media for environmental remediation and drinking water pollution, which often require the removal of trace molecules of interest among natural constituents.

Electrochemical processes are an attractive modular platform for water purification, as they do not require solvents that change solution chemistry or impart environmental effects downstream, as well as heat and pressure inputs like conventional techniques.[1] Electrochemical separations can be further designed to selectively target certain species where methods such as distillation or chromatography may be inefficient.[1,2] Redox-active materials such as metallopolymers have previously been shown to be able to electrochemically separate targeted inorganic oxyanions[3] and organic anions in aqueous[4,5] and organic[6] solution with high selectivity over competing species via the activation of specific chemical binding. The adsorption and desorption processes are reversible and switched via an electrochemical stimulus.

We report a new asymmetric dual-functionalized Faradaic system that leverages the redox-active iron centers of a metallopolymer at the anode and tunable inorganic mixed-valence crystalline intercalating species at the cathode to carry out enhanced electrochemical separation of transition metal oxyanions in aqueous media. The process uses heterogeneous conductive composite electrodes containing these species to suppress solution pH changes from the parasitic water splitting side reactions, provide narrower potential windows to perform the separation, facilitate increased anionic adsorbate removal via cathodic cation intercalation, and also prevent oxidation state changes of homogeneous redox-active species during electrosorption.

References:

(1) Anastas, P. T.; Zimmerman, J. B. *Environ. Sci. Technol.* **2003**, *37*, 94A.

(2) Sholl, D. S.; Lively, R. P. *Nature*, **2016**, *532*, 435.

(3) Su, X.; Kushima, A.; Halliday, C.; Zhou, J.; Li, J.; Hatton, T.A. *Nat. Commun.* **2018**, *9*, 4701.

(4) Su, X.; Kulik, H. J.; Jamison, T. F.; Hatton, T. A. *Adv. Funct. Mater.* **2016**, *26*, 3394.

(5) Su, X.; Tan, K. J.; Elbert, J.; Rüttiger, C.; Gallei, M.; Jamison, T. F.; Hatton, T. A. *Energy Environ Sci.* **2017**, *10*, 1272-1283.

(6) Achilleos, D.; Hatton, T. A. *ACS Appl. Mater. Interfaces* **2016**, *8*, 32743-32753.

560159 Amine-Based CO_2 Capture Aided By Acid/Base Catalyst: Advancement of Amine Regeneration Using Fe-MCM-41 Catalyst

Amine-based CO_2 capture aided by acid/base catalyst: Advancement of amine regeneration using Fe-MCM-41 catalyst

Xiaowen Zhang [1], Hongxia Gao [1], Zhiwu Liang [1, *], Paitoon Tontiwachwuthikul [1], Maohong Fan [2]

[1] Joint International Center for CO_2 Capture and Storage (iCCS), Provincial Hunan Key Laboratory for Cost-effective Utilization of Fossil Fuel Aimed at Reducing CO_2 Emissions, College of Chemistry and Chemical Engineering, Hunan University, Changsha 410082, PR China

[2] College of Engineering and Applied Science, University of Wyoming, Laramie, WY 82071, USA

Keywords: CO_2 capture; catalytic CO_2 desorption; Brϕnsted acid site; basic site, heat duty reduction.

* Author for correspondence, Email: zwliang@hnu.edu.cn

Abstract

Carbon dioxide (CO_2) has become the focus of much attention because it is the most important human caused greenhouse gas, contributing more than 60% of the enhanced greenhouse effect. Amine-based CO_2 capture process is the most developed and the most commercial mature technology that can be used for CO_2 capture. 5 M monoethanolamine (MEA) solvent is the benchmark solvent and has been the most frequently studied. However, commercial implementation of this technology suffers from significant challenges, particularly the enormous energy consumption for the rich amine solvent regeneration, resulting in its immensely high operating costs of CO_2 capture.

Our previous work demonstrated that the introduction of solid acid catalysts to the CO_2 desorption process is a potential approach to decrease the heat duty [1]. Several solid acid catalysts including, H-ZSM5, SAPO-34, and $SO_4^{2-}/ZrO_2/$ gamma-Al_2O_3 have been reported and presented superior catalytic desorption performance for this purpose [2, 3]. To further advance this promising method, in the present work, a composite catalyst MCM-41 (MFe) was synthesized using a post synthesis of the combined impregnation ultrasound method and used for the first time as a catalyst in a rich MEA solution regeneration process. MCM-41 has the advantages of highly ordered porous structure, suitable Lewis acid sites (LAS), larger surface area, and thermal stability, but encounters lower B/L ratio and weak Brϕnsted acid sites (BAS). The MCM-41 is modified by iron oxide (Fe_2O_3) will generate more BAS and basic active sites. Potentially, the MFe catalyst may show a superior catalytic CO_2 desorption performance for the CO_2-loaded amine solvent regeneration process.

In this work, MFe was prepared and used for the rich 5 M MEA solvent regeneration process to increase the CO_2 desorption kinetics. Three MFe catalysts with different Fe_2O_3 loading contents (5%, 10% and 15%) were synthesized, and marked as MFe5%, MFe10% and MFe15%. The procedure of catalyst preparation is based on Lan et al. [4], with a modification. All the catalysts were measured with XRD, FT-IR, ICP, N_2 absorption, Py-IR, and CO_2/NH_3-TPD techniques.

The regeneration experiment was conducted using a batch reactor, which is reported in previous work [3]. However, the mechanical agitator was replaced by a magnetic stirrer with a stirring speed of 500 rpm in this work. Typically, 1 L rich MEA solutions with the desired CO_2 loading (0.5 mol CO_2/mol amine) were added into a 2 L four-necked flask, which was immersed in the oil bath. 12.5 g catalyst was put into the MEA solution. After the desorption experiment started (initiating from 0 min after reaching the desorption temperature of 333 K, and the final temperature was kept at 371 K), the regenerated amine solution was taken to determine the CO_2 loading at 30, 60, 90, 120, 240, 360 and 480 min.

The heat duty (HD, kJ/mol) for the CO_2 desorption process was monitored by an electric energy meter, which was calculated by equation (1).

$$HD = \frac{\text{Heat input / time}}{\text{Amount } CO_2 \text{ / time}} = \frac{\text{Electricity(kJ)}}{\text{Amount of desorbed } CO_2 \text{(mol)}} \quad (1)$$

The relative heat duty (RH, %) was calculated by equation (2).

$$RH = \frac{HD_i}{HD_{baseline}} \times 100\% \quad (2)$$

where $HD_{baseline}$ is the HD of MEA without catalyst regeneration in first 90 min, while HD_i is the HD of MEA with different catalysts regeneration in first 90min.

The CO_2 desorption rates (DR, mol/min) are determined by using a linear fitting the curves of CO_2 loadings against its corresponding times (at the first 90 min), and the slope of the fitted curve is defined as the desorption rate.

Desorption factor (DF, mol^3/(kJ.min)) was applied to fully assess the catalytic CO_2 activity of the prepared catalysts. Where, CC (mol) refers to the amount of desorbed CO_2 at the first 90 min.

$$DF = \frac{DR \times CC}{HD} \quad (3)$$

The XRD patterns and FT-IR spectra of MFe10% and MCM-41 are shown in Figure 1, which certify that the MFe catalysts were successfully synthesized. The ICP test results also indicate that the actual Fe loading of the MFe catalyst is close to the normal value. The N_2 physisorption isotherms for all catalysts are illustrated in Figure 2 (d). All the isotherms are type IV with a specific hysteresis loop, representing typical mesoporous material characteristics. The BET surface area and pore volume of the MFe catalysts are slightly decreased with metal loading increasing up to 15%. However, the acid sites, BAS, B/L ratio, and basic sites of the MFe catalysts are higher than that of the parent MCM-41.

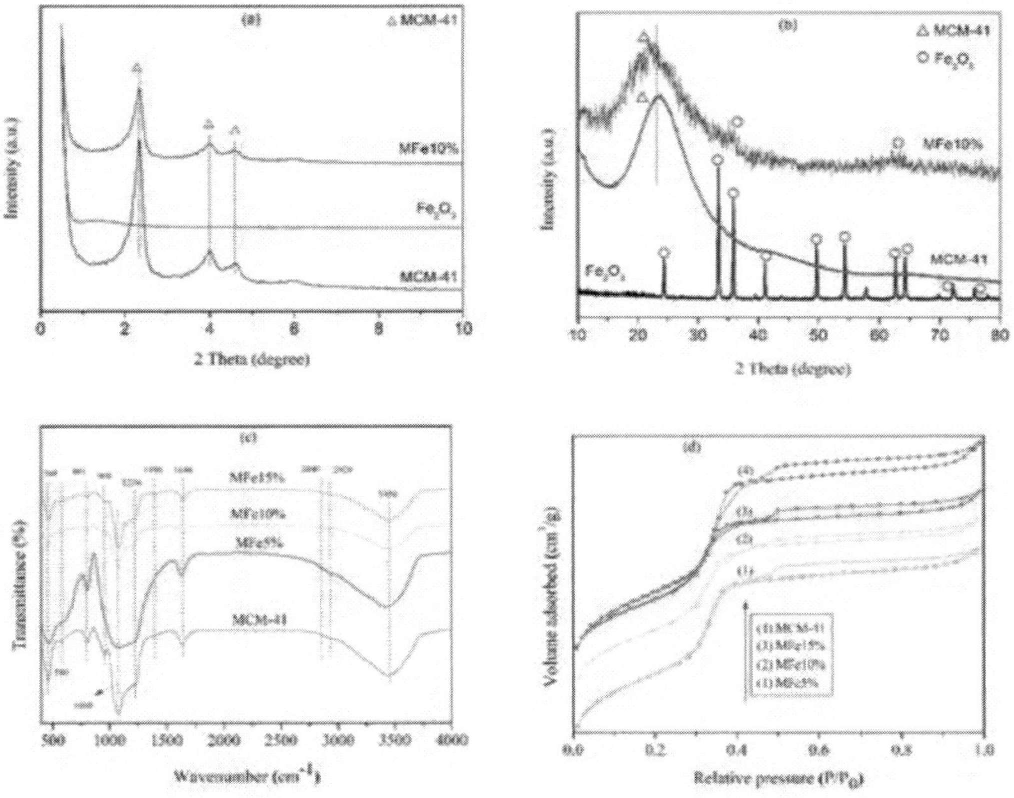

Figure 1. Characterization results of various catalysts: (a) low angle XRD patterns, (b) high angle XRD patterns, (c) FT-IR spectra, (d) N_2 physisorption isotherms.

The CO_2 desorption curves of the 5 M MEA solution with different catalyst are displayed in Figure 2. These curves manifest the catalytic performance of five catalysts in MEA solution for the whole 480 min duration conducted at 371 K. It is noticed that the CO_2 loading decreased rapidly in the first 90 min after which the extent of CO_2 loading change was smaller. Thus, all analyses for the catalytic performance of each MEA-catalyst system were performed based on the first 90 min of the desorption processes. As shown in Figure 2(a), all the catalysts quickened the CO_2 desorption rate compared with the blank run. Three MFe catalysts demonstrated better catalytic activity than the parent catalysts MCM-41. It is important to highlight that the lean CO_2 loading of all the MEA-catalyst systems tended to the same value at about 0.35 mol CO_2/mol amine, implying that the catalyst only accelerated the desorption rate but did not change the thermodynamic equilibrium. With regard to the RH, MFe10% offered the best catalytic activity, which decreased the heat duty by 32.5% as compared with the blank run (Figure 2(b)). The desorption factor of MFe10% catalyst is 337.3% higher than the catalyst-free test (Figure 2(c)). Moreover, it can be seen from Figure 2(d) that the MFe10% presented good catalytic stability. Therefore, the MFe10% catalyst can be considered as a potential catalyst for reducing the energy consumption for CO_2 desorption reaction and for use in industrial CO_2 capture processes.

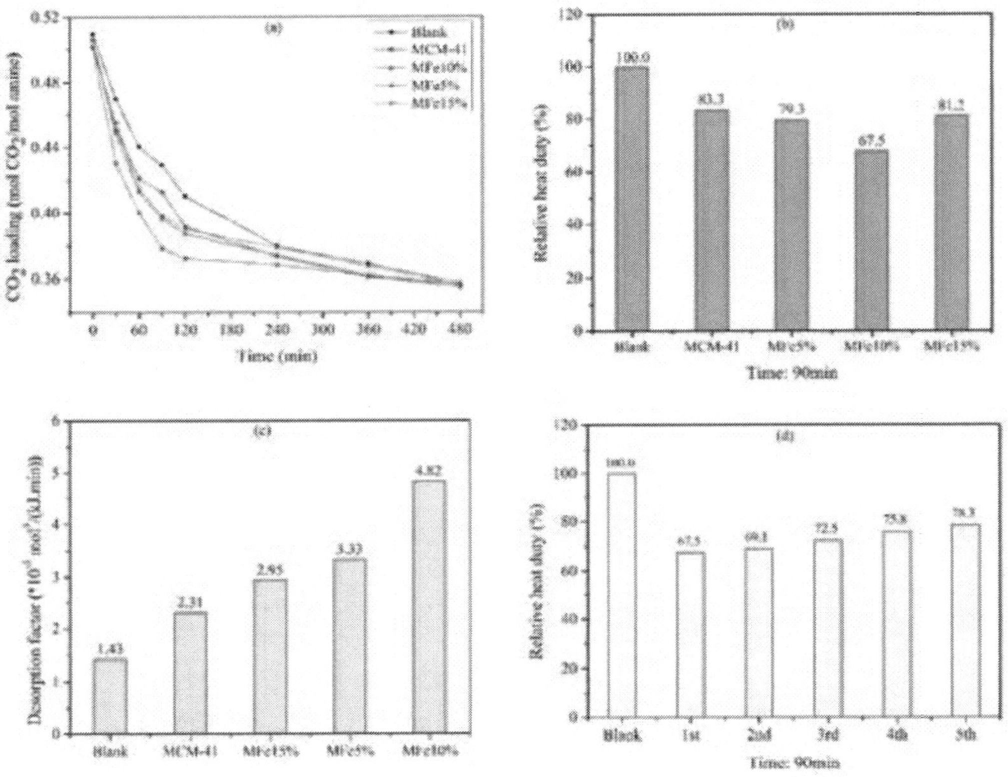

Figure 2. Catalytic CO_2 desorption performance: (a) CO_2 loading change curves, (b) relative heat duty, (c) desorption factor, (d) relative heat duty of cyclic tests.

The structure-activity correlation of the catalyst is investigated. The relationships between the BAS, basic sites and the catalytic activity (i.e. DF) were explored, and the result is shown in Figure 3. Note that the catalytic performance increased with increasing BAS. This confirms that the BAS plays a significant role in its catalytic performance. With regard to the basic sites, it is noticed that there is no linear relationship between the basic sites and the catalytic performance (Figure 3(b)). However, it is inferred that the enhanced catalytic performance of composite catalysts is in relation to their improved basic sites compared with the MCM-41 alone. These results demonstrate that the BAS and basic sites play a predominant role in the catalytic CO_2 desorption activity. Therefore, the enhanced catalytic performance of composite catalysts can be mainly attributed to their improved BAS and basic sites. The surface area also has an important effect on the catalytic performance. However, due to the large surface area of these composite catalysts, the effect of this factor should be similar in the catalytic amine solvent regeneration processes with these MFe catalysts.

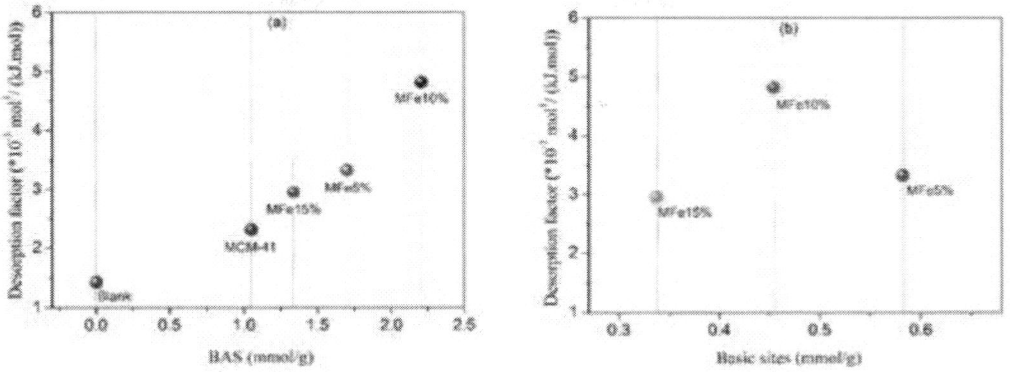

Figure 3. The structure-activity correlation of the catalyst: (a) BAS, (b) basic sites.

Acknowledgment

The financial support from the National Natural Science Foundation of China (NSFC-Nos. 21536003, 21706057, 21606078 and 51521006), China Scholarship Council (201806130065) and Hunan Provincial Innovation Foundation for Postgraduate (CX2018B154).

References

1. Liang, Z., et al., AIChE Journal, 2016. 62(3): p. 753-765.

2. Zhang, X., et al., Applied Energy, 2018. 218: p. 417-429.

3. Zhang, X., et al., AIChE Journal, 2018. 64(11): p. 3988-4001.

4. Lan, B., et al., Chemical Engineering Journal, 2013. 219: p. 346-354.

563468 High Regeneration Efficiency and Low Viscosity of CO_2 Capture By Functionalized Ionic Liquid Organic Solution

To reduce the high regeneration energy consumption of traditional amine or ionic liquid aqueous solution and the high viscosity of the pure ionic liquid or saturated amine non-aqueous solution, a novel functionalized ionic liquid combined with organic solvents is innovatively proposed for CO_2 capture in this project. In this novel system, the functionalized ionic liquid tetraethylenepentamine 2-methylimidazole ([TEPAH][2-MI]) as the main absorbent which ensures high CO_2 absorption rate and loading, and the organic solvent can significantly reduce the regeneration energy consumption of the system compared with water. In addition, ethylene glycol (EG) was added as a regulator to reduce the volatility of the solvent and prevent the absorption product from being precipitated in the saturated solution. The results showed that the [TEPAH][2-MI]/NPA/EG solution had a high absorption loading of 1.72 mol/mol (28 wt% and 257 g/L). The viscosities before and after absorption were 3.47 and 6.20 mPa·s respectively, which were far less than traditional pure ionic liquids and amine non-aqueous solutions. By heating for 45 min at 120 ^oC, the regeneration efficiency of the solutions could reach 90.7% after the 5th regeneration cycle. Moreover, the reaction mechanism was elucidated based on the results of ^{13}C NMR. Polycarbamate and alkyl carbonate formed in the [TEPAH][2-MI]/NPA/EG solution during the absorption process, and then the reaction products could decompose after desorption.

567715 Understanding the Behavior of Degraded Amine Products on CO_2 Capture in Aqueous Amines By Molecular Simulations

In the context of sustainable development and clean energy production, one of the most important alternatives to mitigate anthropogenic CO_2 emissions is to capture and separate CO_2 (CCS) from diluted sources, such as gases emitted from fossil fuel combustion and other industrial processes. The most standard process used in industry today is the separation by absorption in aqueous amine solvents. Among them, monoethanolamine (MEA) is the benchmark molecule because of its properties such as high CO_2 cyclic capacity, high kinetic at low CO2 partial pressure, low viscosity, high water solubility, low price, etc. Nevertheless, it has some disadvantages such intensive energy required for desorption and solvent degradation, the last one counting for around 10% of the total cost of CO2 capture. Furthermore, byproducts cause a noteworthy decrease in process efficiency with solvent losses, corrosion, fouling, foaming and an increase in viscosity.

The degradation of MEA may arise thermally in the presence of CO2, which occurs at high temperature and high CO2 partial pressure in the stripper, or through oxidative degradation, mainly due to large amount of O2 in flue gases. Thermal degradation causes dealkylation, dimerization and cyclisation. First, MEA, as primary amine, reacts with CO2 to form carbamate in the absorber and the carbamate can be transformed into 2-oxazolidone (OZD). Subsequently, OZD reacts with the MEA molecule to form HEEDA (N-(2-Hydroxyethyl) ethylenediamine), precursor of HEIA, which is known to thermally degrade quickly. As a result, HEEDA cannot accumulate in the process and HEIA is formed by the reaction between HEEDA with CO2, and followed by intramolecular cyclisation. The three major thermal degradation products of MEA are, hence, HEIA, the dominant one, HEEDA and OZD. HEIA is the largest degradation product found in solution that do not react to form further polymeric products. Its presence increases with time, which is an indication of its relative stability, and it does not affect the solution's corrosivity.

We will present and discuss results concerning the effect of the presence of degradation amine products into the microscopic behavior of the CO2 capture process with aqueous amines, focused on HEIA. Molecular simulations in the NVT and NPT ensemble have been performed for this purpose. The study has been done in a systematic way. We first studied the pure component (HEIA), validating the force field with experimental density data, followed by the study of the binary (HEIA+H2O, HEIA+MEA, MEA+H2O), ternary (HEIA+H2O+MEA, HEIA+H2O+CO2, MEA+H2O+CO2) and the quaternary system (HEIA+MEA+H2O+CO2). It is observed that the HEIA shows no specific interactions with either the MEA or water, and it forms aggregates to avoid the contact with them. We will discuss the implications of this microscopic behavior into the final absorption performance of the aqueous amine solvent, with and without HEIA. This work shows how molecular simulations can help to elucidate the absorption mechanism and to find optimal absorbents for CO2 capture.

This work has been partially funded by ADNOC Gas Processing and its stakeholders (Total, Shell and Partex) through a Gas Research Center project (GRC2018-003) and Khalifa University of Science and Technology (project RC2-2019-007).

574814 Ptmgfe/Ni-Based Al_2O_3-CeO_2 Catalysts for Dry-Reforming of Methane

The dry reforming of methane (DRM) converts two major greenhouse gases into a synthesis gas with H_2:CO ratio of close to 1 and, thus, can be used readily for Fischer-Tropsch and carbonylation reactions.[1][2] In recent years, CO_2 reforming of methane (dry reforming of methane, DRM) has become an attractive research area because it converts two major greenhouse gasses into syngas (CO and H_2), which can be directly used as fuel or feedstock for the chemical industry.[3] The promotional effect of Fe, Mg, Mo, and Pt on Ni-based catalyst supported on Al_2O_3-CeO_2 (Ni/$A_{12}O_3$-CeO_2) have been investigated in dry reforming of methane (DRM) reaction. To elucidate the influence of metal promoters on the Ni/Al_2O_3-CeO_2 materials, the catalytic reaction was conducted in a continuous fixed-bed flow reactor at various reaction temperatures (550-700 °C), space velocity of 12,000 ml^{-1} h^{-1} and different feedstokcs (pure CH_4 or 50% CH_4 and CO_2 mixture) under atmospheric pressure for 10 h time on stream. The combination of material characterization and catalytic test reveals that Fe, Mg, and Mo-doped Ni/Al_2O_3-CeO_2 are excellent catalyst for this process helping to achieve high degrees of CO_2 conversions. The co-doped Ni/Al_2O_3-CeO_2 catalysts with Mg and Fe significantly enhanced the activity, selectivity and stability of the obtained Mg-Fe/Ni/Al_2O_3-CeO_2 composite catalyst. Further, addition of 0.05wt% Pt to the Mg-Fe/Ni/Al_2O_3-CeO_2 composite catalyst leads to an excellent catalytic activity/stability at 700 oC during the dry reforming reaction

References

[1] H. Wu *et al.*, "Bi- and trimetallic Ni catalysts over Al2O3and Al2O3-MOx(M=Ce or Mg) oxides for methane dry reforming: Au and Pt additive effects," *Appl. Catal. B Environ.*, vol. 156–157, pp. 350–361, 2014.

[2] M. Ocsachoque, F. Pompeo, and G. Gonzalez, "Rh-Ni/CeO2-Al2O3 catalysts for methane dry reforming," *Catal. Today*, vol. 172, no. 1, pp. 226–231, 2011.

[3] Z. BIAN, S. Das, M. H. Wai, P. Hongmanorom, and S. Kawi, "A review on bimetallic Ni-Based catalysts for CO2 reforming of methane," *ChemPhysChem*, vol. 18, pp. 3117–3134, 2017.

562786 Synthesis and Performance Evaluation of New Bleach Activators

Tetraacetylethylenediamine (TAED) is used as a bleach activator in the detergent market. TAED is only used in powder formulations due to its instability in aqueous formulations and its reaction with bleaching agents (peroxide source) under storage conditions. To address the hydrolysis issue, two imide derivatives of TAED are synthesized. Due to the increased steric hindrance close to the imide group, these two molecules exhibited improved hydrolytic stability but still had similar perhydrolytic activity compared with TAED, suggesting their potential as bleach activators.

567378 Sustainable Product Design: A Life-Cycle Approach

A systematic framework is proposed for sustainable chemical product design. The product technical requirements are first identified as design constraints. Then, a base-case product is generated as a reference on top of which a more sustainable product is designed. To do so, the life cycle of base-case product (in particular, whether and how it should be recycled) is decided by using life cycle sustainability assessment (LCSA) or rule-based methods depending on the availability of life cycle inventory data. In addition, the hotspots (i.e., life cycle stages with major impact) are identified. Afterwards, to reduce the impact on hotspots, product design targets and design alternatives are generated using knowledge-base and heuristics. Lastly, LCSA or rule-based methods is applied to decide the most sustainable product from the generated product design alternatives. Two case studies – composite bumper beam and lithium ion battery – are provided to illustrate the framework.

571392 Combined Life Cycle Analysis and Environmental Risk Assessment As a Tool to Evaluate the Circular Economy of Plastics

There is currently over 7.3 billion tons of plastic in the Earth System, and it will grow to be 40 billion tons by 2050. Approximately 10 mega tons of plastics reach the oceans every year; 12 billion tons are projected to be in landfill by 2050. As these are released, the flux of plastics to surface waters, coastlines and oceans is amplified, with significant environmental, social and economic cost. In the global north, plastic ends up in the environment through leakages from waste collection, recycling and disposal, driven by the growth in single-use plastic, with a significant amount of egregious litter and informal disposal. Current understanding of the pathways and fate of plastic in the environment is poorly constrained, with only fragmentary evidence restricted to parts of the cycle. The problem is now part of public consciousness and is on the political agenda nationally and internationally. The work presented here is part of the £1m Redefining Single Use project at the University of Sheffield which aims to stimulate creative thinking across disciplines and explore novel ideas and solutions with potential to deliver more circular economic approaches to plastics manufacture and utilisation.

Overcoming the negative environmental impacts of our current use of plastics requires both a step change in thinking and innovative and transformative actions. To ensure we are not inadvertently causing more problems by solving one, sustainability analysis is essential. Here we combine the two analytical approaches which dominate environmental policy and decision-making to create a tool suitable for analysing the circular economy of plastics: environmental risk assessment (ERA) and life cycle assessment (LCA). Environmental risk assessment (ERA) couples information on the fate and behaviour of chemicals in the environment with information on the effects of chemical exposure on ecological systems to predict the probability that adverse environmental effects will occur (i.e. risk). Life cycle assessment (LCA) provides an integrative assessment of a product and process by considering the environmental and health effects from the extraction of raw materials to their recycling or disposal ('cradle to cradle/grave'). The framework of the new analysis tool will be discussed, along with an example of its use to analyse the use of plastics, and alternatives, in the supply chain of milk from farm to café.

577108 Sustainability Potential of Melamine Non-Woven Prepared As a High-Performance Material for Thermal and Acoustic Applications

Melamine Etherified Resin (MER) fibers, commercially known as smartMELAMINE, are modified melamine resin fibers obtained by melt-blown technology. A combination of organic, lightweight, fire-resistant, acoustic and thermal insulating properties of MER fibers in combination with excellent textile processability offers a wide range of potential industrial use. Results of flammability test of smartMELAMINE compared to commonly used synthetic fibers, showed that smartMELAMINE does not burn, shrink, melt or drip. These properties offer a great advantage for many technological uses, like a fire blocker, construction material, mobility applications (in automobiles, trains, buses, trucks and aircraft), for thermal protective clothing (workwear), heavy-duty military and firefighters gear, dust/hot gas filtration and various other industrial applications.

Various types of smartMELAMINE non-woven material can be produced with different molecular weight, grammage (30 – 600 gsm), material density (60 – 10 kg/m^3) and filament diameter (1 - 25 μm). In addition, obtained melamine non-woven can be chemically modified to obtain material with improved properties. Sustainability potential and recycling opportunities for the above-mentioned material in various stages of its life-cycle will be discussed.

Acknowledgments

Authors would like to acknowledge financial support from the Ministry of Education, Science and Sport of Republic of Slovenia and European Regional Development Fund for the project No. 5442-1/2018/106. This project has also received funding from the European Union's Horizon 2020 research and innovation programme under grant agreement No. 756081.

568330 Shaping Technology Innovation: A Multistage, Multiscale Sustainability Assessment and Decision Support Framework

Industries are experiencing a new wave of technology innovation largely due to the national endeavor for advanced manufacturing in the thrust areas of national importance. The accelerated innovations entail rapid transfer of emerging technologies into design and manufacturing of high performance products. Although the technologies become an engine of change and progress, the net profit brought to the environment and society could be questionable, if sustainability principles are not fully incorporated into technology development and application phases. Unfortunately, there have been nearly no systematic methodologies and effective tools to ensure the sustainability of technology innovations. This renders an urgent research need to develop methodologies and tools for inventors and organizations to evaluate potential sustainability risks of emerging technologies.

In this paper, we will introduce a framework for conducting comprehensive sustainability assessment and impact prediction of emerging technologies. The framework will be composed of a coherent set of concepts, propositions, principles, methodologies, and a tool that could assist organizations, researchers, and policy makers in shaping technology innovation for industrial sustainability. The framework has been tested through investigating new technologies for nanopaint design and nanopaint-based automotive coating manufacturing in highly sustainable process systems. Comprehensive sustainability performance comparison against current painting material and application will be presented.

---

* All correspondence should be addressed to Prof. Yinlun Huang (Phone: (313) 577-3771; E-mail: yhuang@wayne.edu).

566811 Designing an Electrochemical System for Simultaneous Extraction of CO_2 and Rejection of Salts from Seawater

Two of the major challenges of the present times are the rapidly increasing demand for fresh water and the rise of anthropogenic CO_2 in the atmosphere and consequent acidification of the water bodies. Because approximately 70% of the earth is seawater, one of the obvious solutions to meet the freshwater crisis is desalination of seawater. However, seawater is extremely saline and must reject 99% of the salt concentration in order to be accessible for applications like irrigation and drinking. For the commercially available technologies for seawater desalination like Reverse Osmosis (RO) and Multi-Stage Flash Distillation (MSF), there is a trade-off between energy efficiency and salt rejection efficiency. Also, due to the rise in CO_2 concentrations in the atmosphere, more CO_2 gets absorbed in the water bodies leading to the formation of bicarbonates which lower the pH of seawater causing trouble for marine biota. An efficient method to extract this dissolved CO_2 would essentially enable sequestration of CO_2 from the atmosphere without the need to process large volumes of air. To counter both the issues, we propose a stack-type electrochemical cell employing bipolar membranes to desalinate seawater. The cell stack of an acidic solution, bipolar membrane, anion-exchange membrane, cation-exchange membrane, bipolar membrane, and a basic solution divide seawater into a brine stream and a clean water stream by electrodialysis. Due to the high concentration of H+ ions in the acidic chamber of the cell, the equilibrium of the carbonate buffer system shifts so that CO_2 is evolved at the acidic solution and can be sequestered at that point. The feasibility of this design was analyzed using COMSOL Multiphysics. The simulation results show that a stacked arrangement of such cell can successfully reach a desalination efficiency of 99% at near neutral pH. Effect of operating parameters such as the width of the desalination chamber, concentration of salts, and flow rate of seawater is also reported as a part of this work.

566254 Breakdown Products from Perfluorinated Alkyl Substances (PFAS) Degradation in a Plasma-Based Water Treatment Process

Byproducts produced when treating perfluorooctanoic acid (PFOA) and perfluorooctane sulfonate (PFOS) in water using a plasma treatment process intentionally operated to treat these compounds slowly to allow for byproduct accumulation were quantified. Several linear chain perfluoroalkyl carboxylic acids (PFCAs) (C4 to C7) were identified as byproducts of both PFOA and PFOS treatments. PFOA, perfluorohexane sulfonate (PFHxS) and perfluorobutane sulfonate (PFBS) were also found to be byproducts from PFOS degradation. Significant concentrations of fluoride ions, inorganic carbon and smaller organic acids (trifluoroacetic acid, acetic acid and formic acid) were also identified. In addition to PFCAs, PFHxS and PFBS, trace amounts of 43 PFOA-related and 35 PFOS-related byproducts were also identified using a screening and search-based algorithm. Minor concentrations of gas-phase byproducts were also identified (< 2.5% of the F originally associated with the parent molecules) some of which are reported for the first time in perfluoroalkyl substance degradation experiments including cyclic perfluoroalkanes (C_4F_8, C_5F_{10}, C_6F_{12}, C_7F_{14} and C_8F_{16}). The short chain PFCAs detected suggest the occurrence of a step-wise reduction of the parent perfluoroalkyl substances (PFAS) molecule, followed by oxidation of intermediates, perfluoroalkyl radicals and perfluoro alcohols/ketones. Using a fluorine mass balance, 77% of the fluorine associated with the parent PFOA and 58% of the fluorine associated with the parent PFOS were identified. The bulk of the remaining fluorine was determined to be sorbed to reactor walls and tubing using sorption experiments in which plasma was not generated.

566171 Hydrothermal Degradation of Emerging Pollutants from Water

The concentration of emerging pollutants (e.g., hormone, antibiotic) in river water has become an alarming scenario in the United States in last few year. As the half lives of those pollutants are long, runoff of those substance in river water is an unavoidable phenomenon. Hormones are biologically active at very low levels and may disrupt the reproductive endocrinology of fish, while antibiotics facilitates the development and proliferation of antibiotic resistance bacteria (ARB). So, reduction of their concentration in river water is must needed expectation.

Hydrothermal carbonization (HTC) a thermochemical process, where biomass is treated at high-temperature with auto generated high-pressure (mainly water vapor pressure). It is a promising conversion technology that can convert biomass into different value added products. As most pharmaceutical products are vulnerable at high temperature and pressure, HTC technology could break down the emerging pollutants.

The objective of this study was to investigate the degradation phenomenon of emerging pollutants with HTC temperature, residence time, and initial chemical states (acidic or basic). A series of HTC experiments were conducted at different temperatures between 220 and 300 °C, two different residence time of 0.5 hour and 6 hours, and different acidic or basic conditions between pH 3 to 10 by using two model compounds (oxytetracycline and ß-estradiol). HPLC and GC-MS were used to analysis the non-degraded amount of hormone/antibiotic in the sample and the generated product due to their degradations, respectively. The analysis showed that the concentration of hormone and antibiotic reduced significantly at low HTC temperature in presence of highly acidic or basic environment.

(670d) Study the Degradation and Adsorption Processes of Organic Matters from Domestic Wastewater using Chemically Prepared and Green Synthesized Nano Zero-Valent Iron

Rabie S. Farag[1], Maha M. Elshfai[2], Ahmed S. Mahmoud[2], Mohamed K. Mostafa[3], Ahmed Karam[4], and Robert W. Peters[5]*

[1]*Faculty of Science, Chemistry department, Al-Azhar University*

[2]*Sanitary and Environmental Institute (SEI), Housing and Building National Research Center (HBRC), Egypt*

[3]*Faculty of Engineering and Technology, Badr University in Cairo (BUC), Cairo, Egypt*

[4]*Civil, Infrastructure Engineering and Management Department, Nile University, Giza, Egypt*

[5]*Department of Civil, Construction and Environmental Engineering, University of Alabama at Birmingham, Birmingham, AL, USA*

*Corresponding author phone: +201152773314; e-mail: ahmeds197@gmail.com

Abstract

Advanced oxidation processes (AOPs) using chemically prepared and green synthesized nano zero-valent iron (nZVI) has proved to be effective in removing organic contaminants. The green synthesized nano iron (GT-nZVI) was prepared by using extracted black tea reducing agent. The prepared nZVI particles were characterized using X-ray powder diffraction (XRD), scanning electron microscopy (SEM), and Energy Dispersive X-ray Analysis (EDAX) analysis. The main purpose of this study is to compare between chemically prepared nZVI and GT-nZVI in the biological oxygen demand (BOD) removal efficiency from domestic wastewater. Different nonlinear adsorption and kinetic models were also used to describe the mechanism of organic matter adsorption onto differently prepared nano iron particles. The effects of nZVI and GT-nZVI on BOD removal were studied at different pH, adsorbent dose, contact time, stirring rate, and concentrations. The results indicated that nZVI and GT-nZVI are effective in the removal of BOD from wastewater, where the maximum removal efficiencies of 86.4 and 72.5% were achieved for 300 mg/L BOD, respectively. The adsorption and kinetic data of BOD removal using chemically prepared nZVI fitted well to Freundlich and Avrami, respectively. On the other hand, the adsorption data for BOD removal using GT-nZVI fitted well to both Toth and Langmuir adsorption models, while the kinetic data fitted well to Avrami model. Artificial neural network (ANN) adequately predicted BOD removal (R^2 > 90%) for differently prepared nano iron particles.

KEYWORDS: BOD removal; Organic matter removal; GT-nZVI; Artificial Neural Networks (ANN); Isotherm and Kinetic studies

1. Introduction and Background

Wastewater contamination is the most observable environmental problem among many urban areas (Ngoc and Schnitzer, 2009). Increasing world population increases water consumption and produce an enormous amount of wastewater every day (Mahmoud et al., 2018). Nowadays about 2.4 billion people in urban and rural areas don't have contact with adequate sanitation services (Daily and Ehrlich, 1992, Gleick, 2000, Rijsberman, 2006). There are expectations of increasing the demand for sanitation by 2 billion will live in towns and cities, mainly in developing countries within 20 years (Langergraber and Muellegger, 2005). More than 90% of developing countries sewage is released untreated and polluting ponds, rivers and coastal areas (Islam and Tanaka, 2004, Liu and Diamond, 2005, Rabie S. Farag and Peters, 2018). So that, the municipal wastewater has been marked as an international problem as it involves sanitary hazards that containing organic matter and eutrophying substances in nitrogen and phosphorus forms. These hazard materials cause unusual contamination in lakes, seas, and streams (Esrey et al., 2001, Mahmoud et al., 2019a).

Developing nations try to find nontraditional, economic and sustainable techniques for wastewater treatment using active materials and methods to overcome the water deficiency problem (Amin et al., 2014, Ferroudj et al., 2013, El-Shafei et al., 2016). The conventional treatment techniques such as dilution, physicochemical and biological are not adequate to decrease a wide range of wastewater contaminants, highly energy consuming, incomplete pollutant removal and generated toxic sludge (Burkhard et al., 2000, Elawwad et al., 2017). The best treatment process can be achieved by improving the existing methods through some modifications. Nanotechnologies treatment process using active nanoparticles such as nano Zero Valent Iron (nZVI) can be introduced to solve all of these problems in quick and economic ways. Previously, iron was used for degradation of organochlorine pesticides in 1972, also for the degradation of other organic compounds such as trichloroethylene in 1981 from iron powder (Mostafa et al., 2017b). Several studies indicated that nano zero-valent iron (nZVI) is more reactive than micro Zero Valent Iron (mZVI) in both adsorption and degradation process (Mostafa et al., 2017a, Rabie S. Farag and Peters, 2018). Nanoparticles can isolate wastewater contaminants (via adsorption or complexation) making them immobile or can degrade or transform them into innocuous compounds (Farag et al., 2018, Mahmoud et al., 2019b). High production cost, reactivity toward fillers, storage difficulties, and the presence of unwanted materials in the treated water are enough to reduce nano applications in wastewater treatment fields (Jason A. Heberling, 2018, Mahmoud et al., 2019c). Green synthesized nano iron (GT-nZVI) can solve all of these problems because it can prepare using economic and safe materials, and the formation of carbon and oxide outer layers leading to increase its stability (Mahmoud et al., 2019a).

This work attempts to examine the removal efficiency of organic matter represented in BOD from aqueous solution using nZVI and GT-nZVI. The relation between BOD removal efficiency and dosages, pH, contact time, stirring rate and concentration were studied. This study also aims to prepare low cost and high stability nano iron with particle size less than 80 nm. This study act as comparative studies for BOD using nZVI and GT-nZVI at the same operating conditions. Artificial neural network (ANN) technique was also used to find the correlation between the studied parameters and the BOD removal efficiency.

2. Materials and Methods

2.1. Preparations

2.1.1. Preparation of nZVI

Synthesis of Nanoscale Zero Valent Iron (nZVI); 1.0812 g $FeCl_3.6H_2O$ was dissolved in a 4/1 (v/v) ethanol/water mixture (48 mL ethanol + 12 mL deionized water) and stirred well. Sodium borohydride solution 0.1 M was prepared i.e., 0.7564 g $NaBH_4$ was dissolved in 200 mL of deionized water; since for better growth of iron nanoparticles, excess borohydride is needed. The borohydride solution is poured in a burette and added drop by drop (1drop per 2 seconds) into iron chloride solution with a magnetic stirrer, as shown in Figure (1A). The preparation chemical reaction is showing in Eq. 1. After the first drop of sodium borohydride solution, black solid particles immediately appeared as shown in Figure (1B). The magnetic properties of nZVI appear during the preparation process as shown in Figure (1C). Remaining sodium borohydride is added completely to accelerate the reduction reaction and the color of the solution turned to black as shown in Figure (1D). The mixture was left for another 10 min of stirring after adding the whole borohydride solution. The vacuum filtration technique was used to separate the black iron nanoparticles from the liquid phase. Two sheets of Whatman filter papers 42 were used in the filtration process as shown in Figure (1E). The solid particles were washed three times with 50 mL portions of absolute acetone to remove all of the moisture. This washing process is probably the key step of synthesis since it prevents the rapid oxidation of zero-valent iron nanoparticles. The synthesized nanoparticles were finally dried in an oven at 60°C overnight, then collected and covered as shown in Figure (1F). For long-time storage, a thin layer of ethanol was added to preserve the nano iron particles from oxidation.

$$2FeCl_3 + 6NaBH_4 + 18H_2O \longrightarrow 2Fe^0 + 21H_2 + 6B(OH)_3 + 6NaCl \quad (1)$$

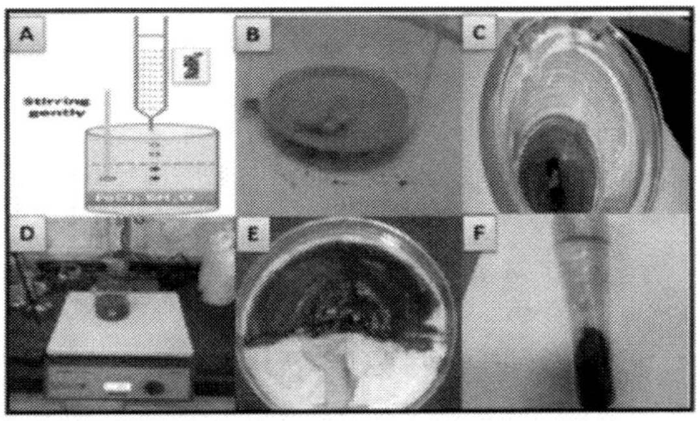

Figure 1. preparation of nZVI.

2.1.2. Preparation of GT-nZVI

Synthesis of green synthesis nano iron using drop by drop method using black tea; About 25 g from black tea was mixed in one liter of deionized water for two hours at 300°C. After cooling, the extracted tea solution was filtrated using Whatman filter paper number 1. About 100 mL of extra pure ethanol solution was added to the infiltrated teat and mixing for 15 min then filtrate again into the extracted tea solution. About 2.1624 g $FeCl_3.6H_2O$ was dissolved in a 4/1 (v/v) ethanol/water mixture (96 mL ethanol + 24 mL deionized water). The extracted tea solution was poured in a burette and added at a rate of one drop per two seconds into ferric chloride solution and stirring at a magnetic stirrer at 150 rpm. After complete reaction between phenolic compounds in black tea and $FeCl_3.6H_2O$, the black precipitate appeared. The black precipitate was collected by centrifuge at 5000 rpm. After washing and drying the precipitate using ethanol and dryer with fixed temperature 105°C the black precipitate was ready to use for any environmental application (Rabie S. Farag and Peters, 2018).

2.2. Characterizations

The prepared nZVI and GT-nZVI were characterized by using X-Ray Diffraction patterns (XRD), Scanning Electron Microscope (SEM), EDAX analysis and Uv-Vis spectrophotometer.

2.3. Batch adsorption studies

The effect of different operating parameters for BOD removal efficiency was studied in an acidic, neutral and alkaline media at pH (3, 5, 6, 7, 8, 9 and 11) for standard BOD solutions (100, 200, 300, 400 and 500 mg/L), using adsorbent doses (0.2, 0.4, 0.6, 0.8 and 1 g) at different contact times (5, 10, 15, 20, 25, 30 and 60 min), and stirring rate (100, 200, 300, 400 and 500 rpm) at room temperature. The removal percentage were calculated using Eq. 2.

$$\text{Sorption } [\%] = [C_o - C_e/C_o] \times 100 \qquad (2)$$

where C_o is the initial BOD concentration, (mg/L) and C_e is the equilibrium BOD concentration (mg/L)

The amount of BOD sorbed by weight of nZVI and GT-nZVI were calculated using Eq. 3.

$$q_e\ [\text{mg/g}] = [[C_o - C_e]V]/\ m \qquad (3)$$

where qe is the equilibrium adsorption capacity (mg /g), V is the volume of aqueous solution (L), and m is the dry weight of the adsorbent (g)

2.4. Isotherm studies

Different isotherm studies were conducted for monolayer and multilayer adsorption processes using nonlinear equations including Freundlich, Langmuir, Redlich–Peterson, Hill, Sips, Khan, Toth, Koble–Corrigan, Jovanovich.

2.5. Kinetic studies

To determine the suitable time for reaching the equilibrium state, standard solutions were sited in contact with adsorbent materials at different times at a fixed temperature. The quantity of sorbed compounds at time t is known Q_t (mg/mg), was calculated using Eq. 4.

$$Q_t = (C_o - C_t)x\frac{V}{m} \qquad (4)$$

where C_o is the initial concentration (mg/L), C_t is the initial concentration at time t (mg/L), V is the volume of the solution (L), and W is the weight of the sorbent.

To explore the suitable kinetic model which illustrates the kinetics of contaminants removal mechanism at different times, the pseudo-first, second-order, Avrami, Elovich, and Intraparticle nonlinear models were calculated.

2.6. Validation of adsorption isotherms and kinetics

Different error parameters were calculated in order to identify the most suitable isotherm and kinetic models that describe the treatment process. Due to the ingrained bias that resulting from linear form of isotherm models, the nonlinear form was good alternative parameter that provides an accurate technique to determine isotherm and kinetic parameters from the original equation forms (Malek and Farooq, 1996, Khan et al., 1996). To evaluate the better fit of nonlinear models the error function is required (Ho et al., 2002). The five error functions were used for all models and the ideal model was selected from the summation of these errors.

2.7. Neural network structure

An ANN model was developed with three main layers: an input, hidden, and output layers, to predict the removal efficiency of biological oxygen demand (BOD). The input layer receives records from five experimental factors (pH, adsorbent dose, contact time, stirring rate, and initial concentration), and the hidden layer consisted of ten neurons. The ANN model was structured as 5–10–1 (Figure 2). Equation 5 represents "tansig" transfer function which has been selected for the hidden layer, while Equation 6 represents "purlin" activation function which has been selected for the output layer. A cascade-forward backprop algorithm was used for data classification (Mahmoud et al. 2018). A comparison was conducted between the target and the output through calculating the mean squared error (MSE). Equation 7 was employed to adjust the weights and biases, where the MSE was passes back from the output node to the input layer. The network was trained using Levenberg–Marquartd method (trainlm), which is appropriate for small to medium-size networks (Mahmoud et al. 2019).

$$f(x) = x, \quad -\infty < f(x) < +\infty \qquad (5)$$

$$f(x) = \frac{e^{x} - e^{-x}}{e^{x} + e^{-x}}, \quad -1 \leq f(x) \leq 1 \qquad (6)$$

$$MSE = \frac{\sum_{i=1}^{N}(t_i - a_i)^2}{N} \qquad (7)$$

where: N is the number of measured data, and a_i and t_i are the outputs and predicted target, respectively.

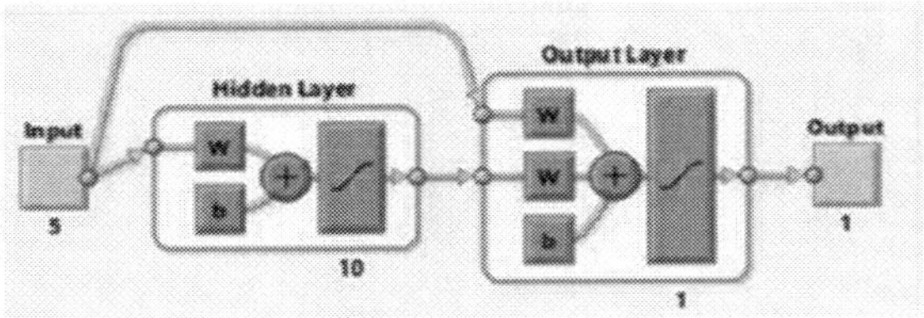

Figure 2. ANN of 5 – 10 – 1 structure used for the prediction of BOD removal efficiency using nZVI and GT-nZVI

3. Results and Discussion

3.1. Characterization of nano Iron carbide

The XRD pattern as shown in Figures (3A) and (3B) show two peaks at 2θ = 44.58° and 64.99° that fit well with the body-centered structure of iron. The XRD analysis in Figure (3A) proved that there is not any formation of oxides or any byproducts in the formation of nZVI, while Figure (3B) proved a formation of different byproducts, such as iron oxides, Carbide, Iron Sodium Phosphate, Magnesium Iron Hydride, Iron-Chromium Oxide, Iron (III) borate and Ferrimolybdite. Figures (3C) and (3D) show the SEM image of the prepared nZVI and GT-nZVI, respectively showing the formation of nano iron with an average diameter of 30 nm for nZVI and 80nm for GT-nZVI.

Figures (3E) and (3F) show EDAX analysis for the selected nZVI and GT-nZVI, respectively show the formation of pure nano iron in the case of nZVI and nano-iron particles with outer oxide and carbon layer in the case of GT-nZVI. Figures (3G) and (3H) show the formation of pure nanoparticles free from oxides and hydroxides in the case of nZVI and GT-nZVI, respectively.

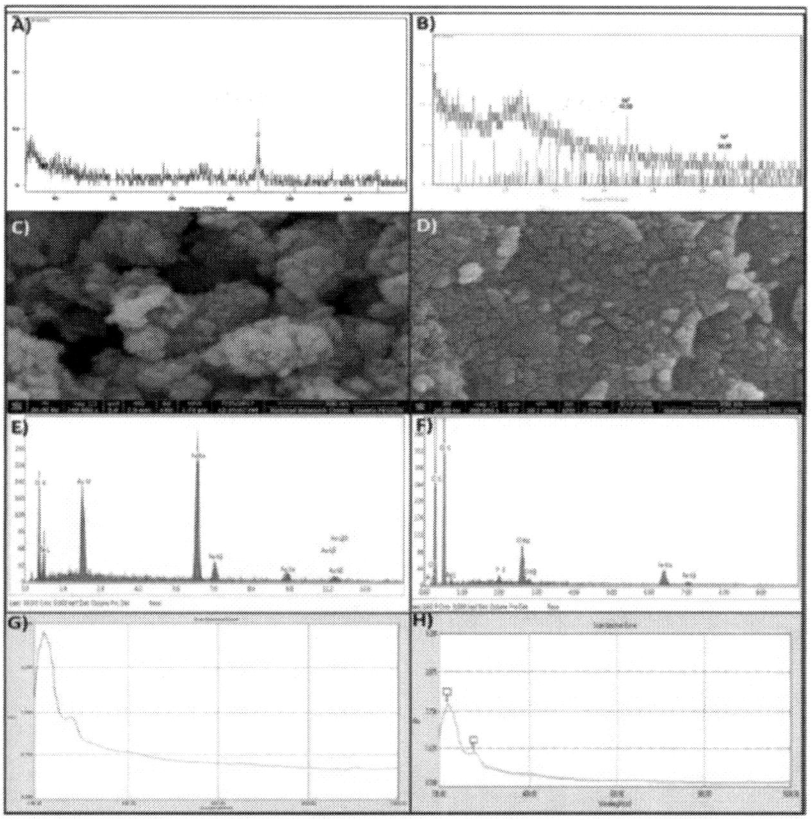

Figure 3. Characterization of nZVI and GT-nZVI.

3.2. Effect of operating parameters

3.2.1. Effect of pH

Figure (4A) shows the effect pH on BOD removal using nZVI and GT-nZVI. The effect of pH was studied in an acidic, neutral and alkaline medium (3-11) for standard BOD solution 300 mg/L, adsorbent dosage of 0.6 g, and stirring rate of 100 rpm and contact time of 10 min for nZVI and stirring rate of 200 rpm and contact time of 20 min for GT-nZVI. The maximum removal efficiency after using nZVI and GT-nZVI, occurs at pH 7 and 8 showing removal efficiencies of 86.4% and 72.5%, respectively. At the optimum pH condition, nZVI has showed a better removal efficiency than GT-nZVI. Low removal efficiencies were observed in both highly acidic and alkaline medias. In highly acidic media, the solubility of some iron nanoparticles particles in acidic solution led to decrease sorbent capacity, while in highly alkaline media, the excess of hydroxyl ions in the solution compete with negative ions on the iron nanoparticles surface which decrease the adsorption process (Kosmulski, 2004, Bezbaruah et al., 2009).

3.2.2. Effect of adsorbent dose

Figure (4B) shows the effect dose on BOD removal using nZVI and GT-nZVI. The effect of dose was studied in the range of (0.2 – 1 g) for standard BOD solution 300 mg/L, pH 7, stirring rate of 100 rpm and contact time of 10 min for nZVI and pH 8, stirring rate of 200 rpm and contact time of 20 min for GT-nZVI. The results showed that the minimum effective adsorbent dose for best removal efficiency is 0.6 g and the removal efficiencies were 86.4% and 72.5% after using nZVI and GT-nZVI, respectively as shown in Figure (4B). An enhancement in BOD removal efficiency was associated with an increase in adsorbent dosage. This result indicated that an increased adsorbent dosage could create a large number of adsorption sites on the nanoparticle surface, which adsorbed significant amounts of pollutants. At the optimum dose of 0.6 g, nZVI has showed a better removal efficiency than GT-nZVI.

3.2.3. Effect of contact time

The effect of contact time was studied at different times (5 - 60 min) for standard BOD solution 300 mg/L, pH 7, adsorbent dosage of 0.6 g, and stirring rate of 100 rpm for nZVI and pH 8, adsorbent dosage of 0.6 g, and stirring rate of 200 rpm for GT-nZVI. Figures (4C) describe the effect of contact time on BOD removal by using nZVI and GT-nZVI, showing that the optimum time for removal of BOD was 10 and 20 min for nZVI and GT-nZVI, respectively, with a removal efficiency of 86.4% and 72.5%, respectively. No improvement in the removal efficiency was observed with a further increase in contact time since portion of the voids was occupied causing a reduction in the adsorption pattern.

3.2.4. The effect of stirring rate

The effect of stirring rate was studied at the different stirring rate (100 - 500 rpm) for standard BOD solution 300 mg/L, pH 7, adsorbent dosage of 0.6 g, and contact time of 10 min for nZVI and pH 8, adsorbent dosage of 0.6 g, and contact time of 20 min for GT-nZVI. Figures (4D) describes the effect of stirring rate on BOD removal by using nZVI and GT-nZVI as sorbent materials showing that the minimum effective stirring rate for best removal efficiency is 100 and 200 rpm, respectively showing removal efficiencies 86.4% and 72.5%, respectively. Previous studies indicated that there is a little increase in removal percent according to the stirring rate relative to chemisorption reaction (Ho and McKay, 1999). The main reason of using nanoparticles for degradation and adsorption of different biological contaminants at little stirring rate and time is high surface area and reactivity of the selected nanomaterials (Machado et al., 2013, Mahdy et al., 2012, Markova et al., 2013, Yang et al., 2014, Yang et al., 2013).

3.2.5. Effect of initial concentration

The effect of initial concentration was studied at the different BOD concentrations (100 - 500 mg/L), pH 7, adsorbent dosage of 0.6 g, stirring rate 100 rpm, and contact time of 10 min for nZVI and pH 8, adsorbent dosage of 0.6 g, stirring rate 200 rpm, and contact time of 20 min for GT-nZVI. Figures (4E) describes the effect of initial BOD concentration on the removal efficiency. The removal efficiency of BOD decreased with the increase in the initial BOD concentration because the iron nanoparticles reach their maximum adsorption capacity.

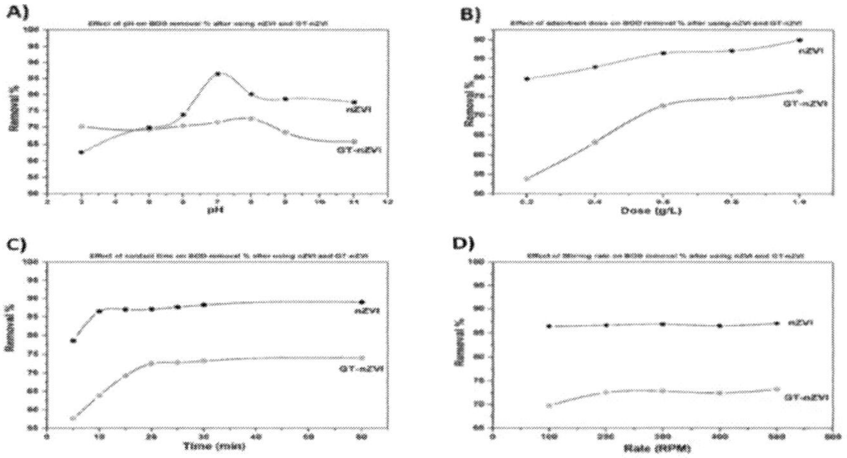

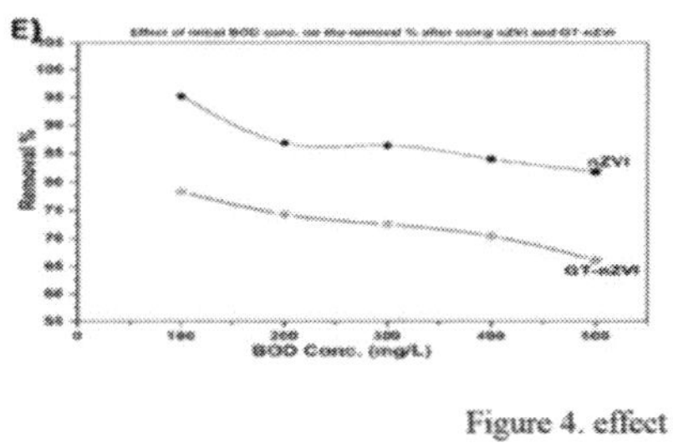

A) Effect of pH on BOD removal using nZVI and GT-nZVI.
B) Effect of dose on BOD removal using nZVI and GT-nZVI.
C) Effect of contact time on BOD removal using nZVI and GT-nZVI.
D) Effect of stirring rate on BOD removal using nZVI and GT-nZVI.
E) Effect of concentration on BOD removal using nZVI and GT-nZVI.

Figure 4. effect of operating parameter.

3.3. Adsorption studies

Figure (5A and 5B) describes nonlinear relations between different adsorption isotherm models. The adsorption isotherms were studied by applying nonlinear equations of Redlich- Peterson, Hill, Sips, Khan, Toth, Koble-Corrigan, Javanovic, Freundlich and Langmuir models. The obtained results indicated that nZVI obey the both Koble–Corrigan and Freundlich with the lowest summation of errors 0.5992. Koble-Corrigan model combines Langmuir and Freundlich adsorption isotherm models along with the adsorption isotherm for pure BOD. Table 1 describe the variation between maximum uptakes between experimental work and each calculated isotherm models after using nZVI. The adsorption isotherm for adsorption of BOD onto GT-nZVI indicating that the most preferred models which can describe the adsorption process are Toth and Langmuir models with the same lowest summation of errors 0.1587. Langmuir assumes that the monolayer coverage of adsorbate and adsorption occurs at fixed adsorbent sites without any steric hindrance between adsorbed BOD molecules. This can be attributed to the similarity between activation energy and enthalpy of this active sites. Also, Langmuir model describes adsorption process over a homogeneous adsorbent surface and the equilibrium saturation point can take place when the adsorbent (GT-nZVI) was fully saturated with adsorbed BOD molecules. Toth model improved Langmuir isotherm model using experimental data, also described heterogeneous adsorption process in high and low concentrations through asymmetrical distribution of quasi-Gaussian energy (Toth, 1971). Table 2 describe the variation between maximum uptakes between experimental work and each calculated isotherm models after using GT-nZVI.

Table 1. Shows the results of experimental and calculated Qe (mg/mg) for all nonlinear isotherm models after using nZVI for BOD removal.

Exp. Qe	Calculated Redlich-Peterson Qe	Calculated Hill Qe	Calculated Sips Qe	Calculated Khan Qe	Calculated Toth Qe	Calculated Koble – Corrigan Qe	Calculated Javanovic Qe	Calculated Freundlich Qe	Calculated Langmuir Qe
0.15867	0.113	0.094	0.121	0.122	0.073	0.123	0.068	0.123	0.074
0.28933	0.331	0.301	0.330	0.331	0.319	0.330	0.313	0.330	0.320
0.432	0.427	0.404	0.426	0.425	0.432	0.425	0.431	0.425	0.432
0.55967	0.554	0.550	0.553	0.553	0.565	0.553	0.569	0.553	0.565
0.68133	0.676	0.698	0.674	0.677	0.671	0.678	0.672	0.678	0.672

Table 2. Shows the results of experimental and calculated Qe (mg/mg) for all nonlinear isotherm models after using GT-nZVI for BOD removal.

Exp. Qe	Calculated Redlich-Peterson Q_e	Calculated Hill Q_e	Calculated Sips Q_e	Calculated Khan Q_e	Calculated Toth Q_e	Calculated Koble – Corrigan Q_e	Calculated Javanovic Q_e	Calculated Freundlich Q_e	Calculated Langmuir Q_e
0.1305	0.125	0.097	0.124	0.144	0.125	0.124	0.122	0.144	0.125
0.2477	0.258	0.252	0.258	0.256	0.258	0.258	0.256	0.256	0.258
0.3625	0.363	0.373	0.364	0.351	0.363	0.364	0.365	0.351	0.363
0.4693	0.457	0.467	0.458	0.447	0.457	0.458	0.460	0.447	0.457
0.5508	0.557	0.549	0.556	0.568	0.557	0.556	0.554	0.568	0.557

3.4. Kinetics Studies

Figure (5C and 5D) describes the nonlinear relations between different kinetic models. The kinetic analysis was carried by practicing nonlinear equations of Pseudo First Order (P.F.O), Pseudo Second Order (P.S.O), Elovich, Avrami, and Intraparticle kinetic models. The obtained results indicated that nZVI and GT-nZVI accept Avrami model with the lowest summation of errors equal 0.058. Avrami model describes the kinetics of crystallization, such as the solid transformation from one phase to another at constant temperature. This model can also take into consideration other changes of phase in materials, such as chemical reaction rates (Avrami, 1940). Table 3 shows the results of the experimental and calculated Q_t (mg/g) for all nonlinear kinetic models after using nZVI and GT-nZVI.

Table 3. Shows the results of experimental and calculated Qt (mg/mg) for BOD removal using nZVI and GT-nZVI.

Material	Exp. Q_t	Calc. P.F.O Q_t	Calc. P.S.O Q_t	Calc. Elovich Q_t	Calc. Avrami Q_t	Calc. Intraparticle Q_t
nZVI	0.393	0.393	0.398	0.407	0.393	0.414
	0.432	0.434	0.423	0.420	0.429	0.421
	0.435	0.438	0.432	0.428	0.437	0.426
	0.435	0.439	0.437	0.433	0.439	0.431
	0.438	0.439	0.440	0.437	0.440	0.434
	0.441	0.439	0.442	0.441	0.440	0.438
	0.445	0.439	0.447	0.454	0.440	0.455
GT-nZVI	0.288	0.276	0.285	0.316	0.285	0.311
	0.319	0.341	0.328	0.332	0.327	0.325
	0.346	0.356	0.346	0.341	0.346	0.335
	0.3625	0.360	0.355	0.348	0.356	0.344
	0.364	0.360	0.361	0.353	0.362	0.351
	0.366	0.361	0.365	0.357	0.366	0.358
	0.3701	0.361	0.375	0.373	0.373	0.391

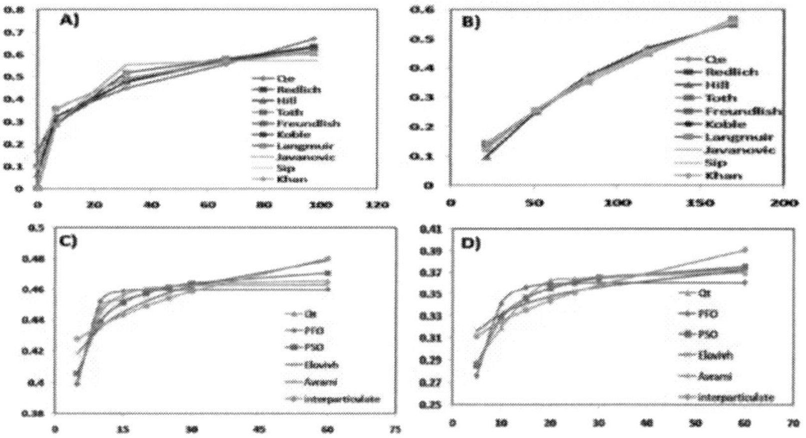

Figure 5. Isotherm and kinetic studies for BOD removal using nZVI and GT-Nzvi: (A) isotherm after nZVI treatment; (B) isotherm after GT-nZVI treatment; (C) Kinetic studies after using nZVI; (D) kinetic studies after using GT-nZVI.

3.5. Artificial neural network

3.5.1. Training and validation performance

The magnitudes of the gradient for BOD was recorded as 1.3793 and 1.3243 for nZVI and GT- nZVI, respectively (Figure 6). Since, these values exceeded the least error level of (1×10^{-5}), the training process was terminated after 6 epochs. For all tested parameters, the MSE of the training phase decreased with the increase in epochs (Figure 7). The MSE of the validation phase increased after epoch 0, indicating the data overfitting. The test and validation curves increased in an analogous way with the number of iterations. The best validation performance was 0.11588 and 1.4634 for nZVI and GT- nZVI, respectively.

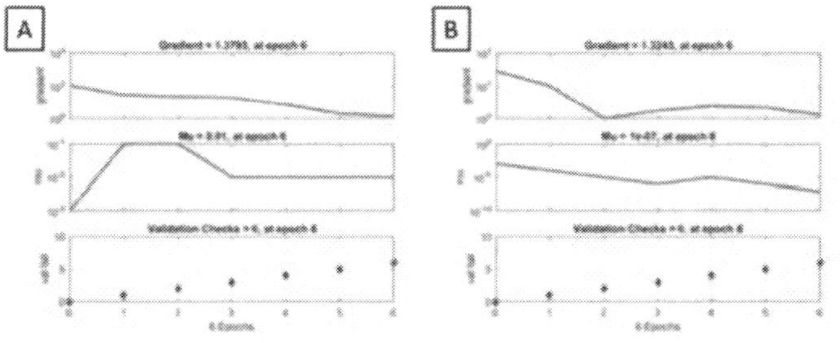

Figure 6. Training performance: (A) nZVI and (B) GT-nZVI.

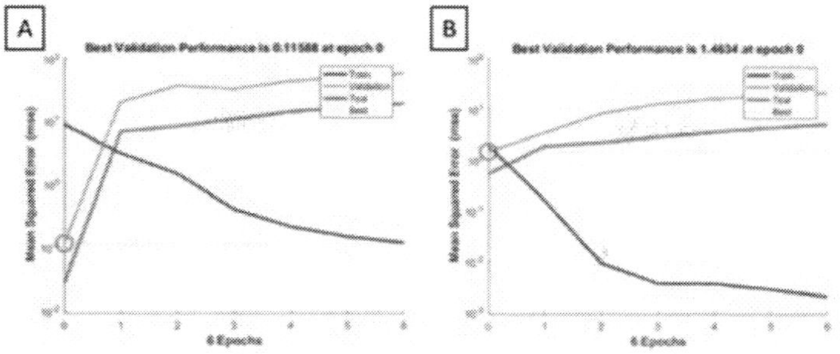

Figure 7. Best validation performance: (A) nZVI and (B) GT-nZVI.

3.5.2. Regression plot

Figure 8 shows a relation between the coefficient of determination (r^2) and outputs and targets of the generated network. A good fit was detected between the output and the target data for the three phases: training, validation and test, with r^2-values of 0.90, 0.99, and 0.99, respectively for nZVI and 0.94, 0.97, and 0.95, respectively for GT-nZVI. The overall r^2-value was 0.91 and 0.95 for nZVI and GT-nZVI, respectively, indicating the model reliability.

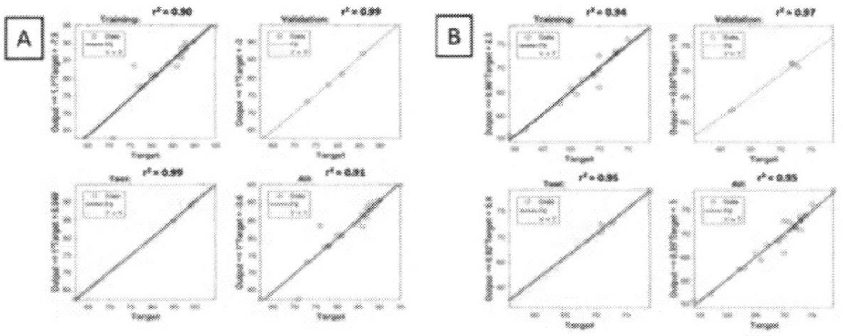

Figure 8. Regression plot between target and output for BOD removal efficiency: (A) nZVI and (B) GT-nZVI.

4. Conclusions

At the optimum conditions of pH 7, adsorbent dosage of 0.6 g, stirring rate of 100 rpm, and contact time of 10 min for nZVI, the BOD removal efficiency was reported as 95.2, 86.8, 86.4, 83.9 and 81.7% for initial BOD concentrations of 100, 200, 300, 400 and 500 mg/L, respectively. At the optimum conditions of pH 8, adsorbent dosage of 0.6 g, stirring rate of 200 rpm, and contact time of 20 min for GT-nZVI, the BOD removal efficiency was reported as 78.3, 74.3, 72.5, 70.4 and 66.1% for initial BOD concentrations of 100, 200, 300, 400 and 500 mg/L, respectively. From the obtained results we can conclude that nZVI and GT-nZVI are an effective adsorbent for removal of BOD. Although, nZVI has shown to be more effective than GT-nZVI in the removal of BOD, GT-nZVI was more cost effective than nZVI since GT-nZVI was prepared from natural material (black tea). Additionally, the GT-nZVI was more stable than nZVI because it contains outer carbon and oxide layers. The obtained results indicated that nZVI preferred Koble–Corrigan and Freundlich isotherm models and GT-nZVI preferred Toth and Langmuir models. The obtained results indicated that nZVI and GT-nZVI preferred Avrami kinetic model. Under the studied conditions, ANN was able to accurately predict the removal efficiency of BOD with r^2 value of 0.91 and 0.95 for nZVI and GT-nZVI, respectively.

5. References

AMIN, M., ALAZBA, A. & MANZOOR, U. 2014. A review of removal of pollutants from water/wastewater using different types of nanomaterials. *Advances in Materials Science and Engineering,* 2014.

AVRAMI, M. 1940. Kinetics of phase change. II transformation-time relations for random distribution of nuclei. *The Journal of chemical physics,* 8**,** 212-224.

BEZBARUAH, A. N., KRAJANGPAN, S., CHISHOLM, B. J., KHAN, E. & BERMUDEZ, J. J. E. 2009. Entrapment of iron nanoparticles in calcium alginate beads for groundwater remediation applications. *Journal of hazardous materials,* 166, 1339-1343.

BURKHARD, R., DELETIC, A. & CRAIG, A. 2000. Techniques for water and wastewater management: a review of techniques and their integration in planning. *Urban water,* 2, 197-221.

DAILY, G. C. & EHRLICH, P. R. 1992. Population, sustainability, and Earth's carrying capacity. *BioScience,* 42, 761-771.

ELAWWAD, A., KARAM, A. & ZAHER, K. 2017. Using an algal photo-bioreactor as a polishing step for secondary treated wastewater. Pol. J. Environ. Stud., 26(4), 1493-1500.

EL-SHAFEI, M., MAHMOUD, A., MOSTAFA, M. & PETERS, R. Effects of entrapped nZVI in alginate polymer on BTEX removal. AIChE Annual Meeting, 2016. 13-18.

ESREY, S. A., ANDERSSON, I., HILLERS, A. & SAWYER, R. 2001. *Closing the loop: ecological sanitation for food security.*

FARAG, R. S., ELSHFAI, M. M. & MAHMOUD, A. S. 2018. Adsorption and kinetic studies using nano zero valent iron (nZVI) in the removal of chemical oxygen demand from aqueous solution with response surface methodology and artificial neural network approach. *Journal of Environment & Biotechnology Research,* 7, 12-22.

FERROUDJ, N., NZIMOTO, J., DAVIDSON, A., TALBOT, D., BRIOT, E., DUPUIS, V., BÉE, A., MEDJRAM, M. S. & ABRAMSON, S. 2013. Maghemite nanoparticles and maghemite/silica nanocomposite microspheres as magnetic Fenton catalysts for the removal of water pollutants. *Applied Catalysis B: Environmental,* 136, 9-18.

GLEICK, P. H. 2000. *The world's water 2000-2001: the biennial report on freshwater resources*, Island Press.

HO, Y.-S. & MCKAY, G. 1999. Pseudo-second order model for sorption processes. *Process biochemistry,* 34, 451-465.

HO, Y., PORTER, J. & MCKAY, G. 2002. Equilibrium isotherm studies for the sorption of divalent metal ions onto peat: copper, nickel and lead single component systems. *Water, Air, and Soil Pollution,* 141, 1-33.

ISLAM, M. S. & TANAKA, M. 2004. Impacts of pollution on coastal and marine ecosystems including coastal and marine fisheries and approach for management: a review and synthesis. *Marine pollution bulletin,* 48, 624-649.

JASON A. HEBERLING, Y. G. A., AHMED S. MAHMOUD, MOHAMED K. MOSTAFA, ROBERT W. PETERS. AOP Performance at Wastewater Treatment Plants. ". Proc. 2018 Annual AIChE Meeting, Pittsburgh, PA, (October 28 – November 2). 2018.

KHAN, A., AL-WAHEAB, I. & AL-HADDAD, A. 1996. A generalized equation for adsorption isotherms for multi-component organic pollutants in dilute aqueous solution. *Environmental technology,* 17, 13-23.

KOSMULSKI, M. 2004. pH-dependent surface charging and points of zero charge II. Update. *Journal of Colloid and Interface Science,* 275, 214-224.

LANGERGRABER, G. & MUELLEGGER, E. 2005. Ecological Sanitation—a way to solve global sanitation problems? *Environment international,* 31, 433-444.

LIU, J. & DIAMOND, J. 2005. China's environment in a globalizing world. *Nature,* 435, 1179.

MACHADO, S., STAWIŃSKI, W., SLONINA, P., PINTO, A., GROSSO, J., NOUWS, H., ALBERGARIA, J. T. & DELERUE-MATOS, C. 2013. Application of green zero-valent iron nanoparticles to the remediation of soils contaminated with ibuprofen. *Science of the Total Environment,* 461, 323-329.

MAHDY, S. A., RAHEED, Q. J. & KALAICHELVAN, P. 2012. Antimicrobial activity of zero-valent iron nanoparticles. *International Journal of Modern Engineering Research,* 2, 578-581.

MAHMOUD, A. S., FARAG, R. S. & ELSHFAI, M. M. 2019a. Reduction of organic matter from municipal wastewater at low cost using green synthesis nano iron extracted from black tea: Artificial intelligence with regression analysis. *Egyptian Journal of Petroleum*.

MAHMOUD, A. S., ISMAIL, A., MOSTAFA, M. K., MAHMOUD, M., ALI, W. & SHAWKY, A. M. 2019b. Isotherm and kinetic studies for heptachlor removal from aqueous solution using Fe/Cu nanoparticles, artificial intelligence, and regression analysis. *Separation Science and Technology*, 1-13.

MAHMOUD, A. S., MOSTAFA, M. K. & ABDEL-GAWAD, S. A. 2018. Artificial intelligence for the removal of benzene, toluene, ethyl benzene and xylene (BTEX) from aqueous solutions using iron nanoparticles. *Water Science and Technology: Water Supply,* 18, 1650-1663.

MAHMOUD, A. S., MOSTAFA, M. K. & NASR, M. 2019c. Regression model, artificial intelligence, and cost estimation for phosphate adsorption using encapsulated nanoscale zero-valent iron. *Separation Science and Technology,* 54, 13-26.

MALEK, A. & FAROOQ, S. 1996. Comparison of isotherm models for hydrocarbon adsorption on activated carbon. *AIChE Journal,* 42, 3191-3201.

MARKOVA, Z., I KOV , K. N. M., FILIP, J., UDA, J., KOL , M., AF OV , K. R., MED K, I. & ZBO IL, R. 2013. Air stable magnetic bimetallic Fe–Ag nanoparticles for advanced antimicrobial treatment and phosphorus removal. *Environmental science & technology,* 47, 5285-5293.

MOSTAFA, M. K., MAHMOUD, A. S., SARYEL-DEEN, R. A. & PETERS, R. W. Application of Entrapped Nano Zero Valent Iron into Cellulose Acetate Membranes for Domestic Wastewater Treatment. 2017 Annual AIChE Meeting; Minneapolis, MN, October 29 - November 3, 2017, 2017a.

MOSTAFA, M. K., MAHMOUD, A. S., SARYEL-DEEN, R. A. & PETERS, R. W. Application of entrapped nano zero valent iron into cellulose acetate membranes for domestic wastewater treatment. Environmental Aspects, Applications and Implications of Nanomaterials and Nanotechnology 2017 - Topical Conference at the 2017 AIChE Annual Meeting, 2017b. 27-34.

NGOC, U. N. & SCHNITZER, H. 2009. Sustainable solutions for solid waste management in Southeast Asian countries. *Waste management,* 29, 1982-1995.

RABIE S. FARAG, M. M. E.-S., AHMED S. MAHMOUD, MOHAMED K. MOSTAFA, & PETERS, R. W. Green Synthesis of Nano Iron Carbide: Preparation, Characterization and Application for Removal of Phosphate from Aqueous Solutions. Proc. 2018 Annual AIChE Meeting, Pittsburgh, PA, (October 28 – November 2). 2018.

RIJSBERMAN, F. R. 2006. Water scarcity: fact or fiction? *Agricultural water management,* 80, 5-22.

TOTH, J. 1971. State equation of the solid-gas interface layers. *Acta chim. hung.,* 69, 311-328.

YANG, J., WANG, X., ZHU, M., LIU, H. & MA, J. 2014. Investigation of PAA/PVDF–NZVI hybrids for metronidazole removal: synthesis, characterization, and reactivity characteristics. *Journal of hazardous materials,* 264, 269-277.

YANG, Y., GUO, J. & HU, Z. 2013. Impact of nano zero valent iron (NZVI) on methanogenic activity and population dynamics in anaerobic digestion. *Water research,* 47, 6790-6800.

571656 Formic Acid Oxidation By the Fenton and Fenton-like Reaction: Experimental Study and Kinetic Modelling

Formic acid oxidation by the Fenton and Fenton-like reaction: experimental study and kinetic modelling

The Fenton reaction is considered to be an efficient purification technique for the treatment of wastewaters from a wide range of industries [1]. The Fenton reaction, known as one of the advanced oxidation processes, is the mixing of ferrous iron with hydrogen peroxide to form hydroxyl radicals:

$$Fe(II) + H_2O_2 \rightarrow Fe(III) + {}^{\bullet}OH + OH^-$$

These hydroxyl radicals are non-selective and powerful in the oxidation of various organics. The Fe(III) formed can react in a reaction know as the Fenton-like reaction:

$$Fe(III) + H_2O_2 \rightarrow Fe(II) + {}^{\bullet}OOH + H^+$$

The Fenton reaction is known to be faster than the Fenton-like reaction (rate constants: 53 and $2{\cdot}10^{-3}$ M^{-1} s^{-1} for the Fenton and Fenton-like reaction respectively). The Fenton and Fenton-like reactions start a sequence of radical combination and propagation reactions, as well as the reactions with the organic compounds of interest.

The large number of reactions that have to be taken into account in the Fenton reaction makes the kinetic modelling challenging. In addition, the radical reactions that have to be included in the model are usually extremely fast and it is therefore difficult to determine the reaction rate constants experimentally. Furthermore, the organic compounds can form various intermediate compounds, which add to the reactions describing the oxidation in the Fenton reaction. For example, Duesterberg et al. described the oxidation of formic acid with a set of 20 reactions [2]. However, more reactions have been suggested in literature. For instance, it is known that Fe(III) can form complexes with water and H_2O_2, such as $[Fe^{III}]^{3+}$, $[Fe^{III}OH]^{2+}$, $[Fe^{III}(OH)_2]^{+}$ and $[Fe_2^{III}(OH)_2]^{4+}$. In addition, it has been shown that ferric-formate complexes could be formed in the Fenton and Fenton-like reaction [3] [4].

In this study it was chosen to study the oxidation of formic acid to limit the amount of intermediate compounds that can be formed in the oxidation process of the organic. In contrast to previously reported studies on the formic acid oxidation by the Fenton reaction [2][5], our study mainly focuses on relatively high concentrations of formic acid (0.01M) and low ratios of iron and H_2O_2 to formic acid. The aim of our work was to extend the current knowledge on the oxiation of formic acid by the Fenton reaction and to find a kinetic model that describes the formic acid oxidation in the both the Fenton and Fenton-like reaction.

In the experimental study, batch experiments were perfomed at constant temperature. The results of these experiments show that the oxidation rate of formic acid in both the Fenton and Fenton-like reaction increases with increasing H_2O_2 concentrations (Figure 1). The most important differences between the Fenton and Fenton-like reactions are observed in the first minutes of the reaction. In the Fenton reaction a steep decrease in formic acid is observed in the first minute of the reactions, which is independent of the H_2O_2 concentration. Afterwards the reaction rate is constant. In contrast, the reaction rate is relatively constant for the Fenton-like reaction for the entire course of the reaction.

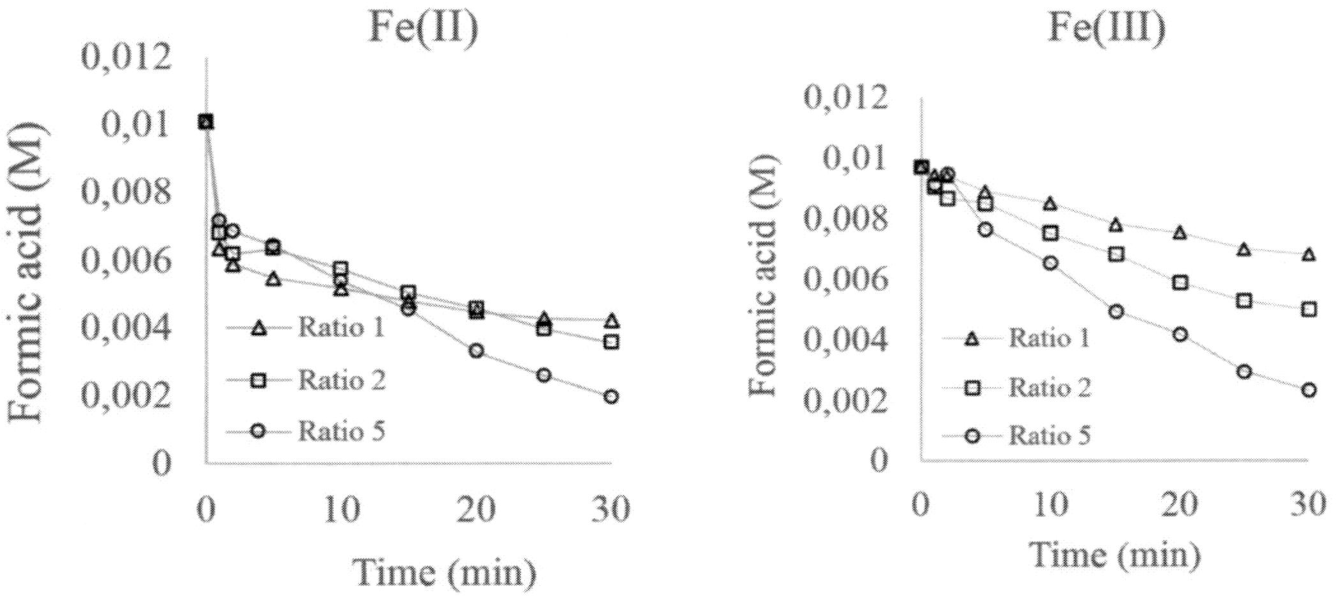

Figure 1. Effect of hydrogen peroxide concentration on formic acid oxidation by Fe(III)/H_2O_2 and Fe(III)/H_2O_2. Conditions: 0.01M formic acid and 0.001M Fe at 25°C. Datasets represent different concentrations of H_2O_2 added to the reaction solution: ▲: 0.01M, ■: 0.02M and ●: 0.05M.

It was decided to study the reactions spectrophotometrically to obtain more information on the different iron species in the reaction solutions. Figure 2 shows the absorbance spectra during the Fenton and Fenton-like reaction. Previous studies have shown that he absorbance at 290 nm could be attributed to $Fe^{III}]^{3+}$, $[Fe^{III}OH]^{2+}$, $[Fe^{III}(OH)_2]^{+}$ and $[Fe_2^{III}(OH)_2]^{4+}$ [6] [49] . $[Fe_2^{III}(OH)_2]^{4+}$ and $[Fe^{III}(OH)_2]^{+}$ are unlikely to form below pH 3 [6] and therefore not taken into account. Although a fraction of the absorbed iron species can be regarded $[Fe^{III}]^{3+}$, its decadic molar absorptivity (315 M^{-1} cm^{1}) is relatively low compared to $[Fe^{III}OH]^{2+}$ (2005 M^{-1} cm^{-1}) and the absorbance at 290 nm is mainly attributed to $[Fe^{III}OH]^{2+}$ [28].

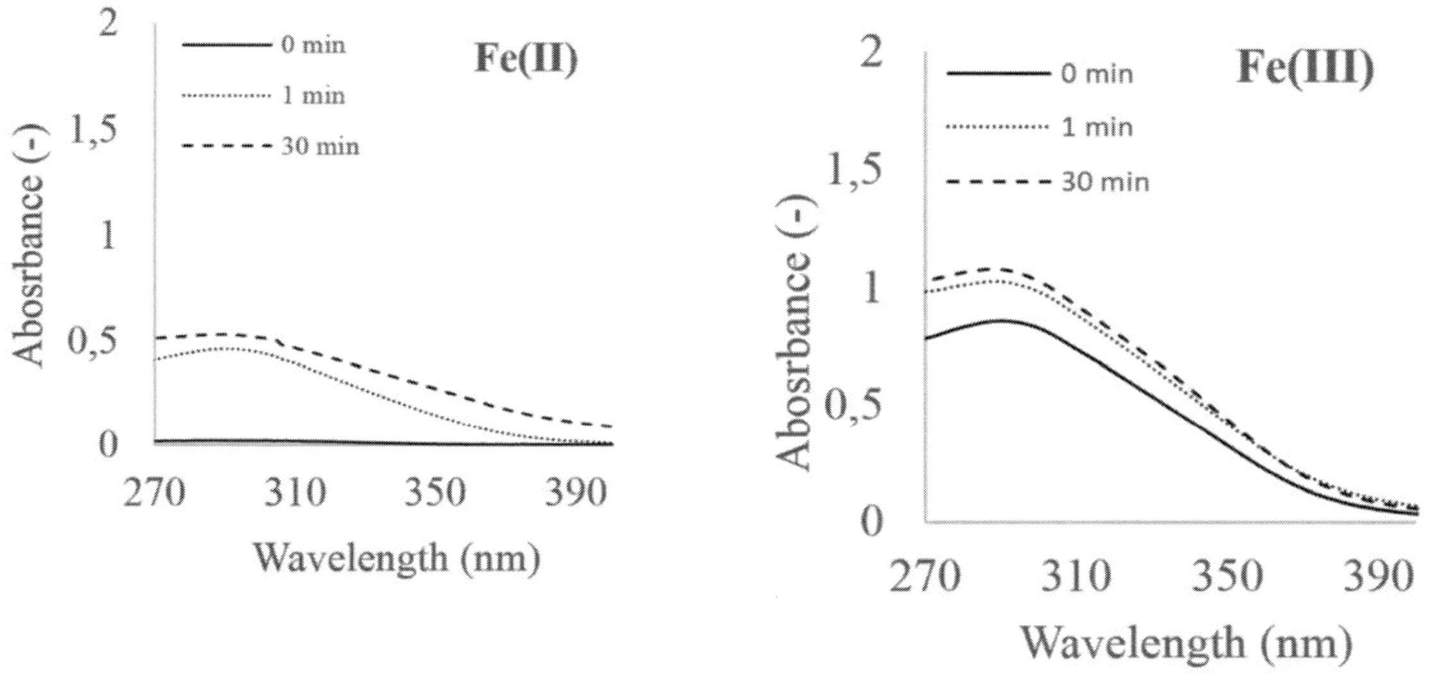

Figure 2. Absorption spectra for formic acid oxidation by Fe(III)/H_2O_2 and Fe(III)/H_2O_2. Conditions: $1 \cdot 10^{-3}$ M Formic acid, $2 \cdot 10^{-3}$ M H_2O_2 and $1 \cdot 10^{-4}$ M Fe at 25°C. Datasets represent different sampling times (minutes after reaction started) — 0 minutes, ··· 1 minutes and --- 30 minutes.

Kinetic models previously presented in the literature are unable to describe our experimental results. For example, the model introduced by Duesterberg et al. shows a fast oxidation rate of formic acid in the first minutes of the reaction, which can be ascribed to the excess of Fe(II) and H_2O_2 to formic acid in their study. The use of this model for our experimental results gives a predicted initial decrease in formic acid in the Fenton reaction that is proportional to the ratio of Fe(II) to formic acid, which is shown in Figure 3 (A). However, the initial decrease observed is 30% for a ratio of formic to Fe(II) of 10:1. The fit to the Fenton-like data is reasonably well although significant differences are observed for the highest concentration of H2O2 (Figure 3B). To improve the fit of the model to the experimental data, the reactions describing the formation of $[Fe^{III}OH]^{2+}$ and the ferric-formate complex were added to the model suggested by Duesterberg et al.

Figure 3 (C and D) shows that the addition of these reactions describing the formation and decomposition ferric complexes improved the fit of the kinetic model to the experimental data. This could suggest that the formed complexes influence the availability of free iron in the solution.

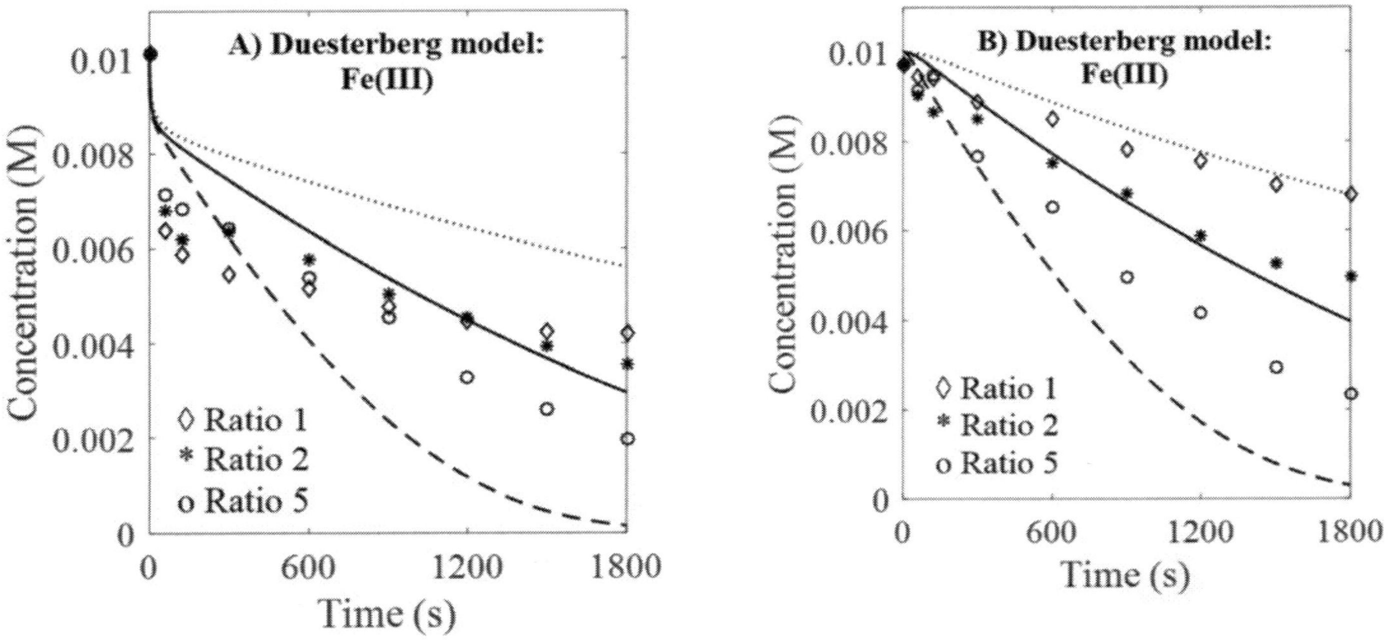
A) Duesterberg model:
Fe(III)
Concentration (M)
Time (s)
◊ Ratio 1
* Ratio 2
o Ratio 5
B) Duesterberg model:
Fe(III)
Concentration (M)
Time (s)
◊ Ratio 1
* Ratio 2
o Ratio 5

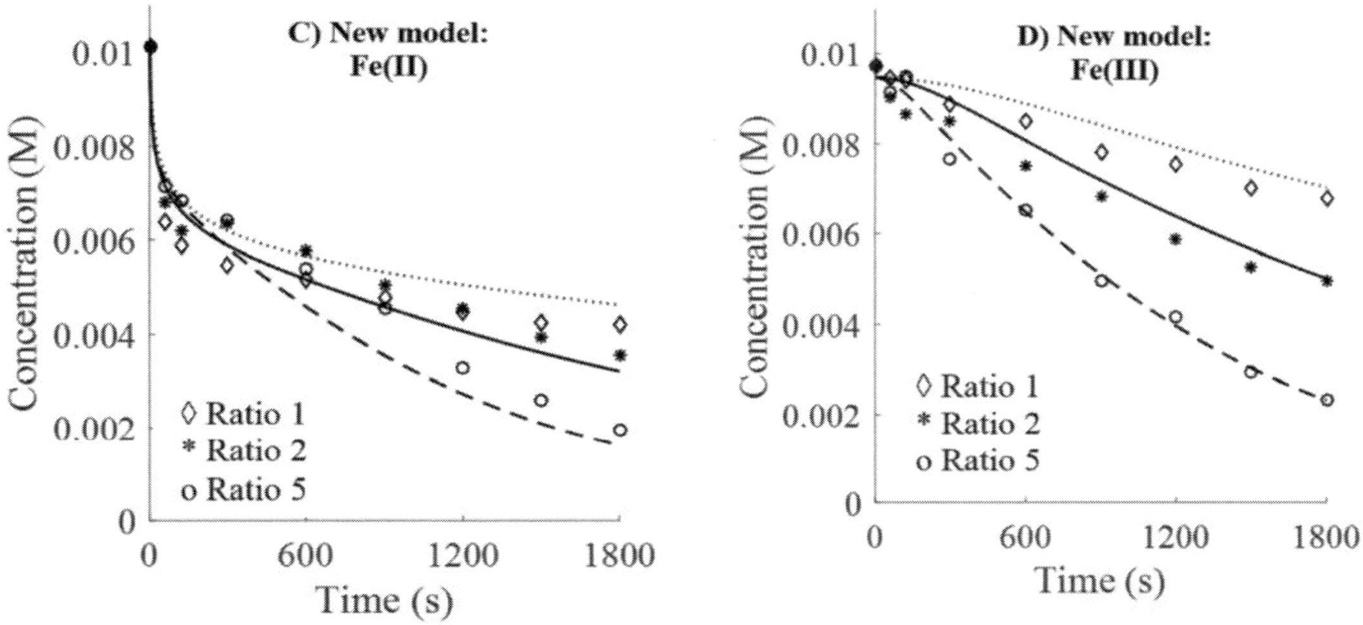

Figure 3. Experimental data and model prediction for formic acid oxidation by Fe(III) /H_2O_2 and Fe(II)/H_2O_2; 0.01M formic acid and 0.001M Fe. A and B: model as described by Duesterberg et al., C and D: new model. Datasets represent different concentrations of H_2O_2 added to the reaction solution. Experimental data: ◊: 0.01M, *: 0.02M and ○: 0.05M.

The limitation regarding the kinetic modelling of the Fenton reaction is the number of reactions and kinetic constants that have to be included to obtain an accurate fit to the experimental data. Nonetheless, the suggested reactions and kinetic constants could give a reasonable fit to our experimental data without using fitting routines to obtain new kinetic constants.

Our work has led us to the conclusion that the initial oxidation of formic acid in the first minutes of the reaction is significantly different for the Fenton and Fenton-like reaction. These experimental results could not be described by kinetic models that are proposed in literature. The addition of the formation of ferric-complexes (both inorganic and organic) improved the fit of the model to the experimental data, especially for the initial decrease in formic acid in the first minutes of the Fenton reaction. However, a perfect match with the experimental data was not achieved.

[1] E. Neyens and J. Baeyens, "A review of classic Fenton's peroxidation as an advanced oxidation technique," *J. Hazard. Mater.*, vol. 98, pp. 33–50, 2010.

[2] C. K. Duesterberg, W. J. Cooper, and T. D. Waite, "Fenton-mediated oxidation in the presence and absence of oxygen," *Environ. Sci. Technol.*, vol. 39, no. 13, pp. 5052–5058, 2005.

[3] M. Dopar, H. Kusic, and N. Koprivanac, "Treatment of simulated industrial wastewater by photo-Fenton process. Part I: The optimization of process parameters using design of experiments (DOE)," *Chem. Eng. J.*, vol. 173, no. 2, pp. 267–279, 2011.

[4] M. Simunovic, H. Kusic, N. Koprivanac, and A. L. Bozic, "Treatment of simulated industrial wastewater by photo-Fenton process: Part II. The development of mechanistic model," *Chem. Eng. J.*, vol. 173, no. 2, pp. 280–289, 2011.

[5] J. D. E. Laat, H. Gallard, S. Ancelin, and B. Legube, "Comparative study of the oxidation of atrazine and acetone by H2O2/UV, Fe(III)/UV, Fe(III)/H2O2 and Fe(II) or Fe(III)/H2O2," vol. 39, no. 15, pp. 2693–2706, 1999.

[6] H. Gallard, J. De Laat, and B. Legube, "Spectrophotometric study of the formation of iron(III)-hydroperoxy complexes in homogeneous aqueous solutions," *Water Res.*, vol. 33, no. 13, pp. 2929–2936, 1999.

577404 Generation of Wastewater Treatment Networks: Integrating Process Efficiency, Economics and Sustainability

The rise in world population and industrialization in developing nations has tremendously increased the demand for water and has resulted in wastewater contaminated with several pollutants. Thus, wastewater treatment for reuse and safe disposal have become crucial for sustainable existence. The treatment methods must vary based on the properties of the inlet wastewater stream entering a treatment plant, such as the number of contaminants, their amounts, toxicity, shape, size, etc. To this end, we believe that generation of a superstructure comprising of all possible treatment methods and flow patterns using a systems approach, followed by elimination of inapplicable methods based on certain physical constraints, will make the designing of wastewater treatment network more efficient. In this work, the technologies/methods involved in wastewater treatment such as sedimentation, filtration, membranes, adsorption, advanced oxidation processes, rotating biological containers, and activated sludge are modeled using the material and energy balance, equipment design, costing as well as environmental impact, which are in the form of linear and non-linear mathematical models. We identify the qualitative aspects and driving force involved in each technology and based on this information as well as typical concentrations of wastewater streams, generate a stage-wise wastewater treatment scheme. Utilizing systematic programming methods, we frame the wastewater treatment network selection as an optimization problem for cost minimization, energy minimization along with sustainable goals. We use the GAMS (General Algebraic Modeling Systems) programming language and the mixed-integer nonlinear programming (MINLP) solvers such as BARON, to solve the optimization problem.

In our analysis, we demonstrate case studies for municipal and pharmaceutical wastewater treatment and determine the best possible strategy for meeting the requirements of the 1972 Clean Water Act (initiated and amended by the United States Environmental Protection Agency). In the next step, we use the P-graph approach for solving the same problems as this tool provides a ranked list of solutions, thus offering insights into non-intuitive solutions which guarantee global optimality. The comparative analysis provides information about the key cost contributors and treatment bottlenecks that lead to insights for future research and development such as the need for efficient methods for wastewater characterization, stream segregation/aggregation analysis, environmental impact assessment and provide information about the process scales at which wastewater treatment can prove cost-effective and sustainable.

Keywords: water, technology model, optimization, sustainable design

571937 Estimating True Cost of Ionic Liquids Using Monetization

Due to the attractive properties of ionic liquids, they have found their way into many applications where they showed high potential in replacing existing chemicals and systems. However, rising concerns over their ecological impacts, e.g. toxicity and biodegradability, and high cost have limited their use. To address this issue, researchers have started to adopt the life cycle thinking for the design of greener and less expensive ionic liquids. However, having multiple metrics like cost and environmental impacts to compare the sustainability of ionic liquids can be challenging for decision making. This is especially true when different weighting factors can lead to different results in a multi-objective optimization problem. In this work, we apply monetization to estimate the true cost of ionic liquids by converting environmental units into a single monetary unit. Our results show that the monetized cost can be higher than or as high as the production cost. In this work, we also investigate the impact of atom economy, size of molecule and cation group on the monetized cost and its environmental contributions. We show how two ionic liquids of different atom economy can have the same cost of production but different monetized costs. We expect this to be a starting point for future studies targeting the design of more sustainable ionic liquids. For example, the impact of the different synthesis procedures of ionic liquids on both the cost and environmental impacts can be investigated. Also, how different anions and cation families affect the sustainability of other ionic liquids qualitatively and quantitavely is a question that need to be answered.

573592 Optimised Actions to Engage Cooperative Partnerships for an Effective Climate Change Mitigation

Climate change is one of the main challenges facing our generation which calls for urgent mitigation actions since otherwise global warning will continue to worsen. World leaders through international agendas have boosted the climate action since the earliest 80s and 90s, such as for example, with the establishment of United Nations Framework on Climate Change Conference and the Intergovernmental Panel on Climate Change, or the adoption of the Kyoto Protocol and more recently the ratification of the Paris Agreement. Unfortunately, despite the efforts made so far, carbon dioxide emissions continue to rise and perhaps more important, the overall pledged mitigation efforts presented by the countries fall short of meeting the long-term Paris goals.

Coordinating global efforts to fight against climate change remains a priority challenge, hindered by the need of satisfying so many conflicting interests and competing priorities in a fair manner. Traditional negotiations schemes have failed so far, leading to an impasse reflecting differences in the willingness to act with other parties. The backsliding of the United States Administration on climate change, with the rescission of the US Clean Power Plan and its intention to withdraw from the Paris Agreement, might trigger a dangerous domino effect among other countries following its intention to inaction. This situation calls for new initiatives and mechanisms to engage countries in a collective action shaping the transition towards a more sustainable future.

In this contribution, we present an energy system model, referred to as ERCOM as acronym of Emission Reduction Cooperation Model, which allows quantifying the benefits of cooperation by automated screening millions of alternatives for differing levels of cooperation, ultimately identifying the most cost-effective solutions for meeting a given CO_2 emission reduction target in the power sector. In essence, ERCOM is formulated as a mixed integer linear programming model which is capable of identifying the least cost pathway (i.e. optimal electricity mixes and electricity transmissions) so as to satisfy a given demand of electricity, where the binary variables are introduced to model whether a region cooperate or not. To demonstrate the capabilities of our tool, we employ the U.S. commitment to the Paris Agreement as a test bed, demonstrating that significant benefits can be attained with states collaborating to curb the CO_2 emissions thereby exploiting cost effective and low carbon regional advantages. In particular, our results show that, with all states cooperating, the cost of the electricity generation in the US could be reduced by 16% (i.e. US$46 billion per year) while simultaneously cutting carbon emissions by 70% below 2012 levels (i.e. 66 Mt CO_2 per year). It might be difficult to reach this full cooperative solution in practise, but our results show that, with only half of the states cooperating, these costs and emissions would be reduced by 10% and 68% respectively. Our findings show that pursuing climate change mitigation in cooperation can bring significant benefits not only for the climate but also for the U.S. economy. Unfortunately, the engagement of some regions might be compromised, since joint action entails an uneven distribution of efforts where states may be either benefited or penalised from cooperating. To circumvent this issue, we propose a compensation mechanism based on the fair allocation of collective gains following a equality principle which may be used as one option to spur cooperation since all states would emerge as benefited.

Overall, we argue that the approach we envisioned in this work may provide key ingredients for devising more effective schemes to overcome the current the impasse in climate change negotiations. Even if full cooperation remains elusive, our results demonstrate that the mere cooperation of a few parties may already lead to major environmental and economic benefits. The way forward to establish new partnerships for effective climate change mitigation will not be easy, but the mutual benefits attained will compensate any efforts made to foster deeper cooperation across regions.

576716 Achieving Campus Carbon Neutrality - Application of Sustainable Engineering Methods to Evaluate and Identify Technical and Ecological Solutions

In 2006, the American College & University Presidents' Climate Commitment (ACUPCC) was created as a public contract to meet the goal of carbon neutrality. Carbon neutrality is a term which means having no net greenhouse gas (GHG) emissions related to University activity through emission reductions and offsets. Currently, over 300 active signatories have committed a goal and target date for carbon neutrality before 2100.

Second Nature, the main sponsor of the ACUPCC, supports the Sustainability Indicator Management and Analysis Platform (SIMAP™) as a tool to calculate GHG emissions. SIMAP prepares GHG emissions inventories based on campus-provided inputs. However, the role of land use and carbon sequestration of campus property is disregarded in this tool. Projects which increase sequestration rates are treated as offsets, and according to Second Nature's *Carbon Markets & Offsets Guidance* should be the last priority for achieving carbon neutrality.

Many other models consider flows of carbon, or the carbon cycle, through sources and sinks. If we consider active sequestration, not carbon storage, as a sink, then we include a constant rate of carbon conversion into biomass across geographic landscapes. If our focus is only minimizing source emissions instead of balancing the cycle, we neglect solutions that enhance sinks.

SIMAP categorizes emissions into three scopes, where Scope 1 is the direct emissions from sources owned or controlled on campus. Similar to the on-site natural gas combustion or campus-owned transportation emissions, the sequestration occurs through campus-owned property and should be included as a campus-owned. Therefore, this research proposes that land use should be included into GHG emissions inventories as negative Scope 1 emissions. Currently, methane recovery in waste disposal is considered a negative Scope 3 emission. What does reframing land as carbon dioxide recovery look like in our sustainability analysis? What impacts could this have on campus sustainability and the pursuit of carbon neutrality?

Data has been collected through the process of writing the Climate Action Plan for The Ohio State University. It shows that carbon sequestration capacity can increase substantially through different scenarios of improved land management and land-use change. The potential of enhancing carbon sinks as a viable solution for University climate neutrality is apparent, resulting in land-use change and improved management scenarios which yield 3-6 times more carbon sequestration than the current landscape. Natural sequestration, which increases over time as biomass grows, is a solution which should be explored immediately, to reap the benefits as soon as possible.

However, how do we compare these results with the trade-offs of other technological and behavioral solutions? Many other necessary technological and behavioral options need to be considered such as: combined heat and power plants, increased building efficiency, on and off-site renewable energy, adjusted climate control settings, financed air travel policies, and greener commuting options. Reaching the goal of carbon neutrality requires drastic action, especially for large universities. Yet, navigating the number of options with constrained budgets and multiple stakeholders can cause paralysis in moving forward. To assist in informing University decision-makers of these trade-offs, optimization methods can be applied to compare the estimated carbon reductions and costs for any number of options, presenting multi-solution strategies that meet the needs of the University. Understanding financial, technological, spatial, and social constraints and dynamics of the carbon reduction strategies can inform the optimization problem to achieve the objective of carbon neutrality, assisting the development of the university's action plan.

Prioritizing land use as a last effort offset is a missed opportunity for colleges and universities and including optimization methods to develop plans and drive decisions is essential in realizing carbon neutrality in higher education. This research discusses some of the options and challenges that exist in reaching carbon neutrality for academic institutions, while also demonstrating a potential plan for a large

University.

577318 Attributional to Consequential Life Cycle Assessment – Steps Toward a Unified Framework

Life cycle assessment (LCA) using well established attributional methods is narrowly focused on determining the environmental impact and properties of a technology or product by portraying it as an isolated entity and linearly scaling it up or down in size to match its demand. It doesn't consider the introduction of this technology into society, its acceptance and the ripple effects that might occur due to such a technology in the real world. For example, introduction of a novel "green" technology might be obstructed due to lack of available resources, acceptance of its price by industries and consumers, and how competing technologies and products adapt themselves in response to its introduction into the market. Such a problem can be thought of as being analogous to a chemical engineering design problem in a laboratory (attributional LCA) versus a full-sized industry and its effect on the associated economy (real world impacts). In this sense, there are some gaps between engineering decisions based on technological excellence and market decisions based on the economy and human preferences.

One of the methods currently used to capture such cascading effects into the market and economy is Consequential LCA (CLCA). It accommodates the interaction between competing products in market, how prices affect the demand and supply of goods and also the incorporation of factor constraints. Even though around for two decades, the method has not matured fully. Different groups of researchers practice different flavors of applying CLCA, each with their own unique advantages and disadvantages. Broadly, the three methods are System expansion, Rectangular Choice-of-Technology, and Equilibrium modeling. System expansion and marginal technology based approach, first introduced by Weidema uses a hierarchical set of rules and decisions to determine competing products and technologies of the system in question and tries to determine in a logical fashion, how the consequences of incorporating this novel technology would propagate through the economy and society. Being a step-wise question and answer based approach, it is evident that the methods suffer from the problem of boundary cutoff and has a narrow scope of the extent of consequences considered. The Rectangular Choice-of-Technology (RCOT) framework has been developed to account for market effects as a form of economy and environmental resource (economy factor) constraints and multiple technology uses in analyzing economic systems. The RCOT framework could show gradual and partial adoption of alternative technologies to displace conventional ones as a consequence of market effects. However, the framework does not account for price changes of commodities due to changing market conditions. Equilibrium modelling is a different approach to CLCA, using the economic general equilibrium models to capture price effects, through supply demand elasticity values and microeconomic relations. They include microeconomic concepts such as price elasticities of supply and demand, market clearing constraints, production and consumption functions, maximization of profits of different agents in the model, such as government, investment agents, producers, household and foreign markets. However, the major drawback of using these models is that they are highly aggregated, making it difficult and uncertain to study the impact of a specific product or technology.

In this research work, we perform a broad comparison of these methods using the same case study and highlight the differences in scope, results and insights obtained from each method. We explore their advantages and disadvantages in detail. Such an exploration and comparison provides insight into the characteristics and benefits of each method. With the knowledge obtained from such an exploration, a CLCA based optimization framework that combines all the methods is suggested to give a novel multiscale solution. The equilibrium models in this framework will help in system expansion which reduces errors in system boundary selection while the marginal-based and RCOT approaches can be focused on processes where greater resolution and details are required. Along with that, the presence of third finer engineering scale enables designing of processes while performing CLCA, which is a novel approach in the Sustainable Process Design field that has typically used ALCA for determining environmental impacts.

Ref:

Ekvall, Tomas, and Bo P. Weidema. "System boundaries and input data in consequential life cycle inventory analysis." *The International Journal of Life Cycle Assessment* 9.3 (2004): 161-171.

Noise Abatement That Substantiated to be Fiscal for Ammonia Complex

Sara Ahsan, Arsalan Naeem

Engro Fertilizers Limited, Pakistan

Abstract:

Engro Fertilizers Limited, one of the leading fertilizers company in Pakistan, commissioned one of the largest single train Ammonia-Urea manufacturing site with a capacity of 2194 MTPD and 3835 MTPD of ammonia and urea respectively. At its early operational age, the complex faced a number of startups and shut downs due to gas curtailments but finally sustained operations after two and half years. This document explicates one of the major persistent issues faced by our site i.e. higher noise index as compared to the anticipated index. Engro, a company known for its HSE standards then launched a study to overcome this concern. During RCA of this problem, it was noticed that the top contributor to high noise index is the Primary Reformer's induced draft fan. A separate study then started to look into the high noise generated by the fan. It was noticed that during normal operation ID fan speed varies by ~10 RPM translating to speed variation in primary driver i.e. steam turbine by ~70 RPM. Moreover, it was detected that ID fan had been operating very close to its surge line with 80% (Pareto rule) of the points falling within 15% of the surge line in unstable zone. The issue was then taken up with the vendor with some probable solutions. The paper encompasses the detailed RCA of the issue, proposed solutions and the most techno-economically viable change implemented on site.

Keywords: Noise Index, Induced draft fan, Octave Frequency Analysis

Introduction to Efert and Ammonia plant

In 1957 vast deposits rich in Natural gas in Mari (Sindh, Pakistan) were explored. It was then decided to put up a Urea manufacturing site in Daharki (about 10 km from Mari field). A plant with an annual capacity of ~173,000 MeT started its production in December 1968. Later, in 2007, the company undertook its largest urea expansion project in 2007. The state-of-the-art plant EnVen has an annual capacity of ~1.3 Million MeT and was put into production in 2010.

EnVen plant is based on Haldor Topsoe and Snamprogetti technologies for Ammonia and Urea plants respectively. It is one of the world's largest single train ammonia-urea complex with name plate capacities of 2194 MTPD ammonia and 3835 MTPD urea.

Following block flow diagram shows the basic process of Haldor Topsoe Ammonia technology plants.

Block flow diagram of ammonia plant:

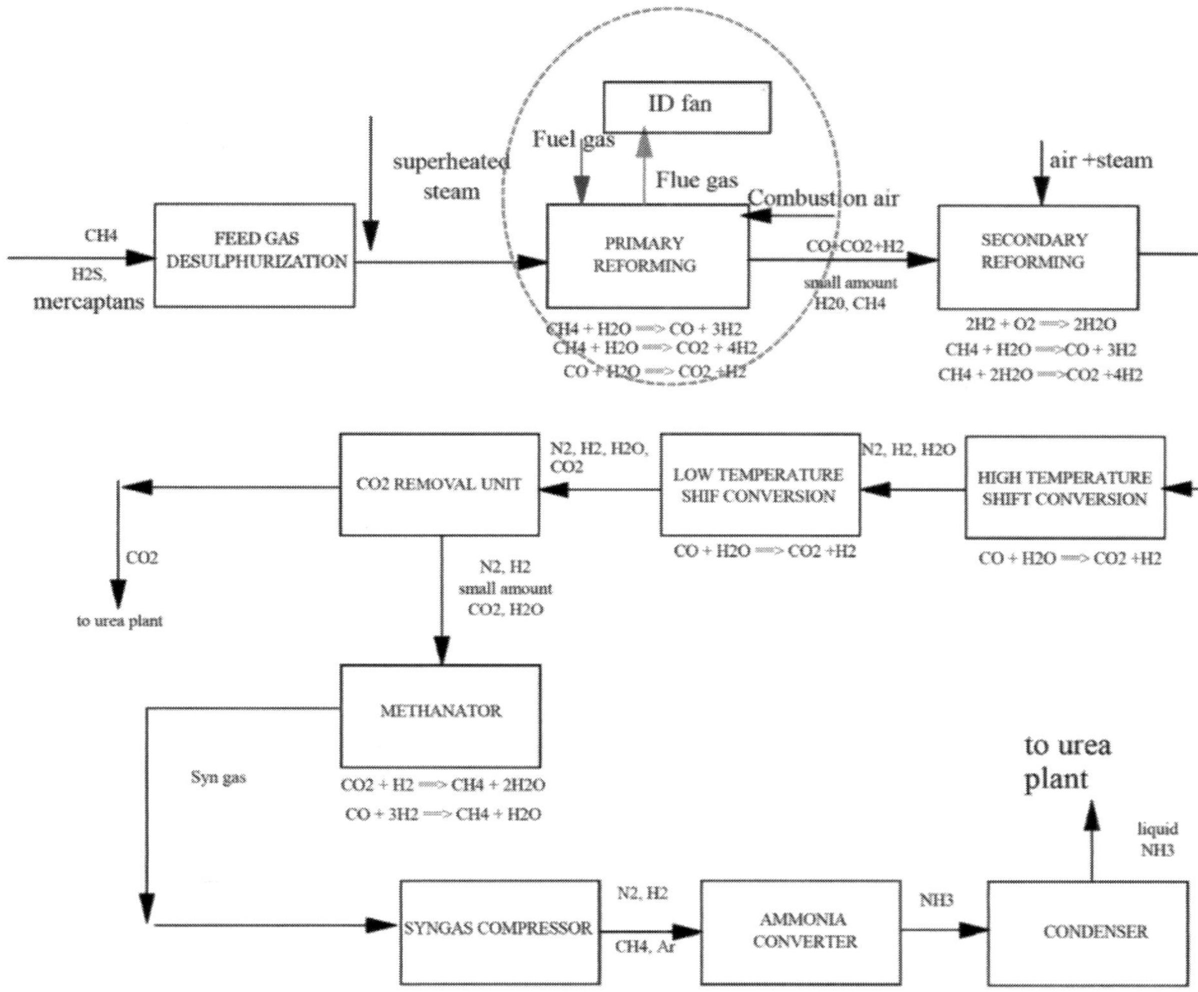

Description of Problem:

After commissioning of EnVen 1.3, site noise index increased exponentially. This issue became more evident when Engro started to receive noise complaints from neighboring industry, Mari gas- the raw feed gas supplier for Engro (Refer to below area layout for clarity regarding site location).

1- Engro Production site
2- Location of Induced Draft fan
3- Mari gas residential area
4- Engro residential area

Hence, an investigation was started to find out the root cause of such high noise. Therefore, an in-house noise survey was conducted to further determine the major source of noise at site. During this survey, the highest noise contributor was measured to be the Primary Reformer Induced draft fan with an average of ~102 dB during day and ~98 dB in night. This survey was carried out through Octave Band Frequency meter to get a detailed profile of site noise. The survey further revealed that the noise constitutes of low frequency bands. Results of the survey are as under:

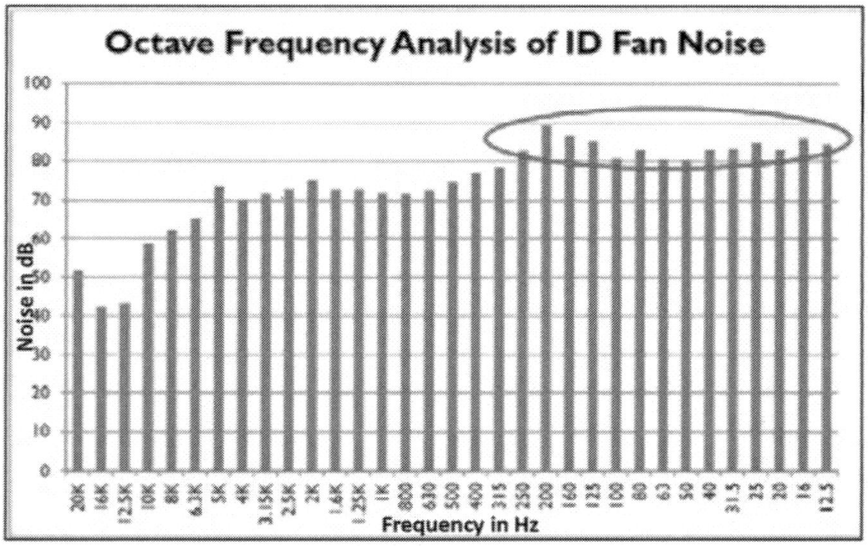

Since the low frequency noise has a tendency to travel to distances, therefore, high noise complaints were received from neighboring sites. Now, the challenge was to study the root cause behind high noise emissions from the induced draft fan compared to design value which was ~87 dB.

Root Cause Analysis:

ID fan is connected with a steam turbine and a motor with a dog-clutch arrangement. Among the two drivers, the one with the higher speed will be engaged through the dog clutch arrangement. During the startup, a VFD motor operates the blower at 600 rpm. After this, steam turbine is started which operates the blower at 990 rpm during normal operation; whereas, VFD motor keeps on running at 600 rpm. In case of any upset, if steam turbine rpm reduces, the motor cuts in and ramps up to 990 rpm to ensure safe operation of Primary Reformer.

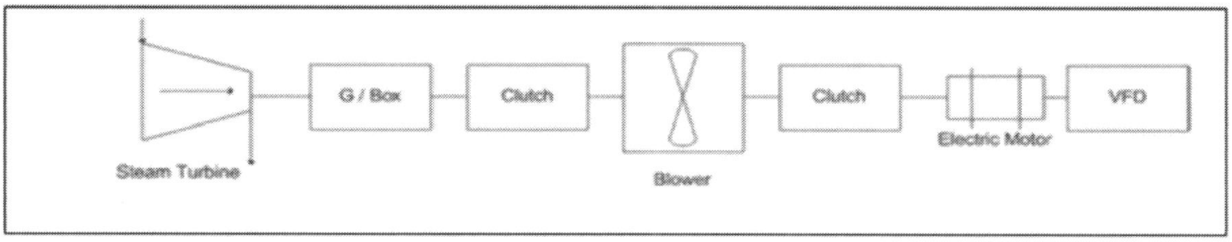

During the investigation, following issues were noticed with the ID fan operation:

- During normal operation, Variation in speed of ID fan (~10 RPM) was observed. This translates into speed variation of ~70 RPM of its primary drive i.e. steam turbine. This speed variation in ID fan caused hunting of steam turbine governor. These also resulted in breakage of capacity pocket of steam turbine
- Machine's thrust vibration also had an increasing trend. After the machine's overhaul in May 2014, these vibrations subdued for some time but the increasing trend started again. This indicated operation of machine in an unstable zone. Trends and location of probes are as below:

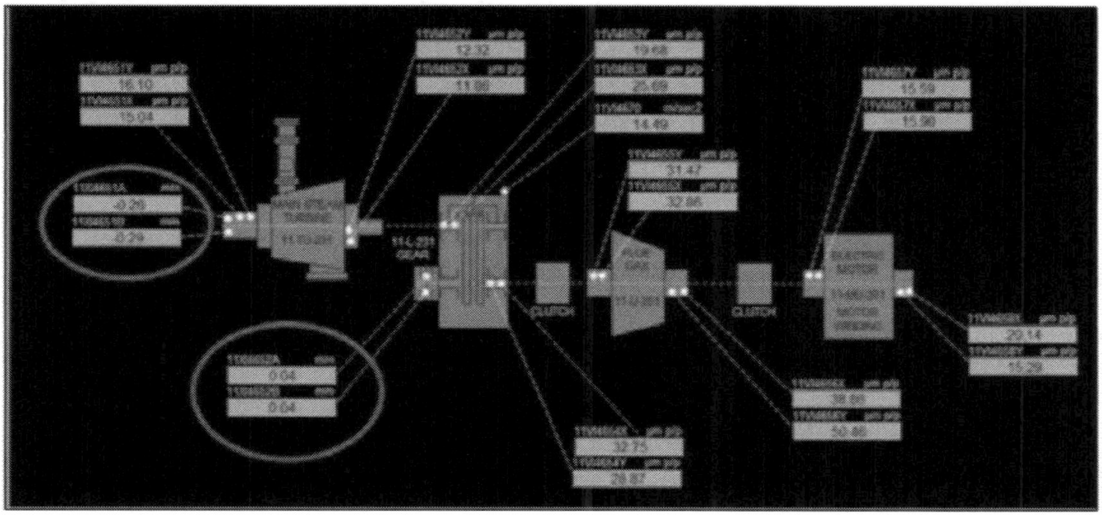

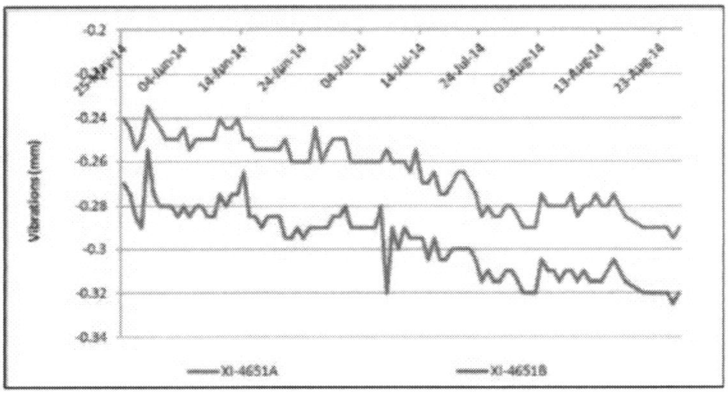

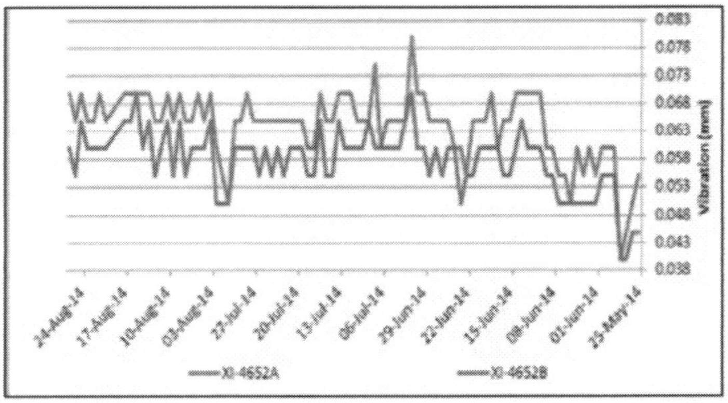

Here, XI-4651 and XI-4652 indicate turbine's and gear box thrust vibrations respectively.

This led us to investigate whether machine was being operated in an unstable zone or not. However, as per original design fan suction pressure indicator was installed at upstream of dampers rather than at ID fan suction. This made precise evaluation of ID fan very difficult; therefore, a new Pressure Indicator (PI-2519) was installed (Range: 0 to -1800 mmH20) at suction of ID fan downstream of the dampers, as shown in figures below:

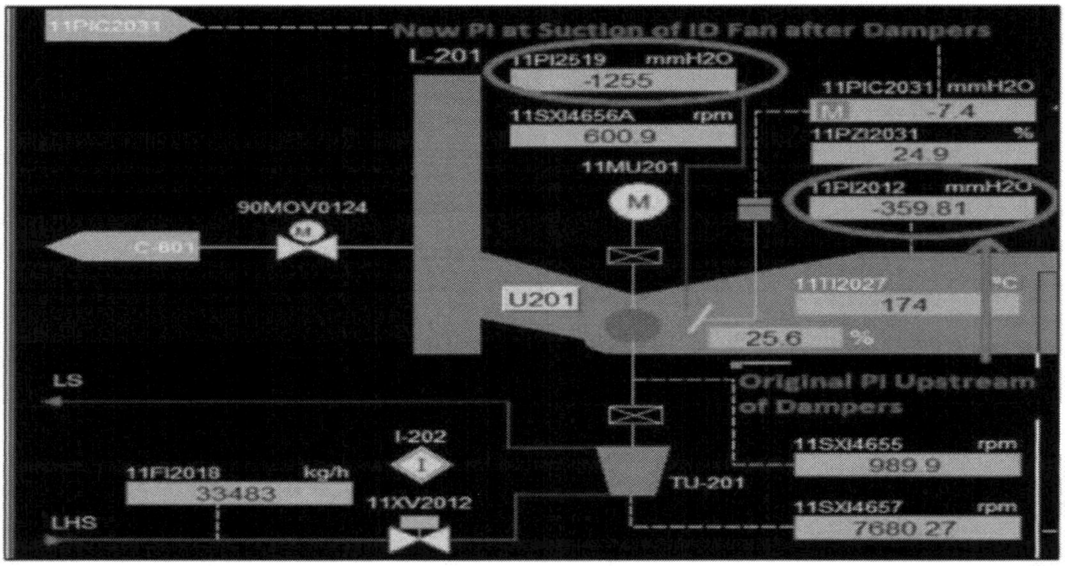

The suction pressure of ID fan from new PI was then mapped on its performance curve. Below graph shows performance of ID fan at operating speed of ~990 RPM.

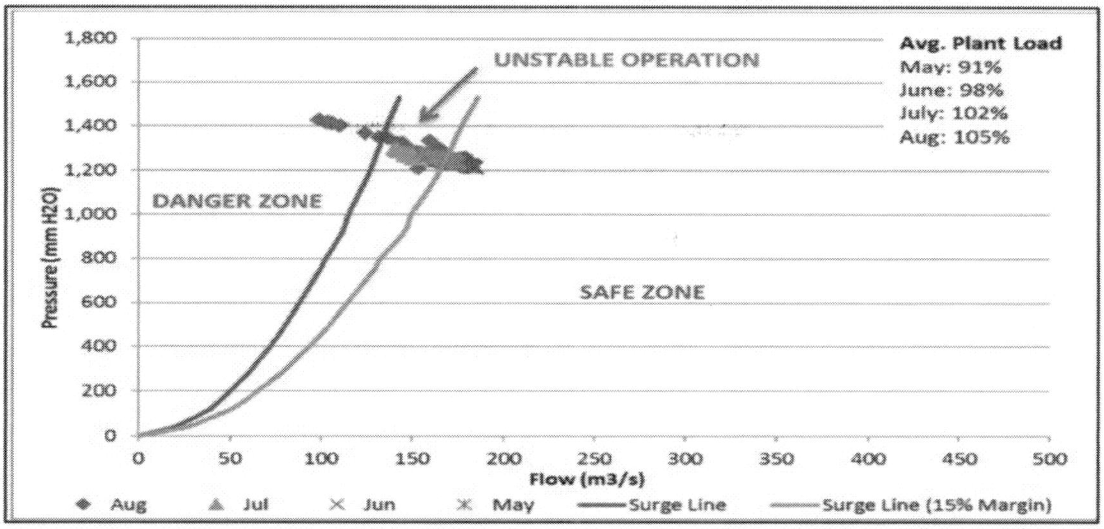

This graph indicates that ID fan has been operating very close to its surge line and 80% of the operating points are within 15% of surge line indicating operation in an unstable zone. These analyses lead us to take this issue up with the fan vendor Boldrocchi. It was mutually concluded that the installed fan is quite large compared to the requirements leading to all of the above mentioned issues.

Options studies for Noise Abatement:

To overcome this long pending issue, following options were explored:

a) Reduction in fan speed
b) Re-sizing ID fan motor
c) Installation of stack silencers
d) Re-sizing of ID fan

a) Reduction in fan speed:

A test run was carried out during plant outage and the noise survey was again carried out. This involved the following steps:

1. Bypass of motor cut-in at 960 rpm
2. Reduction of turbine speed from 990 to 800 rpm
3. Speed was decreased in small steps of 20 rpm per 20 min
4. I-1 actuation security at high draft remained in service for plant safety

Following results were noticed after this test run:

1. Vibrations of ID Fan turbine remained normal.
2. Steam consumption of ID Fan Turbine reduced by ~6 Ton/hr.
3. Blower suction pressure reduced from 1290 to 847 mmH2O.
4. F-201 stack damper opening increased from 21.5% to 32.6%.
5. Noise levels at the site boundary wall reduced from 69 dB to 64 dB.

Following conclusion was drawn after this test run:

In order to increase damper opening to required level of 80%, (for reduction in noise), we need to bring the fan to its minimum continuous speed of 600 rpm. However, this cannot be done with existing driver and major driver revamp will be required to decrease fan speed to desired level (motor revamp)

b) Re-sizing ID fan motor:

It was proposed to re-size the ID fan motor to reduce the speed of driven equipment. This did not require any modifications to the ID fan. However, the modification required extensive testing and long outage; therefore, based on low IRR the option was rejected.

c) Installation of Stack Silencers:

Acoustic Engineering contractors were contacted regarding the high noise emissions. All the companies strongly suggested installing a stack silencer at F-201 stack. However, during discussion with Haldor Topsoe experts, it was suggested that it is very difficult to mitigate low frequency noise and in their experience, installation of stack silencers is not a very good solution to the problem. Moreover, it was suggested to rectify the issue at source and minimize the noise from ID Fan rather than taking any remedial measures.

d) Re-sizing of ID fan:

The most effective option based on the payback period came out to be re-sizing the ID fan itself as it involved low cost as well as steam saving of ~10 Met/hr during operation . Payback was calculated to be less than 1 year. Fan vendor Boldrocchi also gave their consent for this option.

Following basis were taken for resizing of F-201's ID fan:

Parameter	Units	Min	Normal	Max
Plant Load (Un-corrected)	%	78	105	DBN Case (122%)
Flue Gases Flow Rate	m^3/hr	557,000	675,000	775,000
Pressure upstream of Dampers (measured value)	mm H_2O (g)	-340	-410	-520*
Temperature upstream of Dampers (measured value)	°C	174	170	167
Damper Opening	%	22.9	28.3	-
FLUE GAS COMPOSITION (Wet Basis)				
O_2	wet mol %	1.91	1.75	1.65
N_2	wet mol %	67.52	64.25	69.66
CO_2	wet mol %	8.74	8.39	8.38
Ar	wet mol %	0.81	0.8	0.84
H_2O	wet mol %	21.02	24.82	19.47
*estimated value based on measured pressure drop using: $DP_1/DP_2 = (Q_1/Q_2)$^2				

To reach the above performance vendor proposed installing a new impeller type 35LA – 300/1700 M DWDI, having special backward curved blades (airfoil shape) in order to obtain a diameter of about 3100 mm Vs. already installed of 3500mm.

Expected Performance as provided by the vendor is summarized below:

Parameters	Units	Min	Normal	Max
Capacity	m3/h	557,000	675,000	775,000
Static Pressure at dampers inlet	mm H2O	520	410	340
ID Fan Speed	RPM	990	990	990
Damper Opening	%	35	80	90
Power Absorbed	kW	1.2	1.4	1.8
Eq. Steam Consumption	MeT/hr	18	20	24

As per vendor the noise levels will reduce to originally guaranteed numbers. The new Impeller design will allow a more opening percentage of the existing dampers in order to reduce the high noise emissions by the Fan, from the actual 90-93 dBA (with damper opening less than 40%) to foreseen 85-87 dBA (with damper opening at 80%). Hence, the reason for high noise was lower damper opening and operation of ID Fan in unstable zone.

Results:

ID fan modification was implemented in 2016 outage. Following impact on noise and steam savings were measured.

a. Noise Survey Post Modification:

The below pictorial view and table shows the before and after modification noise levels at different locations.

Sr. No	ID fan		Plant Gate		Mari Gas colony		Efert Colony	
	Before	After	Before	After	Before	After	Before	After
1	98	88		59	66	52	63	50

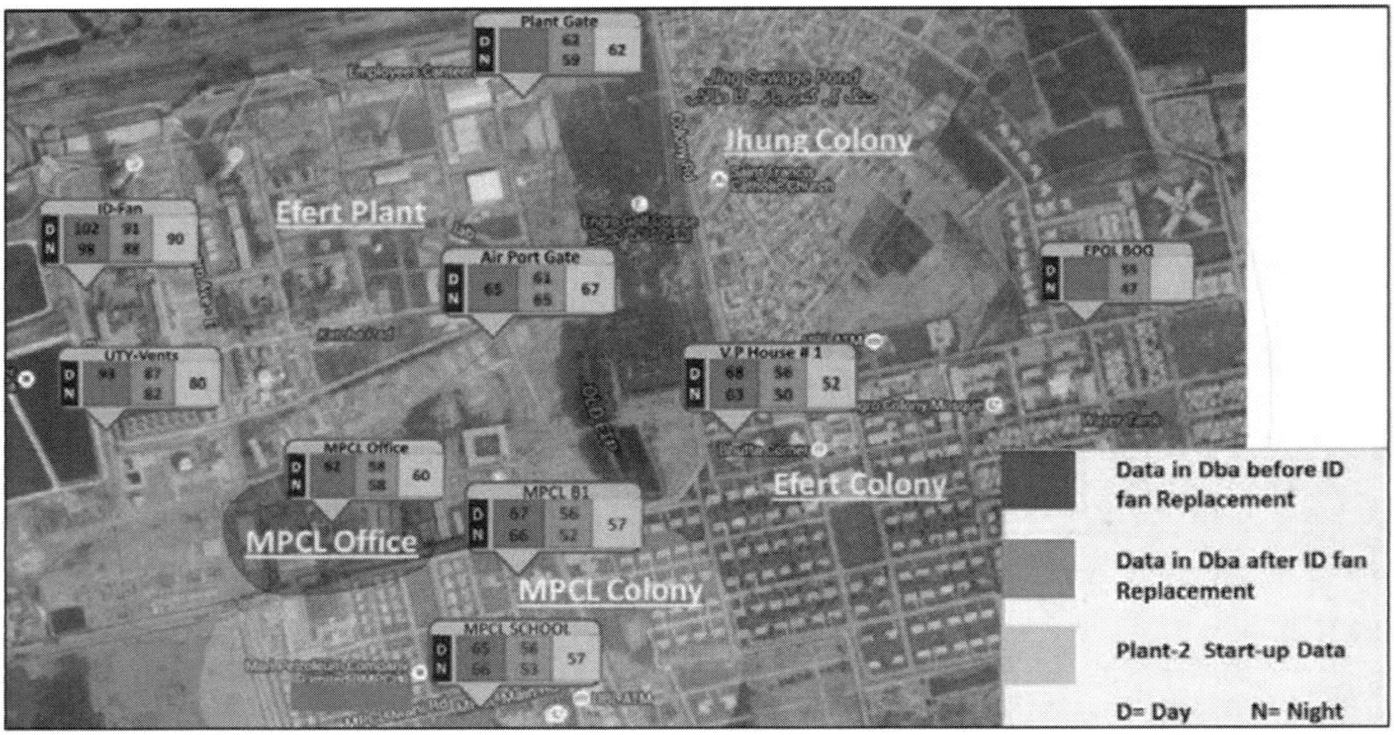

b. Steam Balance Post Modification:

Ammonia plant detailed steam balance (before/after) has been attached on the following pages

AMMONIA-III STEAM BALANCE

Parameter	4/12/2016	5/30/2016	Difference
FEL Load	88.0	104.0	16.01
S/G	2.54	2.55	0.01

KS HEADER

GENERATION (MeT/h)				CONSUMPTION (MeT/h)			
Date	4/12/2016	5/30/2016	Δ	Date	4/12/2016	5/30/2016	Δ
11-E-208, SECONDARY REFORMER OUTLET W.H.B	337.01	388.3	51.3	11-TK-431, SYNTHESIS GAS COMPRESSOR	208.7	258.6	49.9
11-E-210, HTSC OUTLET W.H.B				11-TK-441, REFRIGERATION COMPRESSOR	30.8	31.5	0.7
11-E-501, BACK-END W.H.B				10-TK-01, CO2 COMPRESSOR	104.5	108.6	4.1
11-E-219 A/B, DE-SUPERHEATERS	10.0	16.5	6.5	11-PV-7101, LED DOWN TO 'HS' HEADER	2.5	5.5	3.0
				LOSSES	0.3	0.4	0.1
TOTAL	**347**	**405**	**58**	**TOTAL**	**347**	**405**	**58**

HS HEADER

GENERATION (MeT/h)				CONSUMPTION (MeT/h)			
Date	4/12/2016	5/30/2016	Δ	Date	4/12/2016	5/30/2016	Δ
11-TK-431	154.9	193.8	38.9	PROCESS STEAM TO F-201	121.5	143.0	21.5
11-E-453, DE-SUPERHEATER	3.0	2.5	-0.5	STEAM TO AIR COIL	7.00	10.3	3.3
11-TK-441,	0	0.0	0.0	11-TK-421	66.2	67.8	1.6
11-E-452, DE-SUPERHEATER	0	0	0.0	11-E-553, ARU	2.5	2.5	0.0
11-PV-7108A	0	0	0.0	10-R-02, UREA HYDROLYSER	5.0	5.0	0.0
11-E-224A, DE-SUPERHEATER	0	0	0.0	EXPORT TO LHS HEADER	0.0	0	0.0
11-PV-7108B, LED DOWN	0	0	0.0	SMALL CONSUMERS	0	0	0.0
11-E-224B, DE-SUPERHEATER	0	0	0.0	11-E-221 LIVE STEAM TO E-302	0	0	0.0
11-PV-7101, LED DOWN	2.5	5.5	3.0	LOSSES	0	0	0.0
11-E-224C, DE-SUPERHEATER	4.0	2.5	-1.5				0.0
IMPORT FROM 'LHS' HEADER	38.0	25.0	-13.0				0.0
TOTAL	**202**	**229**	**27**	**TOTAL**	**202**	**229**	**26**

LHS HEADER

GENERATION (MeT/h)				CONSUMPTION (MeT/h)			
Date	4/12/2016	5/30/2016	Δ	Date	4/12/2016	5/30/2016	Δ
IMPORT FROM URUT-1	141.7	113.4	-28.3	11-TP-1403A/B CW PUMP	32.1	30.4	-1.7
IMPORT FROM AMMONIA PLANT	0	0	0.0	11-TP-1301A, BFW PUMP	22.3	22.0	-0.3
				11-TP-1202A, PW PUMP	2	2.0	0.2
				11-TU-201, ID FAN	32.2	24	-8.2
				11-TP-402A LO PUMP	5.0	4.2	-0.8
				11-TP-401A, TC PUMP	4.2	4.0	-0.2
				10-PV-1048,	0	0	0.0
				EXPORT TO 'HS' HEADER	38.0	25	-13.0
				13-PV-6661A/B	0	0.0	0.0
				90-E-807, CDR RECLAIMER	4.2	0.0	-4.2
				TP-1474, TP-1475, TP-1473A	1.3	1.0	-0.3
				LOSSES	1.0	0	-1.0
TOTAL	**142**	**113**	**-28**	**TOTAL**	**142**	**113**	**-28**

LS HEADER

GENERATION (MeT/h)				CONSUMPTION (MeT/h)			
Date	4/12/2016	5/30/2016	Δ	Date	4/12/2016	5/30/2016	Δ
11-TP-1301A, BFW PUMP	22.3	22.0	-0.3	11-TK-421, PAC	0.0	0.0	0.0
11-TP-1202, PW PUMP	1.0	2.0	1.0	13-V-1301, DE-AERATOR	18.1	21.3	3.2
11-TU-201, ID FAN	32.2	24.0	-8.2	10-TK-01	13.0	16.5	3.5
11-TP-402A LUBE OIL PUMP	5.0	4.2	-0.8	90-E-805, CDR REBOILER	15.0	15.0	0.0
11-TP-401A, TC	4.2	4.0	-0.2	SMALL CONSUMERS	4.0	3.0	-1.0
13-PV-6661A/B, LED DOWN	0.0	0	0.0	AMMONIA HEATER, 30-E-3002	0.0	0	0.0
13-E-1301, DE-SUPERHEATER	0.0	0	0.0	LOSSES	17.0	3.0	-14.0
TP-1474, TP-1475, TP-1473A	1.3	1.0	-0.3				
V-201 BLOW DOWN	1.0	1.5	0.5				
TOTAL	**67**	**59**	**-8**	**TOTAL**	**67**	**59**	**-8**

All values are in TPH

- Reduction in Steam Import after turn around
- Reduction in steam consumption of air turbine by ~10 TPH (when LS induction was lined up)
- Reduction in TU-201 consumption to 24 TPH
- Reduction in overall steam losses

Conclusion:

The noise abatement project involving Ammonia plant's ID fan not only helped us in reduction of Site noise Index but also decreased the steam consumption of the said blower, saving ~1200 MMBTU/day of energy

580330 Application of Green-Accounting Methods for the Assessment of the Chicago Metropolitan Area's Sustainability Trends

Assessment of regional sustainability entails an integrated evaluation of trends in economic wellbeing and environmental indicators. We approach the measurement of sustainability, defined as non-declining utility or non-declining wealth, by resorting to green-accounting methods. This study is part of a multidisciplinary research project that deals with the quantitative assessment of urban sustainability at the US Environmental Protection Agency. In this regard, we explore the sustainability trends of the Chicago Metropolitan Area via two metrics rooted in economics, Green Net Metropolitan Product (GNMP) and Genuine Savings (GS). GNMP adjusts the traditional metric of economic wellbeing, the metropolitan Gross Domestic Product (GDP), by subtracting from it the values corresponding to the consumption of manmade capital and natural capital as well as the monetary values of damages caused by environmental pollution. GS measures changes in the total capital stock (or wealth), including manmade capital, human capital, and natural and environmental capital. Regional sustainability requires that GNMP be non-decreasing, and that GS be non-negative over time. We have estimated GNMP and GS for the Chicago Metropolitan Area (CMA) for the period, 1990-2015. Variation of similar methods have been applied in the literature for sustainability measurements of larger geographical entities, e.g. Heberling et al. (2012); Mota and Domingos (2013); and Wu and Heberling (2016). Environmental variables considered in our assessment are damage cost estimates due to air pollution, solid waste, scarcity rent due to water depletion, and value of ecosystem services of urban forests. In addition, we incorporate the value of time to account for the role of technological progress. With proper adaptation, these methods could be applied for the determination of sustainability of industrial processes and/or products. For example, Demeke et al. (2018) have shown that green-accounting methodologies can be coupled with Life-Cycle Assessment (LCA) to measure the sustainability of an industrial product.

579789 Sustainability Metrics for Managing and Processing Food Wastes

Food waste is a significant issue in the United States and globally. About 1/3 of the quantity of biomass that is produced to provide food is not consumed as food. This work addresses sustainability metrics related to the management and processing of food wastes. Food wastes may be used as raw materials to produce ethanol, biodiesel, methane, and fertilizer. Some are used as animal feed. Waste management options include incineration wastewater treatment, landfill disposal, and composting. Sustainability metrics include cost/income energy, carbon dioxide emissions, methane emissions, risk, and nutritional value. Waste minimization options are important as well. Several management options will be considered

563641 Sustainable Soluble Coffee Production Using Dynamic Membrane Processes

A study, in partnership with Nestlé Beverage USA, and U.S. Environmental Protection Agency, is underway to improve the sustainability profile for the manufacture of instant (soluble) coffee products. In this study, the application of membrane processes in the pre-concentration of coffee extract prior to spray/freeze drying has been found to be a promising alternative to energy-intensive separation methods such as evaporation. Such approach opens opportunities in the intensification of soluble coffee production in terms of energy-efficiency, water recovery, and waste reduction, thereby reducing the carbon footprint of the overall operations. In addition, the use of a novel dynamic vibratory separation process to address the limitations of traditional crossflow filtration, by imparting shear at the membrane surface, reducing gel formation and fouling, and improving flux was investigated. In this study, we highlight the benefits and limitations of the technology, and how laboratory investigations help guide industrial use and scale-up considerations; and the overall assessment of environmental impacts. Using reconstituted coffee extract solutions, we tested and screened various types of membranes to determine their performance in concentrating the samples. Parametric studies were also conducted in order to determine the effects of varying feed concentrations, transmembrane pressure, and surface shear (vibrational displacement) on permeate flux, and selectivity of filtration in terms of organic/inorganic rejection efficiencies. Scale-up trials from laboratory scale system were correlated to estimate design parameters for a commercial membrane filtration unit. Different cases involving the integration of membrane processes are compared with the conventional evaporative pre-concentration of coffee extract. As criteria for sustainability, the potential energy savings, operating costs reduction, economic feasibility, and resulting life cycle emissions profiles of the scaled-up operation are also presented. As a promising area for further investigation, our research shows opportunities in making soluble coffee and similar powdered beverage manufacturing operations greener.

570940 Sustainability Assessment: Uncertainty Handling and Impact on Decision Making

Sustainability assessment is a very complex appraisal method. It entails not only multidisciplinary aspects (environmental, economic and social), but also cultural and value-based elements. There have been various types of sustainability indicators for different types of systems and applications, and methodologies for indicator formulation strategy, data scaling, normalization, weighting, and aggregation. The practicing approaches for developing SA frameworks are: (a) a top-down approach – to enable experts/researchers to define and overall structure for achieving sustainability and subsequently indicators, (b) a bottom-up approach – to require participation of various stakeholders to understand the framework and propose key indicators, and (c) a mix of both. Each sustainability dimension (i.e., environmental, economic, or social) requires selection of a number of indicators to measure the performance of different aspects. This has posed a challenge: data/information uncertainty, which is pervasive in almost every phase of sustainability study. Examples include the available data about material or energy utilization, toxic/hazardous waste generation, and market fluctuation; the multifaceted makeup of the inter-entity dynamics, dependencies, and relationships; the prospect of forthcoming environmental policies, and the interrelationship among the triple-bottom-line aspects of sustainability, weighting methods, weights' values and aggregation methods.

In this presentation, we will discuss the characteristics of two classes of uncertainties, i.e., the aleatory and epistemic uncertainties that appear in various types of industrial sustainability problems. We will also describe four uncertainty processing methods: the probability bounds analysis method, the fuzzy arithmetic method, the information gap theory based method, and the interval parameter based method. The applicability of these methods in sustainability assessment at the product and process level will be discussed, and an application of these methods to geographically distributed biofuel manufacturing will be presented.

---

* All correspondence should be addressed to Prof. Yinlun Huang (Phone: (313) 577-3771; E-mail: yhuang@wayne.edu).

567869 Statistical Analysis of Streamlined Process Simulation-Based Gate-to-Gate Energy Use for Chemical Reactions and Separations

Chemical process energy is a key component of chemical life cycle inventory (LCI) data. Energy used in manufacturing of chemicals account for 50%-80% of the environmental impacts associated with chemical production. Current chemical LCIs are typically either dependent on aggregated plant-based data that are not specific to the target chemical, or on estimations based on simplified design calculations and/or use of proxy data. Such estimation techniques miss the process-level details and advanced thermodynamic calculations of chemical process simulation tools commonly used in chemical engineering design. In this work, a streamlined process simulation-based methodology for process energy data generation is used to generate gate-to-gate energy use data for manufacturing of organic chemicals. Chemical process is divided into three major blocks, with the reaction at the core, energy for pretreatment before the reaction, followed by energy required for separation and purification of the product. Chemical process simulator Aspen Plus is used to model and simulate 150 chemical processes based on this methodology. Potential process heat integration is analyzed using Aspen Energy Analyzer, which uses Pinch Analysis to provide an estimate of target heating and cooling utilities. The extracted energy data for each chemical process are used to provide a time-efficient method for predicting process heating and cooling requirements through multi-variate statistical methods such as regression analysis. Process design-based predictor variable such as heat of reaction, type of reaction (addition, substitution, etc.), and unit process (hydrogenation, nitration, etc.), and phase were used in the analysis. Heat of reaction has a strong and positive correlation with reactor heat duty, with a R^2 of 0.8. Separation operations on average require 46% (SEM = 4%) of the heating energy and 57% (SEM = 4%) of the cooling energy. The models from this study can be used for predicting process energy requirements for chemical production in early design stages, and for chemical LCI data generation in the absence of detailed industry data.

577341 A Dynamic Genome-Scale Metabolic Network Model for a Novel Methanotroph-Cyanobacteria Coculture

A Dynamic Genome-scale Metabolic Network Model for a Novel Methanotroph-Cyanobacteria Coculture

Kiumars Badr, Q. Peter He and Jin Wang*

Department of Chemical Engineering, Auburn University, Auburn, AL, 36849, USA. *Email: wang@auburn.edu

Biogas is comprised primarily of methane (50%~70%) and carbon dioxide (30% ~50%), which can be produced from various waste sources, including landfill material, animal manure; wastewater; and industrial, institutional, and commercial organic wastes. EPA estimates that currently US biogas production potential is 654 billion cubic feet per year, which could displace 7.5 billiongallon of gasoline [1]. It is clear that biogas has immense potential as a renewable feedstock for producing high-density fuels and commodity chemicals. However, the utilization of biogas represents a significant challenge due to its low pressure and presence of contaminants such as H_2S, ammonia, and volatile organic carbon compounds. To tap into this immense potential, effective biotechnologies that can co-utilize both CO_2 and CH_4 are needed.

In our previous work, we have demonstrated that metabolic coupling of methane oxidation to oxygenic photosynthesis can be a highly efficient way to recover the energy and capture carbon from biogas [1, 2]. However, development of multi-organism platforms for commercial biogas conversion present significant challenges which center around our ability to control function and composition of species in the Coculture. An essential tool for the optimization, design and analysis of the coculture based biogas conversion is the development and validation of kinetics models that can accurately describe and predict the co-culture growth under different conditions [3]. To this end, using *Methylomicrobium buryatense - Arthrosipira platensis* as the model coculture system, we have developed an unstructured model to capture the growth dynamics. Specifically, Monod-like models were developed to capture coculture growth. Two sources of substrate were considered in the model: gas transferred from gas phase and gas produced *in situ*. In addition, we rely on the fitted maximum cell growth rate for both strains to capture other potential interactions. Using designed experiments and the developed model, we clearly demonstrated that the synergistic effect within the coculture cannot be fully explained by the *in situ* substrate exchange, and there must be other "metabolic links" to explain the significantly enhanced cell growth of both strains in the coculture.

In order to test different hypothesized "metabolic links" between the Methanotroph and cyanobacteria in the Coculture, in the work, we developed a structured, dynamic genome-scale metabolic network model for the Coculture system. These GEMs, especially the dynamic GEMs, offer a comprehensive picture of cellular metabolism and serve as a bridge that can better link the work of microbiologists and engineers in understanding and optimizing complex cellular metabolism. By integrating the available knowledge on each strain with data obtained in our own experiments, we use DFBAlab to implement the dynamic GEM for the Coculture [4]. For the dynamic GEM, besides the GEMs for each individual strain, the key inputs to the DFBA model are the uptake kinetics for different substrates. In this work, the substrate update rates are provided from the unstructured dynamic model we already developed. In addition, the product secretion rates and cell growth rates predicted by the validated unstructured model were used to validate the coculture dynamic GEM. Although the experimental measurements to validate the model are mainly limited to cell growth, substrate consumption and product secretion rates for each organism in the coculture, the overall dynamic trajectory (both measured low-frequency samples and unstructured model predicted high-frequency samples) offers significantly more power in validating the model, compared to a few steady state conditions. The validated coculture GEM enables the testing of different hypotheses on potential "metabolic links" within the coculture.

Reference:

[1] Badr, K., Hilliard, M., He, Q.P., & Wang, J., "Understanding the Stability and Robustness of a Methanotroph-Cyanobacterium Coculture through Kinetic Modeling and Experimental Validation." Annual AIChE Meeting. Pittsburgh, PA (2018).

[2] Badr K, Roberts N, Hilliard M, He QP, Wang J. "Photoautotroph-Methanotroph Coculture – A Flexible Platform for Efficient Biological CO_2 – CH_4 Co-utilization," Dynamics and Control of Process Systems, including Biosystems - 12th DYCOPS. (2019); accepted.

[3] Almquist J., M. Cvijovic, V. Hatzimanikatis, J. Nielsen, M. Jirstrand, "Kinetic models in industrial biotechnology – improving cell factory performance", Metab. Eng., 24 (2014), pp. 38-60.

[4] Gomez JA, Höffner K, Barton PI. DFBAlab: a fast and reliable MATLAB code for dynamic flux balance analysis. BMC Bioinformatics. (2014),15:409.

567780 Kinetics and Reactor Design for Pyrolytic Treatment of Petroleum-Contaminated Soils

Scope and Significance

While marine oil spills from offshore platforms or tankers attract most of the public attention, the vast majority of oil spills (about 98%) occur on land. Every year about 10-25 million gallons of petroleum products are spilled mostly from pipelines and fixed facilities. Petroleum hydrocarbons pose long term threats to groundwater quality, inhibit the germination and root elongation rates of plants and are extremely toxic to important soil microbes. Moreover, polycyclic aromatic hydrocarbons (PAH) present in petroleum can damage the immune systems of aquatic organisms and wildlife.

Current remediation methods for petroleum-contaminated soils are either relatively slow or have unintended consequences in the form of soil damage and high-energy usage. Even worse, some processes such as aerobic bioremediation can activate toxic hydrocarbons and transform them to more noxious byproducts such as PAH derivatives. For all these reasons, there is still a pressing need for more efficient and *sustainable remediation* of petroleum-contaminated soils.

Pyrolysis is receiving increasing attention as an on-site remediation approach because of its potential to rapidly and reliably remove total petroleum hydrocarbons (TPH) with lower energy requirements and better post-treatment soil fertility than other *ex situ* thermal remediation approaches [1-3]. Pyrolytic treatment of contaminated soils at 420°C with only 15-min residence time in a continuous reactor reduced TPH by 99.9%, lowered the total PAH concentration by almost 95%, and restored soil fertility to 98% of the clean soil level [3]. These results clearly suggest that pyrolysis has the potential for improved ecosystem restoration following remediation compared to traditional thermal technologies. However, the same study also revealed potential tradeoffs between pyrolytic treatment intensity, soil detoxification efficacy and soil fertility restoration. Specifically, treatment at 470°C for 15- or 30-min reduced soil fertility to 51% and 39% of the clean soil level, which was only marginally higher that the fertility of the contaminated soil.

The existence of an optimal treatment intensity for contaminated soils underscores the need for mechanistic insight into the pyrolysis process and for a deeper understanding of how the detoxification of contaminated soil and the restoration of soil fertility are affected by (a) the properties of both the soil and the petroleum crude, and (b) the operating conditions of pyrolysis reactors.

Kinetics and Reactor Design for Pyrolytic Treatment of Petroleum-Contaminated Soils

Taking the first steps towards a rational design of pyrolysis reactors, we have combined mathematical techniques and advanced analytical measurements to:

1. Develop a full kinetic model for the pyrolytic treatment of soils contaminated with petroleum hydrocarbons. The new model describes all thermally-induced transformations of soil components (e.g. clays, carbonates, or SOM), the evaporation of light hydrocarbons and the pyrolysis of the heavier hydrocarbon fractions.
2. Develop a mathematical model describing the operation of continuous reactors (rotary kilns) that will be used for *ex situ* treatment of petroleum-contaminated soils. This dynamic model will then be used to estimate the optimal (a) pyrolysis temperature, (b) reactor residence time, and (c) gas flow rates that will achieve a specified TPH reduction at the minimum energy cost for any given throughput of contaminated soil.

Two clean soils with different compositions were spiked with oil crudes to prepare petroleum-contaminated soils. Thermogravimetry with evolved gas analysis (TG-IR and TG-MS) was used first to identify the mineral transformations taking place as the clean soils were heated in an anoxic atmosphere. The same analytical techniques were applied to study the two distinct processes occurring during the pyrolytic remediation of contaminated soils. Desorption of the lighter hydrocarbons is the dominant process for temperatures between 150 and 350°C. Thermal cracking reactions dominate in the 350-500°C range releasing pyrolysis products (like hydrogen, methane, or olefins) and leading to the formation of a carbonaceous solid material (char) that remains in the treated soil [2]. It is important to note that hydrocarbon desorption and pyrolysis overlap with the temperature ranges over which soil mineral transformations take place.

To develop our kinetic model, we assumed that the soil consists not only of inert components but also of "reacting pseudo-components" (e.g. clays, carbonates, SOM) that release either water or carbon dioxide as the soil is heated. Weight losses due to water or carbon dioxide release accounts for more than 95% of the weight loss observed during heating of clean soils. Contaminated soils contained additional pseudo-components that account for the weight losses observed due to the desorption of light hydrocarbons and the pyrolysis of the heavy petroleum fractions to yield gaseous products and char (coke). Using the distributed activation energy approach (DAE) approach, each pseudo-component was assumed to consist of a large number of species that react with activation energies distributed according to Gaussian probability density functions (PDF). To obtain the kinetic parameters, an optimization problem was formulated by fitting model predictions to experimental thermogravimetric data. Solving this nonlinear constrained optimization problem with an interior-point algorithm, the expected activation energies and variances of all pseudo-components were estimated, together with the preexponential factors for all the reactions and the initial mass fractions of all pseudo-components.

Excellent fits to the experimental data were obtained using four reacting pseudo-components for the clean soils. Three pseudo-components were required to model clay dehydration and dehydroxylation processes that occur over a wide range of temperatures and generate different multi-peak water release patterns for each soil. One pseudo-component was enough to model carbon dioxide release from either carbonate decomposition reactions or soil organic matter (SOM) pyrolysis. Two additional pseudo-components were required to model hydrocarbon desorption and pyrolysis.

The kinetic models were then used to simulate the operation of non-isothermal rotary kiln reactors that will be used for *ex situ* treatment of petroleum-contaminated soils. To preserve anoxic conditions, the reactors were heated externally while a stream of nitrogen gas flowed through the rotary kiln to sweep the desorbing hydrocarbons and pyrolysis products. A finite stage or mixing cell model with separate solid and gas phases was used to describe the operation of the pyrolysis reactor.

Model predictions were compared to experimental data we obtained with a continuous rotary kiln reactor [3]. The TPH reductions predicted by the model agreed very well with the experimental data for all the conditions considered in our earlier study: pyrolysis temperatures, 370, 420 and 470°C; residence times 15, 30 and 60 min; solid flow rates 7, 14, and 25 lb/h.

More importantly, however, our model was able to explain the trade-offs observed in our earlier study between pyrolysis treatment intensity, soil detoxification and soil fertility restoration [3]. The clay components of the soil used in [3] undergo significant dehydration between 350 and 450°C, a range that overlaps with that of pyrolysis reactions. As contaminated soil moves through the reactor zone kept at 420°C, the reaction rate for the heavy hydrocarbon fraction is fast enough to achieve almost complete conversion within 15 or 30 min. But, the same conditions are not severe enough to completely remove the water from the clay components of the soil. If the soil is treated at 470°C, on the other hand, the clay components are almost completely dehydrated, causing irreversible soil damage and loss of fertility.

By emphasizing the importance of soil composition and by raising the possibility of trade-offs between detoxification and fertility of treated soils, this study has significantly advanced our fundamental understanding of the *multiple* and *concurrent*physicochemical processes taking place during the pyrolytic treatment of petroleum-contaminated soils. Theoretical kinetic and reactor models are essential for designing pyrolysis reactors optimized for specific petroleum-soil systems, enabling robust operation and facilitating acceptance of this novel technology by regulatory agencies and stakeholders.

References

1. Vidonish, J. E.; Zygourakis, K.; Masiello, C. A.; Gao, X.; Mathieu, J.; Alvarez, P. J., *Environmental Science & Technology* **2016,** *50*, (5), 2498-2506.
2. Vidonish, J. E.; Alvarez, P. J. J.; Zygourakis, K., *Industrial & Engineering Chemistry Research* **2018,** *57*, (10), 3489-3500.
3. Song, W.; Vidonish, J. E.; Kamath, R.; Yu, P. F.; Chu, C.; Moorthy, B.; Gao, B. Y.; Zygourakis, K.; Alvarez, P. J. J., *Environmental Science & Technology* **2019,** *53*, (4), 2045-2053.

573966 Well Placement Optimisation in a Heterogeneous Petroleum Reservoir with Geological Uncertainty

ABSTRACT

Huge investments are required in the development of oil and gas fields; this implies that decisions at any stage of the development process must be aided by sound and proven mathematical-oriented methods, because the use of intuitive engineering judgement alone cannot guarantee sustainable profitability over long periods [1]. This presents an ever-increasing motivation to apply optimisation techniques for the enhancement of petroleum production from both planning and operational/scheduling perspectives. The planning aspect typically involves activities such determining the number and locations of wells, drilling the wells and infrastructure installation. On the other hand, the operational/scheduling aspect comprises of well rate allocation, flow scheduling and routing problems [2]. Incorporating optimisation at the planning stage is a pre-emptive step that mitigates subsequent operational difficulties over the field's lifetime. However, embedding optimisation in a planification problem can be more challenging compared to production and injection scheduling optimisation in an oil and gas field. Although many studies [3, 4] have addressed the operational/scheduling aspect of the field development problem, this work is premised on the proposition that, jointly optimising the planning and scheduling problem better enhances field operability.

The uncertainty of subsurface geology is another constraint to take into consideration when optimising field productivity [5]. To capture the uncertainties in the subsurface geological (reservoir) model, geostatistical realisations of the most accurate reservoir representation are obtained using available information (permeabilities, porosities, fluid saturation) [6]. These uncertainties in the model are often reflected in the diverse outcomes of well productivities and ultimately the field's Net Present Value (the objective function) [7, 8]. Given the numerous number of possible realisations, it is impractical to consider all realisations in the optimisation; hence, it is important to apply classical reduction methods to ensure computational feasibility.

In this work, we specifically address the well placement planning and rate control optimisation problem under geological uncertainty. The difficulty of this problem is characterised by the presence of discrete variables, nonlinear and nonconvex objective function and constraints. Increased complexity arises from the computationally demanding functional evaluations from full-scale reservoir simulation, which is done in ECLIPSE®. We also exploit the MATLAB® Reservoir Simulation Toolbox – MRST® [9] for interfacing with ECLIPSE® and performing derivative-free optimisation tasks. The developed workflow is applied to a synthetic case study, for which it is demonstrated using several operational scenarios that using optimally selected realisations reduces computational time while providing best operating well locations. A comparative analysis of the performance of different derivative-free optimisation algorithms is also presented.

REFERENCES

1. Epelle, E.I. and Gerogiorgis, D.I., Optimal Rate Allocation for Production and Injection Wells in an Oil and Gas Field for Enhanced Profitability. *AIChE J.*, **in press** (DOI: 10.1002/aic.16592).
2. Tavallali, M.S. et al., 2014. Well placement, infrastructure design, facility allocation, and production planning in multireservoir oil fields with surface facility networks. *Ind. Eng. Chem. Res.*, 53(27), 11033-11049.
3. Gunnerud, V., Conn, A. and Foss, B., 2013. Embedding structural information in simulation-based optimization. Comput. *Chem. Eng.,* 53, 35-43.
4. Kosmidis, V.D., Perkins, J.D. and Pistikopoulos, E.N., 2004. Optimization of well oil rate allocations in petroleum fields. *Ind. Eng. Chem. Res.,* 43(14), 3513-3527
5. Li, L., Jafarpour, B. and Mohammad-Khaninezhad, M.R., 2013. A simultaneous perturbation stochastic approximation algorithm for coupled well placement and control optimization under geologic uncertainty. *Comput. Geosci.*, 17(1), 167-188.

6. Wang, H., Echeverría-Ciaurri, D., Durlofsky, L. and Cominelli, A., 2012. Optimal well placement under uncertainty using a retrospective optimization framework. *SPE J.*, 17(01), 112-121.
7. Bellout, M.C. et al., 2012. Joint optimization of oil well placement and controls. *Comput. Geosci.*, 16(4), 1061-1079.
8. Gerogiorgis, D.I., Georgiadis, M., Bowen, G., Pantelides, C.C. and Pistikopoulos, E.N., 2006. Dynamic oil and gas production optimization via explicit reservoir simulation. *Comput. Aided Chem. Eng.*21, 179-184.
9. Lie, K.A., 2014. An introduction to reservoir simulation using MATLAB: user guide for the Matlab Reservoir Simulation Toolbox (MRST). *Sintef ICT*, Trondheim, Norway.

569460 Modeling of Thermal Cracking of Natural Gas Liquids: Optimization with Support Vector Machine-Based Constraint Handling Scheme for Stiff ODEs

Rapid increase in shale gas production in the Appalachian Basin (Marcellus and Utica Shales) for the past 5 years has led to a significant growth in natural gas liquids (NGLs) production as well as a projected increase of these petrochemical feedstocks in the upcoming years [1, 2]. In this perspective, many existing ethylene crackers have expanded capacity and new crackers are becoming online to benefit from this unique opportunity [3]. Hence, the mathematical modeling and optimization of this process emerges as a necessity so as to determine the optimal operating conditions and the reactor configurations for the steam cracker, in such a way that the profit from thermal cracking of NGLs is maximized.

In this work, we are demonstrating (i) the comprehensive first-principles modeling of an adiabatic propane steam cracking process as a 1D plug flow reactor with coking effects [4] and (ii) the data-driven optimization of this reactor model using the ARGONAUT [5-6] framework. The mathematical model for thermal cracking includes the reaction rates expressed with Arrhenius equation, as well as the ODEs for continuity, energy, and momentum balances. The kinetics of thermal cracking of propane is adapted from Sundaram and Froment (1979) and the cracking model is validated using industrial data [7]. This simulation model is later used in the data-driven optimization phase to collect input-output data. However, the stiffness in the reactor models poses an additional complexity in retrieving the optimal operating conditions for the cracking process, especially when the data-driven optimization strategies strongly rely on collecting numerically stable samples from the reactor simulator. To this end, a novel Support Vector Machine (SVM)-based filtering approach is applied to eliminate candidate sampling points that are susceptible to the stiffness in the model and that will result with an abrupt termination of the cracking simulation due to high pressure drop. The SVM classifier identifies and removes the numerically unstable candidate points *a priori* to simulation execution by formulating an optimal separating hyperplane between stable/unstable points. This enables the data-driven algorithm to consistently converge to a profitable and feasible reactor configuration for the propane cracking process with an improved objective function value.

References

[1] Energy Information Administration. Today in Energy: Appalachia, Permian, Haynesville drive U.S. natural gas production growth, August 2018. URL https://www.eia.gov/todayinenergy/detail.php?id=36934. (accessed 04.10.2019).

[2] Energy Information Administration. Short-term energy outlook, April 2019. URL https://www.eia.gov/outlooks/steo/. (accessed 04.10.2019).

[3] Energy Information Administration. U.S. ethane consumption, exports to increase as new petrochemical plants come online, February 2018. URL https://www.eia.gov/todayinenergy/detail.php?id=35012. (accessed 02.26.2018).

[4] O. Onel. (2017). Advances in Modeling, Synthesis, And Global Optimization of Hybrid Energy Systems Toward the Production of Liquid Fuels and Olefins. Princeton University.

[5] F. Boukouvala and C. A. Floudas. (2017). ARGONAUT: AlgoRithms for Global Optimization of coNstrAined grey-box compUTational problems. *Optimization Letters*, 11(5), 895-913.

[6] B. Beykal, F. Boukouvala, C. A. Floudas, N. Sorek, H. Zalavadia, E. Gildin. (2018). Global optimization of grey-box computational systems using surrogate functions and application to highly constrained oil-field operations. *Computers & Chemical Engineering*, 114, 99-110.

[7] K. M. Sundaram and G. F. Froment. (1979). Kinetics of coke deposition in the thermal cracking of propane. *Chemical Engineering Science*, 34, 635-644.

574887 Towards a Novel Energy Price Predictive Framework: The Texas A & M Energy Price Index

The energy markets are sensitive and volatile to technological breakthroughs, changes in monetary policy, or major global challenges. Moreover, various governmental agencies, political and commercial organizations, think tanks as well as researchers and academics around the globe, consider various energy policies and their effects in an attempt to deal with the increasing concerns in energy independence, energy scarcity, energy sustainability, and pollution caused by the utilization of energy. Also, in the current environment in which all strategic political and commercial decisions and policies are assessed in economic terms, it is of outmost importance to determine the cost of energy precisely so as to evaluate their effectiveness. Considering the fact that energy affects every single individual and entity in the world, it is crucial to quantify "the unit price of energy" accurately, and study how it evolves with respect to time, major breakthroughs, challenges, energy and monetary policies. Given the lack of such a tool, we develop a novel framework, the Texas A&M Energy Price Index (EPI) that can be used as a benchmark to calculate the average price of energy to the end-use consumers in the United States.

The complicated energy landscape of the United States is carefully depicted to determine the products that are directed to the end-use sectors of the U.S. economy [1-5]. The total energy demand of these products, together with their prices, serve as the backbone of the Energy Price Index [6-10]. Since several key components (e.g. prices, demands) of the Energy Price Index have a lag between two and three months, we introduce a rolling horizon based parameter estimation model to approximate the current value of the Energy Price Index. The predictive methodology is tested on 120 months to show the accuracy of the developed framework in the prediction of the current value of the Energy Price Index. Energy price sub-indices for the transportation, residential, commercial, and industrial sectors are also developed. Some of the potential applications of the proposed index in the areas of economics, finance, engineering, law, and policy are demonstrated.

References:

- Elia, J. A.; Baliban, R. C.; Xiao, X.; Floudas, C. A. Optimal energy supply network determination and life cycle analysis for hybrid coal, biomass and natural gas to liquid (CBGTL) plants using carbon-based hydrogen production. Computers and Chemical Engineering 2011, 35, 1399-1430.
- Elia, J. A.; Baliban, R. C.; Floudas, C. A. Nationwide energy supply chain analysis for hybrid feedstock processes with significant CO2 emissions reduction. AIChE Journal 2012, 58 (7), 2142-2154.
- Elia, J. A.; Floudas, C. A. Energy Supply Chain Optimization of Hybrid Feedstock Processes: A Review. Annual Review of Chemical and Biomolecular Engineering 2014, 5 (1), 147-179.
- Onel, O.; Niziolek, A. M.; Floudas, C. A. Integrated biomass and fossil fuel systems towards the production of fuels and chemicals: state of the art approaches and future challenges. Current Opinion in Chemical Engineering 2015, 9, 66-74.
- Floudas, C.A.; Niziolek, A. M.; Onel, O.; Matthews, L.R. Multi-Scale Systems Engineering for Energy and the Environment: Challenges and Opportunities. AIChE Journal 2016, 62 (3), 602-623.
- Energy Information Administration (EIA), Annual Energy Outlook, 2019
- Energy Information Administration (EIA), Monthly Energy Review, 2019
- Lazard, Levelized Cost of Energy Analysis Version 12.0, 2018
- Bureau of Labor Statistics (BLS), Producers Price Index, 2018

U.S. Department of Energy (DOE), Alternative Fuels Data Center, 2018

575666 Intensification of Biorefinery Operations Using Building Blocks

Bioenergy is an attractive form of energy since it requires renewable sources that can be produced locally and used as raw materials for its production. One of the possible routes for bioenergy production is through lignocellulosic biorefineries, which utilize various biomass sources, e.g. corn stover, wood, rice straw and forest wastes, to produce biodiesel, steam and power in addition to pulp and paper [1]. Increasing awareness towards sustainable and modular technologies highlights the need for novel processing pathways leveraging on process intensification principles that can yield drastic improvements in several metrics, including economics, energy consumption and chemical footprint [2]. Recent developments in process intensification resulted in development of Integrated biorefinery concept with two new processing pathways: (i) pre-extraction of hemicellulose from biomass and its conversion to ethanol, and (ii) production of syngas from gasification of black liquor [3]. However, there exist further unexplored opportunities for intensification of the biorefineries to enhance the reaction yield and mass/heat transfer within the several parts of the process, e.g. digestion of pulp, evaporation of black liquor, etc. [4].

Recently, a new method for systematic process intensification, which relies on building block-based representation of chemical process flowsheets, has been introduced [5]. These building blocks can be integrated to form a two-dimensional grid embedding numerous non-trivial process alternatives [6-9]. Each block can be assigned with several physicochemical phenomena, e.g. reaction, vapor-Liquid equilibrium, solid-liquid equilibrium, etc., combination of which can yield various intensification pathways. In this work, we extend this novel representation methodology to various operations that take place in a biorefinery. Pulping which occurs in a pulp digester lies at the heart of a pulp mill and, here, we develop a new representation based on fundamental building blocks to represent the pulp digestion process. Mass transfer boundary layer is represented by the boundary shared by two neighboring blocks in different phases and digestion process is represented within the interior of the block. This building-block based representation is modeled using a mixed-integer nonlinear programming (MINLP) problem so as to minimize the energy consumption and increasing the yield of the process while searching for several intensified pathways. The proposed approach is demonstrated through several case studies focusing on the conventional Kraft pulping process, Kraft pulping process with pre-extraction of hemicellulose, Kraft pulping process with pre-extraction of hemicellulose and short fibers.

[1] Huang, H., Lin, W., Ramaswamy, S., Tschirner, U. (2009) Process Modeling of Comprehensive Integrated Forest Biorefinery – An Integrated Approach. *Applied Biochemistry and Biotechnology*, *154*, pp. 205-216.

[2] Tian, Yuhe., Demirel, S.E.., Hasan, M.M.F., Pistikopoulos, E.N., (2018) An overview of process systems engineering approaches for process intensification: State of the art. *Chemical Engineering and Processing - Process Intensification, 133*, pp. 160-210.

[3] Aden, A., Ruth, M., Ibsen, K., Jechura, J., Neeves, K., Sheehan, J., Wallace, B., Montague, L., Slayton, A., Lukas, J., (2002) NREL Report TP-510-32438. Lignocellulosic Biomass to Ethanol Process Design and Economics Utilizing Co-Current Dilute Acid Prehydrolysis and Enzymatic Hydrolysis for Corn Stover.

[4] Huang, H., Ramaswamy, S., Al-Dajani, W.W., Tschirner, U. (2010) Process modeling and analysis of pulp-mill based integrated biorefinery with hemicelluloses pre-extraction for ethanol production: A comparative study. *Bioresource Technology 101*, pp. 624-631.

[5] Demirel, S.E., Li, J., Hasan, M.M.F., (2017) Systematic process intensification using building blocks. *Computers and Chemical Engineering, 105*, pp. 2-38.

[6] Li, J., Demirel, S. E., Hasan, M. M. F., (2018) Process Synthesis Using Block Superstructure with Automated Flowsheet Generation and Optimization. *AIChE Journal, 64(8)*, pp. 3082-3100.

[7] Li J., Demirel S.E., Hasan M.M.F. (2018) Process Integration using Block Superstructure. *Industrial &*

Engineering Chemistry Research, 57(12), pp. 4377–4398.

[8] Li, J., Demirel, S.E., & Hasan, M. M. F. (2019). Building Block-Based Synthesis and Intensification of Work-Heat Exchanger Networks (WHENS). *Processes, 7(1),* 23.

[9] Demirel, S. E., Li, J., and Hasan, M. M. F. (2019). A General Framework for Process Synthesis, Integration, and Intensification. *Industrial & Engineering Chemistry Research.* DOI: 10.1021/acs.iecr.8b05961.

576096 Hyperbolicity of Heat Transport Processes

Heat transport models are ubiquitously present in chemical, petrochemical, manufacturing and process industry as mathematical models of distributed parameter systems given by parabolic partial differential equations (PDEs). The conductive (diffusive) transport over macroscopic length scale is characterized by the well known Fourier's law and a large number of research efforts have been realized in the realm of modelling, control and optimization of such models. However, the salient characteristic of these models is that any initial disturbance in the material body is propagated instantly due to the parabolic nature of the partial differential equation (PDE) model. To eliminate this unphysical feature, we revisit the Fourier law to take into account the thermal inertia, resulting in a hyperbolic heat equation which avoids the phenomenon of infinite speed of propagation of initial data [1,2].

The main conservation laws are embedded in a modelling variety provided by the hyperbolicity of the transport systems which is physically relevant and desired property as action at distance is precluded and physically meaningful finite speed of phenomena propagation is ensured. Furthermore, the hyperbolicity mathematically ensures the well-posedness of local Cauchy problems [3]. In addition, a Stefan problem is considered, which is a specific type of heat transport boundary value problem for a partial differential equation describing the heat distribution evolution in a phase changing medium [4]. Since the moving interface is unknown a priori, the solution needs to take into account determination of the moving boundary position as well as accurate heat propagation. Finally, we provide derivation of the hyperbolic heat transport equations, the stability analysis and the numerical results for simple heat diffusion problem on fixed domain and Stefan problem.

Literature Cited

1. Cattaneo C. On a form of heat equation which eliminates the paradox of instantaneous propagation. *C R Acad Sci Paris*. 1958;pp. 431–433.
2. Vernotte P. Les paradoxes de la theorie continue de l'equation de la chaleur. *C R Acad Sci Paris*. 1958;246:3154–3155.
3. Fischer AE, Marsden JE. The Einstein evolution equations as a first-order quasi-linear symmetric hyperbolic system, I. *Communications in Mathematical Physics*. 1972;28(1):1–38.
4. Gupta S. *The Classical Stefan Problem. Basic Concepts, Modelling and Analysis*. North-Holland: Applied mathematics and Mechanics. 2003.

569568 Constructing Efficient Surrogate Models for High-Dimensional First Principles Systems Under Uncertainty

A plethora of chemical engineering processes are informed by sophisticated numerical simulations based on first-principles models (e.g., consisting of a set of coupled partial differential equations) that exhibit highly nonlinear behavior and are computationally expensive. It is often the case that the parameters in these models as well as the initial and boundary conditions in dynamical systems are uncertain and, thus, this uncertainty is propagated to the model outputs. Therefore, relying on deterministic simulation results could possibly lead to erroneous outcomes for the design, operation, and control of these processes. The computational burden associated with solving the entire set of equations can be prohibitive when one wishes to run the model for the whole spectrum of possible input parameter values, even in this era of high-performance computing, especially in the context of optimization applications. Metamodeling techniques aim at alleviating this problem by approximating first-principles model responses under uncertainty with efficient black-box mappings that can serve as surrogate models for the original high-fidelity models [1]. The primary notion in metamodeling is to calibrate the surrogate models using the fewest possible runs of the original model without compromising their predictive capability.

In this work, we explore the effectiveness of various well-established meta-modeling techniques applied to a basic, but relevant crystallization problem, i.e., the growth/dissolution of a 2D crystal population through cooling/heating cycles. This technique is used, possibly in combination with milling, to manipulate the crystal shape of active pharmaceutical ingredients, and efforts have been recently devoted to optimize and control this process. Even though relatively simple, such a problem represents an interesting test-case for metamodeling since it is highly nonlinear, inherently dynamic, and industrially relevant. [2,3] The crystallization system is described by a system of integro-partial differential equations, whose parameters are inherently uncertain. The parameter uncertainties are modeled by suitable probability distributions selected according to available experimental data and expert knowledge.

We first employ sparse polynomial chaos expansions (PCE) whose coefficients are estimated based on least-angle regression and explore their scalability with an increasing number of uncertain inputs. PCEs have been traditionally used for uncertainty quantification in numerous applications (e.g., see [4]). We then present the fundamentals of Gaussian Processes (i.e., Kriging) [5] and explore their coupling with sparse PCEs in order to create surrogate models with optimal local and global predictive capabilities [6]. Finally, the use of NARX neural networks [7] is demonstrated to deal with the continuous dynamic behavior of the system. The accuracy of these three surrogate modeling approaches in capturing the highly nonlinear spatio-temporal dynamics of the crystallization process at hand is assessed using state-of-the-art cross-validation and validation measures, and their computational aspects are discussed.

[1] Bhosekar A., Ierapetritou M., "Advances in surrogate based modeling, feasibility analysis, and optimization: A review", Computers and Chemical Engineering, 108:250-267, 2018

[2] Salvatori, F., Mazzotti, M., "Manipulation of particle morphology by crystallization, milling, and heating cycles—A mathematical modeling approach", Industrial and Engineering Chemistry Research, 56:9188–9201, 2017.

[3] Bötschi, S., Rajagopalan, A. K., Morari, M., Mazzotti, M., "Feedback control for the size and shape evolution of needle-like crystals in suspension. I. Concepts and Simulation Studies", Crystal Growth and Design, 18:4470–4483, 2018.

[4] Paulson J.A., Mesbah A., "An efficient method for stochastic optimal control with joint chance constraints for nonlinear systems", International Journal of Robust and Nonlinear Control, 1-21, 2017.

[5] Kawai S., Shimoyama K. "Kriging-model-based uncertainty quantification in computational fluid dynamics. 32nd AIAA Applied Aerodynamics Conference, 2014, 2737.

[6] Schöbi R., Sudret, B., Wiart J., "Polynomial-Chaos-based Kriging", International Journal for

Uncertainty Quantification, Begell House Publishers, 5:171-193, 2015.

[7] Mai C.V., Spiridonakos M.D., Chatzi E.N., Sudret B., "Surrogate modeling for stochastic dynamical systems by combining nonlinear autoregressive with exogenous input models and polynomial chaos expansions", International Journal for Uncertainty Quantification, 6:313-339, 2016.

AUTHOR INDEX

AUTHOR INDEX

AUTHOR INDEX

AUTHOR INDEX

AUTHOR INDEX

AUTHOR INDEX